methods in
helio- and
asteroseismology

methods in
helio- and
asteroseismology

frank p pijpers
imperial college london, uk

Imperial College Press

Published by

Imperial College Press
57 Shelton Street
Covent Garden
London WC2H 9HE

Distributed by

World Scientific Publishing Co. Pte. Ltd.
5 Toh Tuck Link, Singapore 596224
USA office: 27 Warren Street, Suite 401-402, Hackensack, NJ 07601
UK office: 57 Shelton Street, Covent Garden, London WC2H 9HE

British Library Cataloguing-in-Publication Data
A catalogue record for this book is available from the British Library.

Cover image: "Concentric Rinds" by M. C. Escher © 2006 The M. C. Escher Company – Holland. All rights reserved.

METHODS IN HELIO- AND ASTEROSEISMOLOGY

Copyright © 2006 by Imperial College Press

All rights reserved. This book, or parts thereof, may not be reproduced in any form or by any means, electronic or mechanical, including photocopying, recording or any information storage and retrieval system now known or to be invented, without written permission from the Publisher.

For photocopying of material in this volume, please pay a copying fee through the Copyright Clearance Center, Inc., 222 Rosewood Drive, Danvers, MA 01923, USA. In this case permission to photocopy is not required from the publisher.

ISBN-13 978-1-86094-755-1
ISBN-10 1-86094-755-7

Printed in Singapore

To my closest colleague
Teresa

and

my young discoverers
Tomás and *Laura*

Preface

It is a fundamental aim of all the physical sciences that its theories must be subjected to ever more rigorous tests until they fail to predict correctly the measurements that are made. The discrepancy between predictions and measurements is then a guide to improvements on such theories. Stellar evolution theory is no different. Because of the complexity of stars, well-tested fundamental physics has to be combined with (semi-)phenomenological descriptions of some processes, such as convection. Helioseismology and its younger cousin asteroseismology are new branches on the genealogical tree of stellar evolution studies. Their aim is to use the oscillation properties of stars to infer interior properties, thereby testing ideas on the internal structure of stars. By applying this technique to the Sun and to other stars over a range of masses, ages, and surface chemical composition, stellar evolution theory is tested. Without these techniques, only surface properties of single stars can be determined.

This book is intended to cover the methodologies used in helio- and asteroseismology. It should be viewed as a resource for post-graduate students and researchers already in the field, with the aim of providing background to, and examples of, the techniques that are necessary to follow the path from data to model parameters, and from model parameters to predictions for observables.

Writing a book such as this involves consulting with a number of colleagues, whose help is gratefully acknowledged. Particular thanks I owe to Jørgen Christensen-Dalsgaard, for his personal support, but also for his excellent lecture notes on stellar oscillations without which parts of this book would have been much harder to write.

F. P. Pijpers

Contents

Preface			vii
1.	Global Helioseismology: The Data		1
	1.1	Intensity Variations	2
	1.2	Velocity Variations	4
	1.3	From Time-Series to Frequencies	9
	1.4	Mode Ambiguities and Cross-Talk	17
	1.5	Line Profile Asymmetries	22
	1.6	Oscillations at High Frequencies	25
2.	Global Helioseismology: Modelling		29
	2.1	Separation of Time Scales	30
	2.2	Equations of Hydrostatic Structure	32
	2.3	Evolution of the Sun and Stars	38
	2.4	Numerical Solution Techniques	41
		2.4.1 Discretisation of a continuous potential	43
		2.4.2 Discretisation in the presence of a discontinuity	47
	2.5	The Resonances of Simple Systems	51
		2.5.1 Resonant properties of some simple systems	54
		2.5.2 Example I: Linear waves in a rectangular membrane	55
		2.5.3 Example II: Linear waves in a circular membrane	57
		2.5.4 Example III: Linear waves in a uniform sphere	58
	2.6	Resonant Frequencies of the Sun and Stars	61
		2.6.1 Derivation of the LAWE	61
		2.6.2 Boundary conditions	68

2.7	Oscillations in the Cowling Approximation	69
2.8	JWKB Analysis	74
2.9	Surface Effects	79
2.10	Excitation and Damping of Modes	81

3. Global Helioseismology: Inverse Methods — 85

3.1	The Relationship between Frequencies and Sound Speed	87
	3.1.1 Derivation of the variational structure kernels	94
	3.1.2 Filtering out surface effects	100
3.2	The Abel Transform	102
3.3	The Relationship between Frequencies and Rotation	106
3.4	Linearisation	112
3.5	Linear Methods	114
	3.5.1 Regularised least-squares	114
	3.5.2 Optimally localised averages	116
3.6	Choosing Regularisation Weighting	120
3.7	Non-Linear Methods	123
	3.7.1 The maximum entropy method	123
3.8	Optimal Mask Design as an Inverse Problem	127

4. Local Helioseismology — 131

4.1	The Data: Reduction and Analysis	131
	4.1.1 Tracking, cross-correlation and filtering	132
	4.1.2 Averaging and masking	139
4.2	The Forward Approach: Modelling	145
	4.2.1 Hydrodynamical models with radiation	145
	4.2.2 The influence of magnetic fields	147
	4.2.2.1 Dynamo action	149
	4.2.3 Turbulence	152
	4.2.4 Wave propagation in inhomogeneous media	156
	4.2.4.1 Example I: Scattering on a sphere in a homogeneous medium	157
	4.2.4.2 Example II: Scattering on an infinite cylinder in a homogeneous medium	163
	4.2.4.3 More general scattering problems	164
4.3	Inverse Methods	166
	4.3.1 Ring diagrams	166
	4.3.2 Time-distance techniques	171

		4.3.3	The ray approximation	172
		4.3.4	The Born approximation	177
		4.3.5	The Rytov approximation	183
		4.3.6	Acoustic holography	187
		4.3.7	Seismology of magnetic loops	190
5.	Asteroseismology			195
	5.1	The Data		196
		5.1.1	Signal and noise in the observations	200
		5.1.2	The mode identification problem	208
		5.1.3	Irregular sampling	220
		5.1.4	Non-seismic constraints	227
	5.2	Modelling		230
		5.2.1	The ages of stars	236
		5.2.2	The composition of stars	238
		5.2.3	Pulsation of stars	243
		5.2.4	Excitation	245
		5.2.5	Rotation	246
	5.3	Inverse Methods		254
		5.3.1	Time-series analysis as an inverse problem	254
		5.3.2	Linear and global methods	261
		5.3.3	Detecting discontinuities and kinks	268
			5.3.3.1 Example: Explicit calculations for a crude model	269
			5.3.3.2 Derivative discontinuities in realistic models	274

Appendix A Useful Vector Formulas   279

Appendix B Explicit Forms of Vector Operations   281

Appendix C Useful Constants   287

Bibliography   291

Index   305

Chapter 1

Global Helioseismology: The Data

Global helioseismology concerns itself with the use of resonant modes of oscillation of the Sun to infer its interior structure. Through the intensity and/or velocity amplitude at the visible surface, and the pattern over the surface, these oscillations can be detected and identified and their frequencies can be measured. A comparison of observed and predicted frequencies then provides the possibility to infer solar interior properties.

Results of such comparisons and inversions for the interior structure and rotation of the Sun can be found in a recent review [Christensen-Dalsgaard (2002)], and also in a number of papers describing results up to 1996 from the Global Oscillations Network Group (GONG): [Gough et al. (1996b)], [Harvey et al. (1996)], [Christensen-Dalsgaard et al. (1996)], [Hill et al. (1996)], [Gough et al. (1996a)], [Thompson et al. (1996)]. Since the emphasis here is on the methods rather than the results, the reader is referred to these reviews for the current state of knowledge of the solar interior.

During an oscillation cycle of any oscillation mode in the Sun, the state variables defining the plasma vary periodically. The time-dependency is normally separated out in terms of an exponential $e^{i\omega t}$. If one incorporates non-adiabatic effects on the oscillations the exponent gains a real part. The discussion of that can be deferred since solar oscillations can be treated as being very nearly adiabatic. The spatial part of the oscillation equations can be written in terms of a variety of pairs of state variables. Common choices are the displacement, the velocity, the density, the pressure, or the temperature. In the Sun none of these quantities can be measured directly. Instead the effect of the oscillations on the emerging radiation is what provides us with the data to be analysed. So far two effects of solar oscillations on radiation have been used as tracers of oscillations: intensity and Doppler shift of spectral lines. These correspond to applying the observational techniques of photometry and spectroscopy respectively.

1.1 Intensity Variations

In white light, but also integrated over finite wavelength ranges the intensity of emerging radiation varies during an oscillation cycle. Although it is true that in the solar interior plasma is compressed and expanded in a (nearly) adiabatic cycle, and therefore the temperature at any given depth increases and decreases, this is not the most important factor in producing intensity variations. The intensity variations that can be observed must arise from those geometrical depths in the photosphere where the optical depth is close to unity, because this is always the location where the emergent flux of the Sun is determined. For the plasma to behave adiabatically it must be at much higher optical depths ($\tau \gg 1$). The plasma is also not optically thin ($\tau \ll 1$), which would produce isothermal behaviour. Certainly the amplitude of temperature variations near the surface during the oscillation is much lower than under fully adiabatic conditions. Furthermore the location where $\tau = 1$ moves outward upon compression, because the density increases, to a region with a lower mean temperature. The depth of formation of the spectrum in any given wavelength range therefore moves periodically to deeper and shallower layers. It is a combination of all of these effects on the propagation of radiation that causes the variations in intensity.

The cause of the intensity variations and more detailed modelling is revisited in section 2.9. For the detection of solar oscillations, it is sufficient to note that the oscillations of the Sun give rise to variations in its emergent flux and therefore it is possible to detect solar oscillations through irradiance variations. Examples of instruments in operation are the 'variability of solar irradiance and gravity oscillations luminosity oscillations imager' (VIRGO-LOI) instrument and the 'Michelson Doppler imager solar oscillations investigation' (MDI-SOI) instrument on board the solar and Heliospheric Observatory (SoHO) satellite. Detailed descriptions of these instruments can be found in [Appourchaux et al. (1997)] and [Scherrer et al. (1995)]. The Global Oscillations at Low Frequencies instrument (GOLF) also on board SoHO, is described in detail in [Gabriel et al. (1989)]. It is no coincidence that these instruments are on-board a satellite. The reason for this is that for ground based observation of intensity variations the dominant noise source is transparency variations of the atmosphere. This scintillation effect is related to the the more familiar 'seeing' which is responsible for smearing out stellar images in particular in telescopes that do not have adaptive optics. Both of these effects are due to atmospheric turbulence, and hence have a wide range in timescales over which they affect observations.

Transparency variations are particularly insidious because they are 'multiplicative noise'. To illustrate this assume that outside the atmosphere there is a time variable signal with a constant amplitude at a single fixed frequency $A\cos(\omega t)$. This is modulated by the transparency variations $T(\omega_a)\cos(\omega_a t)$, so that the measured signal $m(t)$ is:

$$m(t) = \int \mathrm{d}\omega_a\, T(\omega_a) A \cos(\omega t)\cos(\omega_a t) \ . \tag{1.1}$$

The Fourier transform of $m(t)$ is:

$$\begin{aligned} M(\omega') &= \int \mathrm{d}t\, m(t) e^{-i\omega' t} \\ &= \int \mathrm{d}t\, e^{-i\omega' t} \int \mathrm{d}\omega_a\, T(\omega_a) A \cos(\omega t)\cos(\omega_a t) \\ &= \int \mathrm{d}\omega_a\, T(\omega_a) \frac{A}{4} \int \mathrm{d}t\, e^{-i\omega' t} \left(e^{i\omega_a t} + e^{-i\omega_a t} \right) \left(e^{i\omega t} + e^{-i\omega t} \right) \\ &= \int \mathrm{d}\omega_a\, T(\omega_a) \frac{A}{4} \int \mathrm{d}t\, \left(e^{i(-\omega'+\omega+\omega_a)t} + e^{i(-\omega'-\omega+\omega_a)t} \right. \\ &\quad \left. + e^{i(-\omega'+\omega-\omega_a)t} + e^{i(-\omega'-\omega-\omega_a)t} \right) \\ &= \int \mathrm{d}\omega_a\, T(\omega_a) \frac{A}{4} \left(\delta(\omega' - \omega - \omega_a) + \delta(\omega' + \omega - \omega_a) \right. \\ &\quad \left. + \delta(\omega' - \omega + \omega_a) + \delta(\omega' + \omega + \omega_a) \right) \ . \end{aligned} \tag{1.2}$$

This is a convolution integral in the Fourier domain of $T(\omega)$ and $\delta(\omega)$, which means that in the Fourier domain every delta-function peak corresponding to the true frequency is replaced by a copy of the spectrum of transparency variations. The transparency variations are stochastic and dependent on the observation site and season. This means that different observing sites do not measure identical oscillation frequencies, and at a single site one will not measure one and the same frequency from season to season, nor even from day to day because the realisation of the turbulent spectrum causing the transparency variations is different from day to day. In practice one combines measurements from different sites around the world in order to combat the adverse effects of day-night gaps in the data and one will notice that in the Fourier domain the peaks corresponding to solar oscillation frequencies have a width that is increased because of these effects, or even multiply peaked if the observing conditions are poor. In space evidently the effect of an atmosphere is removed. This does not mean that the satellite measurements are noise-free since there are other noise sources as well, but noise reduction is around a factor of 4 compared to ground-based instruments in the frequency range relevant for the solar 5-minute oscillations ([Appourchaux et al. (1997)]).

1.2 Velocity Variations

Spectroscopic measurements show that the line-of-sight component of the velocity of plasma taking part in the oscillatory motion can be measured through the Doppler shift of absorption lines formed in that plasma. Some of the lines useful for helioseismology are: the Ni I 676.8 nm intercombination line (cf. [Bruls (1993)]), and alkali lines such as the K I 769.9 nm line and Na I D lines at 589.00 nm (D_2) and 589.59 nm (D_1) (cf. [Bruls et al. (1992)], [Bruls & Rutten (1992)])

It is this technique that is adopted for the GONG instrumentation (cf. [Beckers & Brown (1980)] and also the GONG web-site http://gong.nso.edu/instrument/) and it is also one of the data products of the MDI instrument on board the SoHO satellite (cf. [Scherrer et al. (1995)]). Both the GONG and MDI instruments observe near the Ni I 626.8 nm line.

The way in which the spectroscopic measurements are turned into Doppler images can be illustrated as follows. Starting point is the line profile of a selected spectral line. From theory the line profile of spectral lines must be determined through detailed radiative transfer calculations taking into account that the solar photosphere is stratified and turbulent. Observationally it is known that some lines have strong centre-to-limb variations over the solar disc, and can have asymmetric profiles, inverted cores, or a variety of other properties which can bias measurements of bulk velocities in the line of sight over the solar disc.

Most of the lines discussed previously have been selected for having as uniform properties as possible over the solar disc and for being most representative of the mean line of sight velocity of the photosphere. However, there is a different reason why in particular the lines of Na and K are attractive, which is that one can use a cell filled with vapour of the relevant element as a spectrometer, which works through resonant scattering of the sunlight passing through it (cf. [Chaplin *et al.* (1996)]). The line profile of such lines can be described to a high degree of accuracy by a function $P(\lambda)$ that is symmetric around a simple minimum, such as a Gaussian, with a full width at half maximum (FWHM) $\Delta\lambda$ that is a few times 10^{-5} times the central wavelength of the line λ. This corresponds to a Doppler velocity width $\Delta v \approx c\Delta\lambda/\lambda$ in the range of several km/s. A spectroscopic instrument samples this line at a finite resolution $R = \lambda/\delta\lambda$, so that there are measured intensities I_i:

$$I_i = \frac{1}{\delta\lambda} \int_{\lambda_i-\delta\lambda/2}^{\lambda_i+\delta\lambda/2} d\lambda \, P\left(\lambda - \lambda_{\text{ref}}\left(1 - \frac{v}{c}\right)\right) M(\lambda - \lambda_i) \,. \quad (1.3)$$

Here λ_{ref} is a reference wavelength for the minimum of P for a source that has a 0 radial velocity relative to the detector. M is the spectroscopic response function describing the sensitivity over the bins/pixels of the detector. M is assumed to be symmetric and independent of i. In writing Eq. (1.3) it is assumed that $v \ll c$ so that the full Doppler shift expression can be expanded to first order. This measurement is repeated at a fixed time sampling interval so that a time series of spectra is obtained:

$$\begin{aligned} I_{i,j} &= \frac{1}{\delta\lambda} \int_{\lambda_i-\delta\lambda/2}^{\lambda_i+\delta\lambda/2} d\lambda \, P\left(\lambda - \lambda_{\text{ref}}\left(1 - \frac{v_j}{c}\right)\right) M(\lambda - \lambda_i) \\ &\equiv R \int_{-1/2R}^{1/2R} du \, M(u) P\left(u + \frac{\lambda_i}{\lambda_{\text{ref}}} - \left(1 - \frac{v_j}{c}\right)\right) \end{aligned} \quad (1.4)$$

in which a dimensionless variable $u \equiv (\lambda - \lambda_i)/\lambda_{\text{ref}}$ is introduced. In principle some finite integration time is involved, and hence an integral over time should be present in this expression with an appropriate temporal response function $M(\lambda - \lambda_i, t - t_j)$, but this is omitted for simplicity. For convenience in what follows the spectrum is assumed to be uniformly sampled. A numbering scheme is adopted for the pixels, so that i runs from $-K$ to $+K$, and pixel 0 is centered on the wavelength $\langle \lambda \rangle$ where, averaged over the time series, P has its minimum. The λ_i then satisfy:

$$\lambda_i = \langle \lambda \rangle + i \frac{\delta\lambda}{f_{\text{over}}} \quad (1.5)$$

where f_{over} is a constant factor, that accounts for oversampling of the spectrum, and $\langle \lambda \rangle$ is:

$$\langle \lambda \rangle \equiv \lambda_{\text{ref}} \left(1 - \frac{\langle v \rangle}{c} \right) . \tag{1.6}$$

The expression for the intensities then reduces to:

$$I_{i,j} = R \int_{-1/2R}^{1/2R} du\, M(u) P \left(u + \frac{i}{R f_{\text{over}}} + \frac{1}{c} (v_j - \langle v \rangle) \right) . \tag{1.7}$$

Now one can calculate the part of the intensity profile that is antisymmetric around line centre at $i = 0$:

$$\begin{aligned}
I_{Ai,j} &\equiv \frac{1}{2} \left(I_{i,j} - I_{-i,j} \right) \\
&= \frac{R}{2} \int_{-1/2R}^{1/2R} du\, M(u) \left[P \left(\frac{i}{R f_{\text{over}}} + u + \frac{1}{c}(v_j - \langle v \rangle) \right) \right. \\
&\qquad \left. - P \left(\frac{-i}{R f_{\text{over}}} + u + \frac{1}{c}(v_j - \langle v \rangle) \right) \right] \\
&= \frac{R}{2} \int_{-1/2R}^{1/2R} du\, M(u) \left[P \left(\frac{i}{R f_{\text{over}}} + u + \frac{1}{c}(v_j - \langle v \rangle) \right) \right. \\
&\qquad \left. - P \left(\frac{i}{R f_{\text{over}}} - u - \frac{1}{c}(v_j - \langle v \rangle) \right) \right] \\
&\approx R \int_{-1/2R}^{1/2R} du\, M(u) \left[u + \frac{1}{c}(v_j - \langle v \rangle) \right] \left. \frac{dP}{du} \right|_{\frac{i}{R f_{\text{over}}}} \\
&\approx \frac{1}{c} \left. \frac{dP}{du} \right|_{\frac{i}{R f_{\text{over}}}} (v_j - \langle v \rangle) .
\end{aligned} \tag{1.8}$$

In the above, first use is made of the symmetry of P, then P is expanded in a Taylor series which is truncated after the first term, and then use is made of the symmetry of M. If P or M are not perfectly symmetric this will, to lowest order, introduce constant offsets, which are easily accounted for. The above results indicate that for small velocities in the line-of-sight, the I_A are a linear measure of that velocity. Since the I_A suffer from measurement error, determining v_j reduces to a standard linear least-squares problem. If the errors on the measurements I are Gaussian distributed and uncorrelated the optimal determination of v_j is a weighted sum of the I_A:

$$v_j - \langle v \rangle = \sum_i w_i \left(I_{i,j} - I_{-i,j} \right) \quad (1.9)$$

with a variance:

$$\sigma(v_j)^2 = \sum_i w_i^2 \left(\sigma_i^2 + \sigma_{-i}^2 \right) \quad (1.10)$$

in which the weights w_i are set as:

$$w_i = \frac{c}{2} \left[\frac{1}{\sigma_i^2 + \sigma_{-i}^2} \left. \frac{\mathrm{d}P}{\mathrm{d}u} \right|_{\frac{i}{Rf_{\mathrm{over}}}} \right] \left[\sum_k \frac{1}{\sigma_k^2 + \sigma_{-k}^2} \left. \frac{\mathrm{d}P}{\mathrm{d}u} \right|_{\frac{k}{Rf_{\mathrm{over}}}}^2 \right]^{-1}. \quad (1.11)$$

From Eq. (1.11) one can see that in the wings of a spectral line $\mathrm{d}P/\mathrm{d}u$ is small and therefore the weights are small, as one would expect since the wings are not very sensitive to Doppler shifts of the line. The weights are also small right at the centre of the line because there too $\mathrm{d}P/\mathrm{d}u$ is small. Furthermore, if the line is a strong absorption line, photon noise would tend to make the σ_i larger in the core of the line, further suppressing the weights. The weighting is thus concentrated around the maxima in $\mathrm{d}P/\mathrm{d}u$ which are the inflection points of the line profile. Since lines are superposed, normally in absorption, on a continuum, it is often better to measure an intensity ratio, so that the data is $(I_{i,j} - I_{-i,j})/(I_{i,j} + I_{-i,j})$. Such a differential measurement helps in eliminating biases. Following the same procedure as above, one can see that these data can be combined in the same manner as in Eq. (1.9), with weights $\widetilde{w}_i$ that are related to the w_i as:

$$\widetilde{w}_i = w_i \left[1 + P\left(\frac{i}{Rf_{\mathrm{over}}} \right) \right] \quad (1.12)$$

in which the 1 arises because of the presence of the continuum, and one should note that P is measured relative to the continuum level and is negative for absorption lines.

The implementation of the above algorithm is straightforward, but in the case of satellite bourne experiments still requires either a large datarate or a relatively fast computer, with an associated high cost. The actual implementation of the above scheme is achieved through hardware. The principle is to use two narrow band filters, centered on the inflection points of the line profile at either side of the line centre. The light passing through each of these is then passed through a circular polariser with opposite handed-ness, after which the light is combined. The polarisation of this combined beam is linear. A second polarising filter is placed such that no light passes through for that linear polarisation direction of the combined light-beam produced when equal amounts of light pass through each of the narrow-band filters. When the line shifts due to motions on the solar surface, the narrow-band filters let through unequal amounts of light which makes the polarisation direction of the combined light-beam rotate. The intensity of the light that then passes through the second polarisation filter is proportional to the line shift. This intensity difference is imaged over the solar disc.

Comparing this hardware solution with the algorithm presented before, the narrow band filters are the component that provide weights w_i. Because of technical limitations these weights may well not exactly match the expression from the algorithm. The result of that is that the measurements are still linear measures of the velocity, for small enough velocities, but somewhat sub-optimal in terms of error propagation. The polariser set-up provides the mechanism whereby the signal from either side of the line can be added or subtracted. Further details of the construction of Doppler imaging instruments, using variations of this basic scheme, can be found in [Brookes et al. (1978)] and [Isaak, et al. (1989)] (instrumentation of the Birmingham Solar Oscillation Network (BiSON)), [Scherrer et al. (1995)] (MDI), [Rhodes et al. (1988)], [Cacciani et al. (1990)] (transmission cells).

There is one further point to note with these Doppler measurements, which is that the measurements are made in a frame of reference that has a radial velocity component with respect to the centre of mass of the Sun. This is true both for satellites and for ground-based data. Evidently this is a known velocity component which changes very slowly compared to the periods of oscillation. It would not be particularly difficult to account for if the entire analysis were carried out in terms of software, since one can replace $\langle v \rangle$ in Eq. (1.6) with this slow drift $v_{\text{drift}}(t)$, and carry through the analysis. The hardware solutions described above may experience difficulties. If the 'frame-of-reference' velocity is large enough that it brings the instrument response into the non-linear regime one needs to take steps to compensate for this. The global solar rotation can cause similar problems since it produces a quite large velocity gradient over the solar surface, compared to the velocity variations caused by the oscillations. A description of the post-processing involved, to take into account higher-order terms in velocity, can be found in [Palle et al. (1992)].

1.3 From Time-Series to Frequencies

For a known function of time $f(t)$ the amplitude as a function of frequency $F(\omega)$ is obtained by carrying out a Fourier integral as in Eq. (1.2):

$$F(\omega) = \int_{-\infty}^{\infty} \mathrm{d}t f(t) e^{-i\omega t} . \tag{1.13}$$

The collection of data, whether it be intensity or velocity, requires a finite amount of time of 'integration' during which the detector, normally a CCD, is exposed to radiation, after which there is a dead time during which the CCD is read out and the digitised signal stored on computer or transmitted to a receiving station. The integration time itself is usually short and is treated in what follows as an instantaneous sample: mathematically the response function in time is treated as a Dirac delta function. The generalisation to a finite integration time is straightforward, and can be postponed at this stage. The time series is therefore not measured continuously but sampled discretely. Normally the detection process is automated to sample the time series equidistantly at a rate that is sufficiently high to resolve the relevant time scales of variation ($\sim$ 5 min for the Sun). A trade-off between the required signal-to-noise of the data on the one hand, and the cost of the detection system on the other (including telescope, enclosure, maintenance, staff, data storage etc.), normally means that the sampling rate is no faster than about once per minute.

In the ideal case the data would therefore be discretely sampled at a constant rate, and one would apply a discrete Fourier transform (DFT) to obtain the signal as a function of frequency within a band.

$$F(\omega_k) = \sum_i f(t_i) e^{i\omega_k t_i} \tag{1.14}$$

The upper limit of that band is set by the Nyquist frequency ω_Nyq which is related to the sampling rate Δ_t as:

$$\omega_\mathrm{Nyq} = \frac{\pi}{\Delta_t} . \tag{1.15}$$

The data stream has a finite length, and the consequence of this is that in the Fourier domain there is a finite frequency resolution:

$$\Delta_\omega \propto \frac{1}{T} . \tag{1.16}$$

To see that this must be the case, one can take the view that a finite length time series is the product of an infinite time series and a tophat function which is equal to 1 for the duration of the measurements and 0 outside of that. The Fourier transform of the product in the time domain of two functions, is the convolution of the Fourier transforms of these two functions. The Fourier transform of the tophat function with length T is the sinc function:

$$\text{sinc}\frac{T\omega}{2} = \frac{\sin T\omega/2}{T\omega/2} . \qquad (1.17)$$

The first zeros of the sinc function on either side of its central maximum are at:

$$\frac{T\omega}{2} = \pm\pi . \qquad (1.18)$$

If one assumes that two peaks in the spectrum can be separated if they are further apart than the zero is from the maximum one obtains a resolution of:

$$\Delta_\omega = \frac{2\pi}{T} . \qquad (1.19)$$

However, this does not take into account that the full complex-valued Fourier transform is not normally used to identify oscillation signals. Instead, the complex phase is normally ignored and only the power is considered:

$$P(\omega) \equiv F(\omega)F^*(\omega) \qquad (1.20)$$

where the $*$ indicates taking the complex conjugate. Ignoring the complex phase is loss of half the information, which effectively means a loss in resolution by the same factor, i.e.:

$$\Delta_\omega = \frac{4\pi}{T} . \qquad (1.21)$$

This result is somewhat dependent on the phase difference between the two closely spaced signals at some relevant reference time. One might be lucky enough to separate signals with a spacing between 2 and 4 times π/T if the phase difference was favourable. Generally however, a time series in which two or more signals are present that are more closely spaced than the resolution of Eq. (1.21) will, in the Fourier domain, not show multiple peaks but instead single or deformed peaks or no peaks at all. This also implies that any signal with a frequency that is below this resolution cannot be detected with any confidence, since it is indistinguishable from a constant offset.

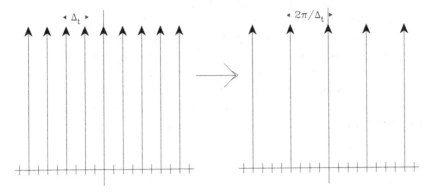

Fig. 1.1 Regular sampling with cadence Δ_t can be represented as multiplying the signal by delta functions with spacing Δ_t (left-hand panel). In the Fourier domain this corresponds to a convolution with a comb of delta functions spaced at frequency interval $\Delta_\omega = 2\pi/\Delta_t$ (right-hand panel).

Signals with a frequency higher than the Nyquist frequency are also not sampled properly and cause spurious peaks within the frequency range $[0, \nu_{\text{Nyq}}]$. This phenomenon is referred to as aliasing. Aliasing can be understood by considering the time series as the product of a continuous time series and a function that is a regularly spaced comb of delta functions, with spacing Δt, which is 0 in between. In the Fourier domain the true spectrum is convolved with the Fourier transform of this comb, which is a comb of delta functions at frequencies which are at all even multiples of the Nyquist frequency (Fig. 1.1). This convolution 'folds' power from outside the interval $[0, \pi/\Delta_t]$ back into this interval.

The Fourier transform of discretely and equidistantly sampled data (discrete Fourier transform or DFT) is achieved most conveniently using what is known as fast Fourier transform algorithms (FFT). FFT algorithms have a particularly advantageous operations count which increases slowly, $O(N \log N)$, with the number of samples N. The primary limitation of standard FFT algorithms (cf. [Press et al. (1992)]) is that they require N to be an integer power of 2. If the number of actual measurements does not meet that requirement, normally the solution is to symmetrically pad out the time series with the mean of the time series or with 0's before the first true measurement and after the final true measurement. The overhead incurred is easily compensated for by the increase in speed compared to other

DFT algorithms. An implementation of an FFT routine can be found in [Press et al. (1992)].

In practice there is a problem, which is that of 'missing data'. Because of technical problems, or poor observing conditions, data can be lost either pointwise or in blocks, which are distributed randomly within the time series. For ground-based telescopes there is also the problem that the Sun can only be observed when it is well above the horizon. This introduces regular gaps in the time series which produces further problems in determining the properties of multiperiodic time series. Together these properties of the data sampling are known as the 'window function'.

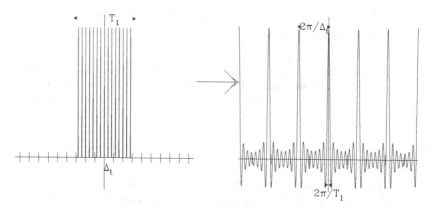

Fig. 1.2 Regular sampling with cadence Δ_t over a length of time T_1 can be represented as multiplying the signal by delta functions with spacing Δ_t and further by a tophat function with width T_1 (left-hand panel). In the Fourier domain this corresponds to a convolution with a comb of sinc-functions spaced at frequency interval $\Delta_\omega = \pi/\Delta_t$. The first zeros of each of these sinc functions are placed at $\pm 2\pi/T_1$ away from the main peak (right-hand panel).

For a finite sequence of discrete samples, in the time domain the window function can be represented as the product of a comb of delta functions spaced at Δ_t, and a broad tophat with length T_1. In the Fourier domain the window function becomes the convolution of a sinc function with the modulated set of peaks obtained before (Fig. 1.2).

The window function describing regular blocks of observations with length T_1, with regular gaps in between so that the blocks are repeated every $T_2 > T_1$, is a convolution in the time domain of the previous window function with a comb of delta functions with spacing T_2. A convolution in the time domain corresponds to a product in the Fourier domain. Therefore in the Fourier domain the window function is represented by a set of delta functions spaced at $2\pi/T_2$ and with an envelope described by the window function shown in figure 1.2. If one finally takes into account that this series of blocks is finite in length with duration $T_3 > T_2$, then in the Fourier domain each of the delta function peaks is again replaced by a sinc function with its first zeros at $\pm 2\pi/T_3$ (Fig. 1.3)

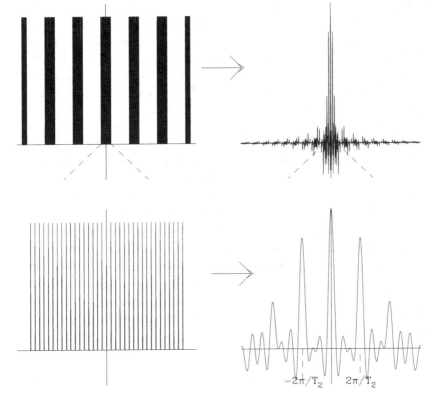

Fig. 1.3 Regular sampling with cadence Δ_t in blocks of length T_1, which are repeated every T_2 over a total length of time T_3. The above figure corresponds e.g. to observing every 15 min. during 9 hours, and repeating this every 24 hours for 6 days. The top left-hand panel shows the window function in the time domain, and the top right-hand panel show the window function in the Fourier domain over the interval $[-\omega_{\text{Nyq}}, \omega_{\text{Nyq}}]$. The bottom left-hand and right-hand panels show blow-ups of the window function in time- and Fourier-domain respectively, where the first main sidelobes in the window function at $\pm 2\pi/T_2$ are clearly identifiable.

Figures 1.1–1.3 give the amplitude in the Fourier domain of a single oscillation signal, for the various sampling scenarios. Evidently a time series in which there are multiple signals present, sampled in the same way, produces a very complicated Fourier spectrum, from which it is extremely hard to extract which frequencies correspond to the true oscillation signals and which are sidelobes generated by the sampling, even in the absence of measurement noise. A number of techniques have therefore been developed to attempt to alleviate the adverse effect of such missing data. A more general discussion of several methods in use in astronomy can be found in [Adorf (1995)] and [Vio et al. (2000)]. Methods designed specifically with helioseismology in mind can be found in [Brown & Christensen-Dalsgaard (1990)] and [Fossat et al. (1999)]. An application of the former to GONG data can be found in [Anderson (1993)], and of the latter to data from the helioseismic telescope network International Research of the Interior of the Sun (IRIS) in [Fierry Fraillon & Appourchaux (2001)].

The method of [Brown & Christensen-Dalsgaard (1990)] uses autoregressive (AR) techniques (see also [Fahlman & Ulrych (1982)]) to interpolate the signal into the gaps, using the measured data after band-pass filtering to reduce its complexity. The latter is necessary because there are very many oscillations present in the signal together with noise, which would require very many coefficients for the AR method. Robust determination of a large number of AR coefficients requires a large number of existing samples, which may not be available. The starting assumption is that the time series is regularly sampled, with missing data, so that for all existing measurements there is an i so that the time of measurement $t_i = t_0 + i\Delta_t$. Within the Nyquist frequency range $[-\omega_{\text{Nyq}}, \omega_{\text{Nyq}}]$ a number of selected band pass filters are applied, and for each band pass with central frequency ω_l the filtered signal y_s is:

$$y_s(t) = \sum_{k=-K}^{K} e^{i\omega_k t} \left[W(\omega_k - \omega_s) \sum_{i=0}^{N} e^{-i\omega_k t_i} y(t_i) \right] \quad (1.22)$$

in which the ω_k satisfy:

$$\omega_k = \frac{k}{2K\Delta_t}. \quad (1.23)$$

The band-pass function W is symmetric and has its maximum at 0 frequency. In order to minimise the influence of the data gaps the inverse Fourier transform w of W needs to have *compact support*, ie. w must be identical to zero outside a given time interval. [Brown & Christensen-

Dalsgaard (1990)] suggest:

$$w(t) = \begin{cases} \frac{1}{2}\left(1 + \cos\frac{\pi t}{m\Delta_t}\right) & \text{for } |t| < m\Delta_t \\ 0 & \text{for } |t| > m\Delta_t \end{cases} . \quad (1.24)$$

Eq. (1.22) expresses a product and shift in the Fourier domain. The same operation can also be expressed in the time domain as a convolution:

$$y_s(t) = e^{i\omega_s t} \sum_{i=0}^{N} y(t_i) w(t - t_i) e^{-i\omega_s t_i} . \quad (1.25)$$

This is computationally efficient because the domain over which w is not equal to 0 is typically small. Using these data the AR coefficients α_j are determined by assuming that the filtered time series satisfies:

$$y_s(t) = \sum_{j=1}^{L} \alpha_j y_s(t - j\Delta_t) + e_t^f$$

$$y_s(t) = \sum_{j=1}^{L} \alpha_j y_s(t + j\Delta_t) + e_t^b \quad (1.26)$$

in which the e^f and e^b are forward and backward prediction errors respectively. The number L is chosen small enough that the coefficients can be determined for any contiguous section of actually measured data: ie. if the length of the smallest block is N' then $L < N'$. The α_j are determined by using a standard least-squares algorithm on Eq. (1.26). With these coefficients the data can then be extrapolated from the measured parts of the time series into the gaps. [Brown & Christensen-Dalsgaard (1990)] also suggest treating these extrapolated values as 'measured' values and iterating to improve the predictor: updating the AR coefficients and hence the extrapolated values of y. If the number of 'missing' samples is as large or larger than the number of measured samples one would not expect such an iteration to converge at all. Even if the number of missing samples is small one should limit the number of such iterations to a few since after a few cycles very little further benefit should be expected.

A slightly different approach, also relying on the autocorrelation of the time series decreasing slowly, is described in [Stoica et al. (2000)]. This has the same limitation that it can correct for smaller gaps in long time series.

The method of [Fossat et al. (1999)], referred to as the method of 'repetitive music', uses the property of solar oscillations that the autocorrelation of the signal has secondary peaks at 4 h and 8 h. One can understand that this is so by considering the autocorrelation in the Fourier domain. The correlation of two functions of time is itself a function of time:

$$\mathrm{Corr}(g,h)(t) = \int \mathrm{d}t'\, g(t+t')h(t')\,. \tag{1.27}$$

The Fourier transform of this is:

$$\mathcal{F}\left[\mathrm{Corr}(g,h)\right] = G(\omega)H^*(\omega) \tag{1.28}$$

and therefore for the auto-correlation ($g = h$) it holds that its Fourier transform is simply the power:

$$\mathcal{F}\left[\mathrm{Corr}(g,g)\right] = |G(\omega)|^2\,. \tag{1.29}$$

The power spectrum $|P(\omega)|^2$ of solar oscillations has a regular comb-like structure, which is mostly due to the fact that the Sun is (nearly) spherically symmetric (cf. section 2.8). From the discussion above a regular comb in the Fourier domain corresponds to a comb in the time domain spaced by $2\pi/\Delta_\omega$ which in the case of the Sun comes out as $\sim$ 4h. This implies that the time series is repeated (almost) exactly every $\sim$ 4h. This fact can then be used to fill in gaps by using the measured time series of 4 hours earlier or later. Evidently the length of gaps that can be filled in this way is less than 8 hours, and it requires the length of the measured data segments to be as large as this.

Helioseismic networks such as IRIS, BiSON and GONG consist of a number of telescopes that are placed at different longitudes on the Earth. Ideally the instruments and telescopes are identical for each of the sites so that the data gathered at each telescope can be combined with all the other telescopes without extensive corrections or cross-calibrations. Also, to keep cost down as well as to obtain a uniform data quality, the operation of the telescopes must be automated as much as possible. The location of each of the telescopes is chosen to try to minimise the loss of data due to poor observing conditions. The distribution over longitude of the observing sites should be relatively uniform, so that the Sun is always sufficiently high in the sky for at least two sites to observe simultaneously, which is necessary to merge the data gathered at the sites. The cost of operation limits the number of sites, so that most networks operate some 6 telescopes. For helioseismic networks such as IRIS, BiSON and GONG this is sufficient to monitor the Sun continuously with a loss of data that is below 20 % over decades. The gap filling techniques discussed above generally suffice to compensate for this loss and to provide frequencies with relative errors in the ppm range or even smaller.

Finally one can note that the estimation of a power spectrum from measured data can also be considered as an inverse problem, and therefore techniques for solving inverse problems can be applied. The discussion of inverse methods is taken up in chapter 3 of this book. The particular case of spectrum estimation from irregularly sampled time series is revisited in the chapter on asteroseismology where this is particularly relevant. Some discussion of inverse methods and other power spectrum estimation methods, such as the Lomb-Scargle periodogram technique (cf. [Lomb (1976)], [Scargle (1982)]) can also be found in [Press et al. (1992)].

1.4 Mode Ambiguities and Cross-Talk

In order to do helioseismology one needs not only the frequencies of the modes but also the structure of node lines over the solar surface. As is demonstrated in section 2.6 the oscillations of the Sun can be considered as a linear superposition of eigenmodes, and associated with each of these is a pattern of node lines over the surface which can be classified in terms of spherical harmonic functions of the co-latitude θ and longitude ϕ:

$$Y_l^m(\theta, \phi) \equiv P_l^m(\cos\theta) e^{im\phi} \qquad (1.30)$$

in which the P_l^m are associated Legendre functions, the properties of which can be found in any number of textbooks on special functions (cf. [Gradshteyn & Ryzhik (1994)]). The spherical harmonic degree $l \geq 0$ and the order m, where $|m|$ has to be $\leq l$, uniquely specify the node line pattern over the surface. For each measured frequency one needs to be able to assign the values of l and m and also the radial mode order n before that frequency can be used to infer solar interior properties.

To perform the identification of modes in terms of l and m several paths can be pursued. One can design the detectors to cover the solar surface in a particular pattern, as is done for the VIRGO-LOI instrument for which the detectors are arranged in such a way as to optimise, as best as possible within the constraints of the instrumentation, the detection of in particular low-l modes. On the other hand one can also use high-resolution images of the solar surface and attempt to decompose the images in terms of the $Y_l^m(\theta, \phi)$ functions.

In either case the intention is to apply filters to the data in order to

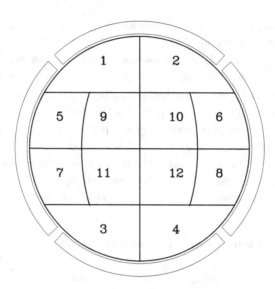

Fig. 1.4 The detector for the VIRGO-LOI experiment, with 12 central 'pixels' designed to be sensitive to modes with $l < 6$, plus an annulus of 4 detectors used for centering on the solar disc (guidance). (figure based on [Appourchaux et al. (1997)]).

isolate signal from every (l, m) pair over a certain range of l and m. An application to the VIRGO instrument can be found in [Appourchaux & Andersen (1990)]. The appendix of that paper details the method which is valid for any imaging instrument. The paper deals with the rather large 'pixels' of the VIRGO instrument (fig. 1.4) but the formulation is sufficiently general that it holds for any pixel size and shape. For any one pixel i with domain D_i on the surface, the signal for a mode with unit amplitude and with spherical harmonic degree l and order m is:

$$S_i(l, m) = \int_{D_i} d\cos\theta \, d\phi \, I(\mu) Z_l^m(\theta, \phi) \mu^q \qquad (1.31)$$

in which μ is the cosine of the angle between the line of sight and the normal to the solar surface at the centre of the area covered by the pixel. The exponent $q = 1$ for observations in intensity, whereas $q = 2$ for observations in velocity because of measuring only the projection of the velocity onto the line of sight. The function Z_l^m depends on whether the mode is a p-mode or g-mode: modes for which gas pressure is the dominant restoring force in the perturbation are referred to as p-modes, modes for which buoyancy is the dominant restoring force are referred to g-modes. A further discussion of these types of modes can be found in sections 2.6 and 2.7. For p-modes the

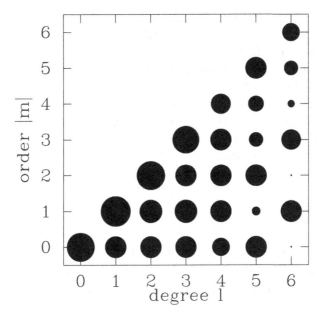

Fig. 1.5 The size of the symbols is proportional to the sensitivity of the VIRGO-LOI detector shown in Fig. 1.4. This is the sensitivity in intensity to a range of p-modes with low l, m. (figure based on [Appourchaux & Andersen (1990)]).

motion near the surface is predominantly radial which leads to the velocity or intensity being proportional to the displacement and therefore there is a simple expression for Z:

$$Z_l^m(\theta, \phi) = Y_l^m(\theta, \phi) \ . \qquad (1.32)$$

For g-modes there is still a proportionality to the Y_l^m but the full expression is more complex since the motion is predominantly horizontal near the surface (see section 2.6). This means that derivatives of the Y_l^m with respect to the angles θ, ϕ enter the expression for intensity fluctuations and the resulting sensitivity is higher near the solar limb than near the centre of the disc. [Appourchaux & Andersen (1990)] give expressions based on the calculations of [Berthomieu & Provost (1990)] for the intensity fluctuations due to g-modes. However, more recent work ([Toutain et al. (1999)]) takes sphericity effects into account in the radiative transfer, whereas [Berthomieu & Provost (1990)] was based on a plane parallel approximation. As one would expect there is little difference for the resulting intensity variations near the centre of the solar disc, but towards the limb there is a substantial difference, which implies that for g-modes the sensitivity calculations need revision. For the detection of g-modes in velocity a recent discussion can be found in [Wachter et al. (2003)].

In Eq. (1.31) I is the intensity distribution over the solar disc, commonly referred to as the limb darkening function . Its argument $\mu \equiv \sin\theta\cos\phi$ is the cosine of the angle between the normal to the surface at location (θ, ϕ) and the line of sight to the observer. The limb darkening function is usually presented as a power series in μ with given coefficients which depend on the wavelength of observation:

$$I(\mu) = \sum a_k(\lambda)\mu^k \ . \tag{1.33}$$

Since the basis functions μ^k are not orthogonal on the domain $[0,1]$, an experimental determination of the a_k is not robust: if more terms are added to the series, all the a_k change value. It would therefore probably be better to use polynomials in μ that are orthogonal (cf. [Meisner & Rast (2003)]) but coefficients for the above expansion can be found tabulated (cf. [Neckel & Heinz (1994)]).

The sensitivities shown graphically in Fig. 1.5 are calculated by summing over all pixels and taking the modulus of the complex quantity:

$$S(l,m) = \sqrt{\left[\sum_i S_i(l,m)\right]\left[\sum_i S_i^*(l,m)\right]} \ . \tag{1.34}$$

In order to construct an optimal mask for the detection of a mode with spherical harmonic degree and order (l,m) and with frequency ω, the sensitivities of the individual pixels can be combined in a weighted sum. [Appourchaux & Andersen (1990)] follow a suggestion of [Christensen-Dalsgaard (1984)] and define the following filter function F:

$$F(l,m,l_0,m_0) = F_+(l,m,l_0,m_0)\delta(\omega - \omega_{nlm})$$
$$+ F_-(l,m,l_0,m_0)\delta(\omega + \omega_{nlm}) \tag{1.35}$$

with

$$F_+(l,m,l_0,m_0) = \left|\sum_i a_i(l_0,m_0)S_i(l,m)\right|$$

$$F_-(l,m,l_0,m_0) = \left|\sum_i a_i(l_0,m_0)S_i(l,m)^*\right| \ . \tag{1.36}$$

The weights a_i are set by minimising:

$$\sum_{l,m} [1 - \delta_{l\,l_0} \delta_{m\,m_0}] F_+^2 \qquad (1.37)$$

in which the δ is the Kronecker delta. The problem of designing an optimal filter can in fact be formulated as an inverse problem, further discussion of which can be found in section 3.9. The above technique for obtaining optimal weights is a version of the Backus-Gilbert method ([Backus & Gilbert (1968)], [Backus & Gilbert (1970)]). The minimisation of Eq. (1.37) produces the appropriate weights for the filter F to match as closely as possible a the product of Kronecker deltas for l and l_0 and for m and m_0:

$$F^{\mathrm{opt}} \approx \delta_{l\,l_0} \delta_{m\,m_0} . \qquad (1.38)$$

This match can never be perfect however. The problem lies that the spherical harmonic functions Y_l^m are only orthogonal when uniformly weighted over the whole sphere. Only half the sphere is observed at any time, and over this the Y_l^m are not orthogonal. The presence of the limb-darkening function in Eq. (1.31) means that the weighting is not uniform which further enhances cross-product integrals between spherical harmonics of different degree and/or order. This has the consequence that even for the optimal filter F^{opt} is not identical to 0 for $l \neq l_0$ and $m \neq m_0$: some signal 'leaks in' from modes with l and m values different from the target values. [Wachter et al. (2003)] also discuss the construction of optimal masks, specifically for g-mode measurements in velocity.

Other issues are encountered when the time series available becomes extremely long. Networks of telescopes have been operating for some decades and also the SoHO satellite has been in orbit long enough to produce time series with lengths in excess of 2000 days, corresponding to some $\sim 3\ 10^6$ samples. The frequency resolution of such a time series is far higher than the natural width of most of the oscillations, which is set by the fact that the modes are damped and their excitation mechanism is stochastic (see next section). To make optimal use of such a data-set [Korzennik (2005)] adapted a multitaper method, and combined both spatial and temporal filtering to obtain per (n, l) multiplet all frequencies simultaneously. As is shown in section 3.3 the frequencies are not dependent on m for spherically symmetric stars, and therefore the m dependence measures departures from spherical symmetry such as caused by rotation. Since in the Sun the departure from spherical symmetry is small, the $(n, l, 0)$ frequency is flanked symmetrically by the frequencies for $(nl, -m)$ and $(n, l, +m)$ in an identifiable multiplet structure. If a measured time series for a set of pixels has been filtered to isolate, albeit imperfectly, the signal $z_{l\,m}(t_i)$ for an l, m-pair, the multitaper filter produces a power spectrum $P(\omega)$:

$$P_{l,m}(\omega; N) = \sum_{k=1}^{N} \mathcal{F}\left(z_{l,m}(t_i) \sin \frac{\pi k t_i}{(M+1)\Delta_t}\right)$$

$$\times \mathcal{F}^*\left(z_{l,m}(t_i) \sin \frac{\pi k t_i}{(M+1)\Delta_t}\right) \quad (1.39)$$

where N is the number of components of the multitaper, and $\mathcal{F}$ indicates the Fourier transform with $\mathcal{F}*$ its complex conjugate. For each successive component $\sin(\pi k t_i/(M+1)\Delta_t)$ of the filter, the multiplication in the time domain corresponds to a convolution or smoothing in the Fourier domain over an increasing width in frequency Δ_ω. This choice of tapers of [Korzennik (2005)] is orthogonal so that the summation for $N \to \infty$ would reproduce the Fourier transform of the untapered time series. In this sense the transformation can also be thought of as a (truncated) discrete wavelet transform.

The summation is stopped at a finite N, such that the final Δ_ω resolves the typical linewidth Γ by a factor of a few. Because of realisation noise due to the stochastic excitation, the full Fourier transform would show a very spiky structure with an envelope given by the damping time of the mode (see next section). The taper smoothes away the spikes and thus achieves an improved recovery of the envelope, which is the quantity of interest. The subsequent step of [Korzennik (2005)] is then to fit to this power spectrum a set of oscillation line profiles, with a given functional form, for all modes with the same n and l but different m taking into account that the spatial filters are imperfect ('spatial leaks') at isolating just a single (l, m) pair, and also including some modes that have different but similar l and m for which the frequency would be located within the frequency window selected for the fitting. For further details the reader is referred to [Korzennik (2005)], and also to [Komm et al. (1999)] in which multitaper analysis is combined with de-noising to improve the extraction of frequencies from time series.

1.5 Line Profile Asymmetries

In the Fourier domain the peaks of the spectrum corresponding to the normal modes of oscillation have a finite width, due to the fact that they are damped modes that are excited stochastically by convection. For a single damped mode the velocity or intensity would behave as:

$$\xi(t) = Ae^{(i\omega - \Gamma)t + i\delta} . \qquad (1.40)$$

With a damping rate Γ. The stochastic excitation means that the amplitude A and phase δ are 'reset' at time intervals following the stochastic properties of the convective excitation. To lowest order there is no correlation between the phase of the oscillation, and the likelihood of excitation, therefore the phase δ is uniformly distributed between 0 and 2π. The amplitude A follows a declining probability distribution function. A Fourier transform of ξ will reflect the stochasticity of the signal. Over long enough times that many excitation events have taken place, the Fourier transform of ξ is the convolution of of $e^{(i\omega_0 - \Gamma)t}$ and of the stochastic variable $Ae^{i\delta}$. The Fourier transform of $e^{(i\omega_0 - \Gamma)t}$ is a Lorentzian:

$$\frac{1/4}{(\omega - \omega_0)^2 + \Gamma^2} . \qquad (1.41)$$

The product of this with the Fourier transform of the stochastic amplitude determines the final shape of profile of the oscillation in the Fourier domain. The Fourier transform of $Ae^{i\delta}$ depends somewhat on the excitation mechanism but it is always a slowly varying function of ω, so the final profile is very similar to this Lorentzian. Frequencies of oscillations could therefore be determined by fitting Lorentzians to the Fourier transform of the time series. However, the 'spectral lines' of the oscillation show asymmetries which are opposite in sign, depending on whether the time series originates from velocity measurements or from intensity measurements ([Duvall et al. (1993)]) (see Fig. 1.6). This asymmetry must be taken into account when carrying out the fitting, in order not to bias the frequency determination (cf. [Rhodes et al. (1997)]).

The Lorentzian clearly is not asymmetric, so the asymmetry must come from the properties of the probability distribution function of $Ae^{i\delta}$ in velocity and intensity respectively. From physical arguments one can argue that the excitation source of the oscillations is located in a thin layer just below the photosphere, and also that one source of noise, the solar granulation, is correlated with the oscillations. More details of the physical picture, and several treatments of this ([Roxburgh & Vorontsov (1995)], [Rosenthal (1998)], [Nigam & Kosovichev (1998)]) are discussed in section 2.9. Based in this physical picture [Nigam & Kosovichev (1998)] provide a fitting function with a few free parameters, which can be used for asym-

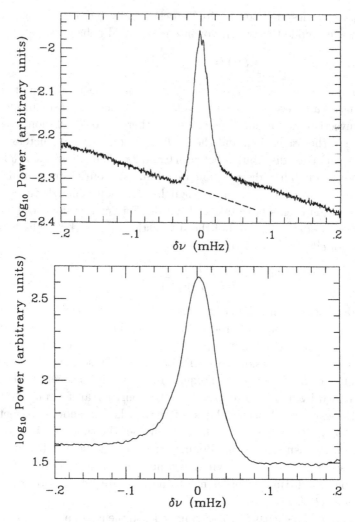

Fig. 1.6 The profile of the acoustic power around the resonant frequency of the oscillation, for $n = 2$ and averaged over $l = 157 - 221$ measured in intensity and in velocity (figure from [Duvall et al. (1993)] reproduced by permission of the AAS).

metric profile fitting. This improves both accuracy and convergence of the fitting and therefore of the fitting parameters. [Nigam & Kosovichev (1998)] provide both a full expression and an approximation which is valid near the resonance frequency ω_{nlm}, ie. for small values of the argument

$x \equiv 2(\omega - \omega_{nlm})/\Gamma$. This approximate fitting function for the power P is:

$$P(x) = A^2 \frac{(1+Bx)^2 + B^2}{1+x^2} + N_{\text{unc}}^2 . \qquad (1.42)$$

The parameters that are fit for are ω_{nlm}, Γ, A, and B. The first two are the real and imaginary components of the complex frequency: i.e. the cyclic frequency and the damping. Within the context of the model of [Nigam & Kosovichev (1998)] the parameter A is interpreted in terms of the location r_s of the source which is assumed infinitely thin, and the amplitude a_ω of the mode. The parameter B also depends on r_s and more importantly, on the ratio of the power in the correlated noise N_{cor}^2 and the power in the mode a_ω^2. This ratio controls the sign of B and therefore the sense of the asymmetry of the profile. It is this ratio which is different in intensity and in velocity and which explains the different sense of the asymmetry between the two. This shape is also the one adopted by [Korzennik (2005)] although the parameters occur in slightly different combinations.

1.6 Oscillations at High Frequencies

The previous sections are all concerned with the global resonant modes of the Sun at low and intermediate degree. These oscillations have sufficiently large wavelength near the surface that their reflection can be regarded as near-perfect. Figure 1.7 shows the acoustic power in grey-scale as a function of l and of frequency ν. At intermediate values of l and of frequency, this power is concentrated in very clearly defined ridges of high power. For sound waves with very short vertical and/or horizontal wavelength this is no longer true. At high resolution, images of the Sun show fine structure on a variety of horizontal spatial scales. Also, from observations at UV and X-ray wavelengths it is known that the density and pressure of the Sun do not disappear outside the photospheric radius. The Sun is surrounded by a chromosphere and corona. The result of this is that the pressure scale height goes through a minimum as a function of radius. In the differential equation describing the propagation of waves (cf. section 2.6 ff.) this gives rise to a term known as the 'acoustic cut-off' frequency.

Acoustic waves with a wavelength that is of the order of the minimum pressure scale height, or equivalently with a frequency that approaches the acoustic cut-off frequency, will no longer be reflected perfectly at the surface. A finite fraction of the energy carried by these waves can be transmit-

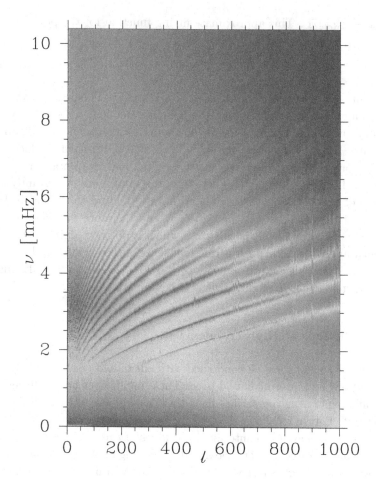

Fig. 1.7 The acoustic power shown as a grey-scale, as a function of degree l and frequency ν, determined from MDI data (figure modified from http://soi.stanford.edu/results/).

ted into the chromosphere and corona in a manner similar to what is known as 'tunnelling' in quantum mechanical settings. The rapid decrease in density in the solar atmosphere will then cause an rapid increase in velocity amplitude, with the result that the waves deform into shocks and radiatively dissipate their energy. Because of the non-linearities, the behaviour of the waves in the chromosphere requires magnetohydrodynamical simulations (cf. [Carlsson & Stein (1999)], [Rosenthal et al. (2002)]). A discussion of these modes within the solar interior, as a high-frequency extension of the resonant modes, can be found in [Roxburgh & Vorontsov (1995)].

As the frequency of acoustic waves increases further to the acoustic cut-off frequency and beyond there is progressively less reflection or no reflection at all. At this point these waves are not proper resonant modes. The acoustic power as a function of frequency does still have structure, and therefore diagnostic potential. The stochastic excitation source of the waves is located near the top of the convection zone where the overturning velocities of the convection are close to the sound speed. This acoustic source, often modelled as being an infinitely thin spherical shell, emits acoustic waves both inward toward the centre and outward towards the photosphere. For purely radial modes the waves emitted towards the centre can be considered as being reflected at $r = 0$ and therefore also travel outward past the source to the photosphere, but with a phase difference compared to the waves emitted directly towards the photosphere. It is then essentially an application of Huygens' superposition principle to see that more power will be detected near those frequencies for which this phase difference ϕ is an integer multiple of 2π. Ignoring effects of sphericity one would obtain for the frequencies of these 'pseudo-modes':

$$2\pi n + \alpha = \omega_n \, 2 \int_0^{r_s} \mathrm{d}r \, \frac{1}{c(r)} \tag{1.43}$$

where $c(r)$ is the sound speed as a function of radius, and α accounts for the phase jump due to the reflection at the centre. With $c(r)$ already determined from the oscillation modes below the acoustic cutoff, the locations ω_n of the peaks of power are a measure of the location r_s of the source. For non-radial modes, emitted at an angle with the normal to the surface, a similar relationship must hold since the same superposition principle applies. In a diagram showing oscillation power as a function of horizontal wavenumber and of frequency, each value of n produces a separate ridge. The above relationship applies to each l and therefore the structure produced by the resonant modes at low and intermediate degrees will persist even in the high frequency regime. A more detailed discussion of the relationship between the location of the acoustic source and the frequencies of the pseudo-modes can be found in eg. [Kumar & Lu (1991)] and [Jefferies et al. (1994)].

Chapter 2

Global Helioseismology: Modelling

In order to use the measured properties of the resonant modes of oscillation of the Sun as diagnostics of its interior one needs also to solve the *forward problem*, which is to determine from a given model for the Sun the frequencies of its resonant modes of oscillation. From the theory of stellar structure, and using perturbation analysis, it is possible to derive a wave equation for the plasma which constitutes the Sun. Appropriately chosen boundary and periodicity conditions then allow for the determination of a discrete spectrum of resonant frequencies: normal modes of oscillation.

The modern theory of stellar structure finds its origins around the turn of the previous century. Evidently the field has seen many advances during this century and the field is mature. Therefore there exist a number of good textbooks that deal with the equations governing models for the structure and evolution of stars (eg. [Clayton (1983)], [Kippenhahn & Weigert (1994)]). Similarly there are textbooks discussing the observation of stars: techniques for measurement of their surface properties through photometry and spectroscopy of the light emitted at ranges ranging from X-ray and ultraviolet through to radio wavelengths (cf. [Mihalas (1978)], [Gray (1992)]).

The particular needs in terms of modelling for the purposes of helio- and asteroseismology are such that it is nevertheless useful to discuss in some detail how modelling is to proceed to satisfy these needs. Without covering all the ground that these other textbooks do, and referring to them for further details, an overview of the physics and the numerical methods is therefore necessary.

2.1 Separation of Time Scales

In the following sections 2.2 and 2.3 the structure and evolution of the Sun are discussed. The discussion of oscillations follows in section 2.6. The physical reason for it to be possible to make this separation is that the time scales over which the Sun adjusts respectively to dynamical changes, thermal changes, and composition changes are very different.

The time scale over which the Sun adjusts to dynamical changes can be estimated to order of magnitude as the solar radius divided by its surface escape velocity:

$$t_{\rm dyn} = \frac{R}{V_{\rm esc}} = \sqrt{\frac{R^3}{GM}} \ . \qquad (2.1)$$

For the Sun $t_{\rm dyn} \approx 30$ min. This is the time scale over which the Sun re-establishes equilibrium if it were to be deformed. Solar oscillations involve displacing material from an equilibrium position which is a deformation. It is no surprise then that the oscillation periods are around 5 min. The radial fundamental mode has a period that is closer to the global dynamical time-scale. The 5 min period modes are of higher order, and hence with shorter length scales which, for modes dominated by pressure effects, reduces their period.

The next important time scale is the time over which the Sun adjusts to thermal changes. It is equal to the time it takes for the total thermal content of the Sun to be radiated away. Because of the virial theorem the thermal energy content and the gravitational potential energy are nearly equal and therefore the thermal time scale can be estimated as:

$$t_{\rm KH} = \frac{GM^2}{RL} \ . \qquad (2.2)$$

This time scale is also known as the Kelvin-Helmholtz time scale, hence the subscript KH. For the Sun this time scale is $\sim 3 \ 10^7$ yr. The ratio of this time scale to the previous is thus $\sim 10^{12}$. This clear separation in time scales implies that for stellar structure calculations the oscillations can be ignored since their effect is averaged out. Conversely the oscillations can be considered under the assumption that the star as a whole is in thermal equilibrium.

There are two effects that change the chemical composition of the Sun. There are diffusive processes, possibly counteracted by turbulent convection or differential radiative pressure on different species. These do not change the overall composition but they do change the distribution of chemical elements within the Sun. The diffusion/settling of heavier elements with respect to Hydrogen is controlled by microphysical plasma transport processes that also determine the diffusive transport of energy. However, there is one important difference between these diffusive processes. Energy transport is mediated by diffusion of photons and therefore the time scale is set by the mean free path of photons. The relative diffusion of elements is set by a much shorter mean free path of atoms. The time scale over which diffusion operates is therefore larger than t_{KH} by a factor of roughly $\sim 10-100$ so that after a few times 10^9 yr effects of diffusion are noticeable in solar models (cf. [Christensen-Dalsgaard (2002)]).

The process that does change the overall composition of the Sun is nuclear fusion, which operates only in the solar core where densities and temperatures are sufficiently high. Outside of the core the conditions very quickly become inadequate for fusion to occur at an appreciable rate. The time scale over which the core composition changes depends very much on the dominant element that is involved in the fusion. The Sun is now in the longest single stage of its evolution, fusing Hydrogen into Helium, which leads to a lifetime of $t_{\text{nuc}} \sim 10^{10}$ years over which half of the nuclear fuel is spent. This is sufficiently long compared to the thermal time scale that one can assume the evolution process is quasi-static: it is a succession of thermal equilibria. As the Sun evolves, the core Hydrogen will eventually be exhausted and the Sun will then need to adjust to the change in energy supply, first because of 'the switching off' of H fusion, and then due to the 'switching on' of He fusion. The stage of core He fusion lasts less long than core H fusion. However it does still last quite long, e.g. some $7\ 10^7$yr for a $3\ M_\odot$ (cf. [Clayton (1983)]). This is in part because the temperatures and densities are sufficiently high just outside the core that H fusion can occur in a shell, simultaneously with the core He fusion. This in fact supplies most of the energy radiated by the star. The nuclear time scale for subsequent elements is progressively shorter, in part because per reaction less energy is freed. One therefore expects that element diffusion is important particularly during the main stages of evolution when H and He are undergoing fusion in the core, but there is too little time for it to make a significant impact in subsequent stages.

In principle it is possible to derive the equations for stellar oscillations, stellar structure, and stellar evolution from first principles (ie. the equations of hydrodynamics) using a formal procedure. One would explicitly introduce the small parameters t_{dyn}/t_{KH} and $t_{\text{dyn}}/t_{\text{nuc}}$ and use 'dominant balance' to separate out three separate sets of differential equations from the terms that are of the same order in these small parameters. This is a rather cumbersome procedure which does not provide additional physical insight, so the more classical approach is presented here.

2.2 Equations of Hydrostatic Structure

The Sun is a high temperature plasma and as such obeys the equations of (magneto-)hydrodynamics: partial differential equations expressing the conservation of mass, momentum, and energy. Although there exist numerical methods and indeed implemented computer codes to solve this set of equations in the form of initial-value problems, it is not very practical to apply those to the problem at hand. As argued in the previous section, there are many orders of magnitude between the time scale of stellar evolution and the time scale associated with dynamical motions such as oscillations. For reasons of speed it is useful to make a number of physically reasonable assumptions which help simplify the problem considerably and speed up computation as well. For the purposes of helioseismology the time-evolution of the Sun is very slow, and at any given epoch during which one measures for instance oscillation frequencies, the Sun is in hydrostatic equilibrium. The equation of conservation of mass is thereby eliminated altogether. The equation of motion reduces to an equation describing the balance of forces:

$$\nabla p = -\rho \nabla \Psi \tag{2.3}$$

in which p and ρ are the gas pressure and density respectively, and Ψ the gravitational potential. The gravitational potential is calculated from the material density using the Poisson equation:

$$\nabla^2 \Psi = 4\pi G \rho \tag{2.4}$$

in which G is the gravitational constant. The previous section also indicates that the Sun is in thermal equilibrium over the time scales relevant to helioseismology. This eliminates the time dependence from the equation of conservation of energy which then becomes:

$$\nabla \cdot F = \rho \epsilon \ . \tag{2.5}$$

Here F is the flux of energy and ϵ is the energy generation rate per unit mass which is non-zero essentially only in the solar core where nuclear fusion takes place. The above equations cannot be solved yet at this point, because the interrelations between ϵ, F, p and ρ have not yet been specified. To do this, astrophysics relies on other branches of physics for input.

The energy generation rate ϵ depends on ρ and on the temperature T of the plasma, and also on the chemical composition: Hydrogen abundance X, Helium abundance Y, and also to some extent the abundance of the heavier elements X_n ($n > 2$). By definition $\sum_{n=2,..} X_n \equiv Z = 1 - X - Y$. A combination of experimental data and theoretical predictions from nuclear physics provides $\epsilon(\rho, T, X_n)$ in the form of tables or parameterised functions. For details, see for instance the European Nuclear Astrophysics Compilation of Reaction Rates (NACRE), which has a web-site at http://pntpm.ulb.ac.be/Nacre/nacre_d.htm (see also [Angulo et al. (1997)], [Adelberger et al. (1998)]).

At this point three thermodynamic variables ρ, p and T have been introduced. Given a chemical composition of the plasma, only two such variables are strictly necessary to describe its thermodynamic state. The third is related to the other two through an equation of state (EOS). Plasma in the solar interior behaves in a way that is very similarly to an ideal gas, except in certain regions where the dominant ionisation stage of a major chemical constituent (H, He) changes. However, even slight departures of the EOS from that of an ideal gas are detectable through helioseismology, so a more sophisticated treatment is called for. This implies calculating particle energy distribution functions for dense, partially ionised plasmas, for a range of astrophysically interesting chemical compositions and temperatures and densities. Furthermore it is necessary to include the pressure of photons, which is not negligible for stellar applications. One widely used calculation of this is the Mihalas-Hummer-Däppen (MHD) equation of state (EOS), the latest update of which is described in [Nayfonov et al. (1999)]. For astrophysics, EOS calculations provide T, ρ or p in terms of the other two variables, for a range of relevant chemical compositions, either in the form of tables or in the form of parameterised functions. Normally derivatives such as the adiabatic exponents Γ_1 and Γ_2 are also explicitly provided:

$$\Gamma_1 \equiv \left(\frac{\partial \ln p}{\partial \ln \rho}\right)_s$$

$$\frac{\Gamma_2 - 1}{\Gamma_2} \equiv \left(\frac{\partial \ln T}{\partial \ln p}\right)_s \equiv \nabla_{\mathrm{ad}}$$

where the subscript s refers to isentropic changes. The symbol ∇_{ad}, related to Γ_2 as shown above, is also referred to quite often. Sometimes the third adiabatic exponent is used, which is related to the first two:

$$\Gamma_3 - 1 \equiv \left(\frac{\partial \ln T}{\partial \ln \rho}\right)_s = \Gamma_1 \frac{\Gamma_2 - 1}{\Gamma_2}. \tag{2.7}$$

For consistency the calculations to determine the EOS are carried out in close collaboration with efforts to determine the opacity of stellar material κ, discussed next, because essentially the same physics needs to be addressed.

To specify the energy flux it is necessary to know what mechanism provides the transport of energy. Neutrinos, generated by fusion processes in the core, carry away some energy. There is so little interaction with matter between core and surface that the flux can be accounted for by reducing ϵ appropriately. In most of the solar interior the mean free path of photons is small compared to the length scales over which density or pressure vary. Photons therefore perform a random walk though the solar plasma. The consequence of this is that the energy flux carried by photons can be described to a high degree of accuracy by a diffusion equation which relates the flux to the temperature gradient:

$$F = -\frac{4acT^3}{3\kappa\rho} \nabla T \tag{2.8}$$

where c is the speed of light, and a is the radiation constant (see appendix C). κ is the opacity of stellar material quantifying the absorption and scattering of photons by the plasma. Only in the outer layers of the Sun does the diffusion approximation break down, in which case a more detailed radiative transfer procedure is required. For details the reader is referred to text books on radiative transfer such as [Mihalas (1978)] and [Gray (1992)]. κ depends on ρ, T and chemical composition, as does ϵ. A combination of experimental data and theoretical predictions from atomic physics provides κ in the form of tables. For details, see for instance the OPAL web-site http://www-phys.llnl.gov/Research/OPAL/ (see also [Iglesias & Rogers (1996)]), or the latest results of the OP project [Badnell *et al.* (2005)]. In addition to this radiative transport of energy, conduction of heat by electrons can also carry some energy. This plays a minor role throughout the solar interior, compared to the flux carried by photons. Heat conduction is a diffusive process which means that the flux carried depends again on the gradient of the temperature as in Eq. (2.8). The two processes can therefore be combined by defining an effective opacity κ_{eff} which takes conduction into account. In order to correspond with the formulation of the differential equations (2.3), (2.4), and (2.5), equation (2.8) is usually re-arranged to have the derivatives on the left-hand side:

$$\nabla T = -\frac{3\kappa_{\text{eff}}\rho F}{4acT^3}. \tag{2.9}$$

Radiation and conduction are not the only mechanism for transport of energy however, and in the outer ~ 25 % by radius of the Sun not even the dominant mechanism. In these layers the temperature gradient ∇T to carry the energy flux, required by solution of Eq. (2.9), is so large that the gas is unstable to overturning motions. Such layers in the Sun and stars are convective. When this occurs hydrostatic equilibrium remains satisfied, with a modified pressure, because there is no net mass flux through any level surface, where level surfaces are defined as surfaces over which $\Psi = $ const.. The pressure needs modification because associated with the overturning motions is an additional pressure, usually referred to as turbulent pressure. Entering the force balance is the gradient of the total pressure which has contributions from normal gas pressure, radiation, turbulence, and possibly magnetic fields as well in the outermost layers.

Any parcel of gas which is displaced adiabatically relative to its surroundings will continue to travel upward or downward, rather than experience a restoring force returning it to its original position. The condition for ∇T at which this transition occurs is known as the Schwarzschild criterion or, in the presence of gradients in the chemical composition, the Ledoux criterion. In textbooks such as [Clayton (1983)] or [Kippenhahn & Weigert (1994)] further details can be found concerning the classical treatment of convection in solar and stellar structure and evolution. For the purposes of solar modelling it is necessary to specify the temperature gradient that convection establishes, or at least, for 1-dimensional models, a mean ∇T over any level surface within a convection zone. Also, an expression for the turbulent pressure needs to be obtained in order to calculate hydrostatic equilibrium correctly. This turns out to be quite problematic. Why this is so can be seen by examining two dimensionless numbers that characterise the flow in convective layers (cf. [Landau & Lifshitz (1987)]). The first of these is the Prandtl number P which is the ratio of the kinematic viscosity ν_v of the plasma, and the heat transport coefficient, which in the case at hand is the term that appears in front of the ∇T term in Eq. (2.8) further divided by ρ and the specific heat at constant pressure c_p. The expression for P is thus:

$$P = \frac{3c_p\nu_v}{4ac}\kappa\frac{\rho^2}{T^3} \,. \tag{2.10}$$

In the solar interior ν_v is very small and therefore also the Prandtl number is very small ($< 10^{-9}$). This implies that motions in the gas once set up, take a long time to damp away viscously. The second dimensionless number is the Grashof number G (or the Rayleigh number $R = GP$):

$$G = \left(\frac{\chi_p k}{m_H \nu_v^2}\right) H_p^3 \nabla T \,. \tag{2.11}$$

Here m_H is the atomic mass unit, and χ_p is the logarithmic isobaric expansivity of the plasma:

$$\chi_p = -\left(\frac{\partial \ln \rho}{\partial \ln T}\right)_p \tag{2.12}$$

where $\chi_p = 1$ for an ideal gas. H_p is the typical length scale over which the mean pressure changes in the convection zone: the pressure scale height. The Grashof and Rayleigh numbers in the solar convection zone are very large ($> 10^{24}$ and $> 10^{15}$ resp.) which implies that there is very strong driving of motions through the convective instability. The combination of high R and low P implies that the motion within convective zones of the Sun (and stars) is highly turbulent, with a very large range of length scales over which overturning motions take place simultaneously.

In the Sun the major contribution to the viscosity is the redistribution on a microscopical scale of momentum through the emission, scattering and absorption of photons. This implies that the mean free path of photons is small compared to the range of scales over which turbulent motion is taking place. Away from the transition boundary between convectively stable and unstable regions, the turbulent motion is therefore to a good approximation adiabatic. The result of this is that only a very small amount of super-adiabaticity is required to set up sufficiently vigorous convection to transport all energy that is not carried by photons. The temperature gradient set up is essentially that of an adiabatically stratified layer:

$$\nabla T = \frac{\Gamma_2 - 1}{\Gamma_2}\frac{T}{p}\nabla p \,. \tag{2.13}$$

This holds throughout the bulk of the convection zone in the Sun, so within this zone Eq. (2.13) can replace Eq. (2.9). However, in the outermost layers, the mean free path of photons increases, and the convection becomes inefficient. In this case it is necessary to determine the relative efficiency of radiative and convective energy transport to be able to calculate the temperature gradient from the flux. One of the first approaches, still widespread in its use, is to replace the turbulent spectrum by a single mixing length scale which is characteristic or representative in terms of the energy flux carried by convection. This is the mixing length theory (MLT) (cf. [Böhm-Vitense (1958)]). One can argue that the mixing length is locally proportional to the pressure scale height $\lambda_{\mathrm{mix}} = \alpha H_p$ but the actual value of the mixing length parameter α is determined for the Sun by demanding that the radius of solar models correspond to the measured value. This is less than satisfactory because there is no reason to assume that the resulting value is valid for any other star. It would be preferable not to have this free parameter or at least to constrain it in some independent manner.

One path that is followed in modern approaches is to attempt to model the turbulence more accurately in terms of Reynolds' stresses (cf. [Canuto & Mazzitelli (1991)], and more recent developments in [Canuto & Dubovikov (1998)], and [Canuto & Minotti (2001)]). What is done is to derive transport equations for correlations between the stochastic fields of density, temperature and velocity variations. This allows one to calculate fluxes of energy and momentum, as well as other quantities of interest, to be determined by solving an extra set of differential equations together with the standard stellar structure equations. A problem with this approach is that each differential equation describing the transport of a particular correlation between n of these quantities involves source terms which are correlations between $n + 1$ (or more) quantities. In other words there is an infinite hierarchy of transport equations to solve. This is normally dealt with by postulating at some level that the higher-order correlations can be expressed algebraically in terms of lower-order correlations, which is referred to as *closure*. There is still some calibration necessary, in the form of these closure relationships for higher order correlations of the various quantities characterising the turbulent velocity field. However, the hope is that the choice of these parameters is more universally valid than a single mixing length, so that for instance in situ measurements in a laboratory or in the Earth's atmosphere can be used to establish such relationships, which are then to be applied universally .

Another path is to perform three-dimensional numerical simulations of convection. Although it is computationally intractable to tackle problems with relevant Rayleigh and Prandtl numbers, here the assumption is that by simulating convection over a feasible range, one can safely extrapolate to higher Rayleigh and lower Prandtl numbers. [Porter & Woodward (2000)] have explored such extrapolations numerically. Also there is significant effort in the modelling of convection in the Sun and stars (cf. [Stein & Nordlund (2000)]), also taking into account the effects of radiation transport and emergent flux (cf. [Asplund et al. (2000)]), rotation ([Brummell et al. (1996)] [Brummell et al. (1998)] [Miesch et al. (2000)] [Robinson & Chan (2001)] [Brun & Toomre (2002)]) and magnetic fields ([Tobias et al. (1998)], [Tobias et al. (2001)], [Brandenburg & Dobler (2002)]), and the behaviour near the edges of convection zones where moving elements can 'overshoot' into initially stable regions ([Brummell et al. (2002)]). These references are merely a selection from a very large body of research. Direct incorporation of such convection simulations in stellar evolution codes is not currently contemplated because the computing requirements are too severe. Instead, progress is likely to occur through parameterisation of the results for the ratio between convective and total energy transport in terms of global stellar parameters. One step in this direction, in which hydrodynamical simulations are used to derive a modified temperature stratification ∇'_{ad} is [Li et al. (2002)].

The differential equations (2.3), (2.4), (2.5), and (2.8) or (2.13), allow for the formulation of a boundary value problem to solve the interior structure of the Sun in terms of all thermodynamic variables. The chemical composition of the plasma needs to be specified as well as appropriate boundary conditions, which include specifying the total stellar mass and the mean radius and shape of the surface. The Sun is rotating sufficiently slowly, and has a sufficiently weak magnetic field that it is to a high degree spherically symmetric in structure, and can be modelled with only the radius as independent variable. For other stars it can be necessary to consider departures from spherical symmetry but this has so far been done relatively little (cf. [Deupree (2001)] [Maeder & Meynet (2003)] [Maeder & Meynet (2004)] [Maeder & Meynet (2005)] [Eggenberger et al. (2005)]). For a review up to 2000 see [Maeder & Meynet (2000)].

2.3 Evolution of the Sun and Stars

The previous section concerns itself with calculating the structure of the Sun and stars for a given chemical composition. However, this composition changes with time through the fusion processes that are responsible for the luminosity of the star, and also because elements heavier than hydrogen have a tendency to sink down with respect to hydrogen. One widely used recipe to derive diffusion coefficients for stellar plasmas is that of [Paquette, et al. (1986)], which is first applied to solar models by [Proffitt & Michaud (1991)]. In those layers that are convectively unstable, this heavy element settling is suppressed by the vigorous mixing of turbulent convection. Differential radiative pressure on different species can also counteract their settling in zones which are convectively stable. Note that even in the convectively stable regions of the Sun or stars, shear flows can be present. This can occur for instance directly due to differential rotation, through meridional circulation induced by rotation, overshooting of gas motion from convectively unstable layers into stable layers etc.. These shear flows easily give rise to turbulence because the viscosity of stellar plasma is so low. This implies that element diffusion is not purely controlled by the microphysics of different chemical elements interacting in a gas or plasma, but by migration of 'tracers' in a turbulent medium. It requires a treatment similar to that of deriving the transport equations for correlations of fields in turbulent convection zones. The calculations of [Paquette, et al. (1986)] do not take into account any modification of diffusion coefficients because of turbulence, which is therefore one area where improvements may well significantly affect solar and stellar models.

The chemical composition of the solar interior and stellar interiors is therefore neither constant nor homogeneous within the star. It is this which causes the Sun and stars to evolve. The modelling of this process is achieved by introducing reaction-diffusion equations for the chemical species. The structure is assumed to adjust very quickly compared to the time scale over which significant composition changes occur. However, in the energy equation (2.5) time dependence must be taken into account on these time scales. This in essence because the gravitational potential energy involved in expansion or contraction of the star is a sufficiently large contribution in the total energy budget that it has a significant influence on evolution. Thus evolution is modelled by introducing time dependence through the chemical composition and the energy equation. Evolution is modelled as a succession of quasi-static equilibria, with a slightly changed mixture and distribution of chemical elements from step to step. To simplify matters further, it is usual not to solve for every individual chemical element, but only for H, He and one or a few generic 'heavy elements' that are representative of a group. One of the most extensive of such treatments thus far is [Turcotte et al. (1998b)]. For any species X_i there is then the additional equation:

$$\frac{\partial X_i}{\partial t} + \nabla \cdot (c_{Di} \nabla X_i) = -X_i \sum_{j \neq i} \sum_k r_{ik}^j X_k + \sum_{j \neq i} \sum_{k \neq i} r_{jk}^i X_j X_k \quad (2.14)$$

where c_{Di} is the (turbulent) diffusivity of element i, and the r_{ik}^j and r_{jk}^i are reaction rates, respectively between elements i and k, with element j as one of the reaction products, and between j and k with element i as one of the reaction products. On the left-hand side of this equation are the transport terms: rate of change and diffusive flux of particles. On the right-hand side are sink and source terms. A sink for element i is any reaction of i with any other element. A source is any reaction between elements j and k which has element i as one of its reaction products.

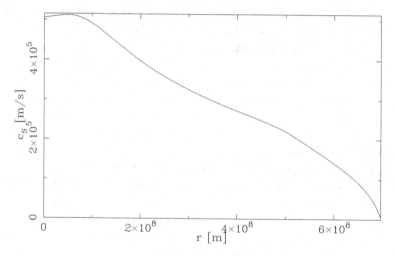

Fig. 2.1 The sound speed as a function of radius in the Sun at current age calculated using the most recent updates of physics as outlined in this chapter (courtesy J. Christensen-Dalsgaard).

The above implies that in order to model the Sun at its current age one needs to know the current distribution of chemical elements within it. One can do this for instance by starting off with the Sun at a notional zero-age at which the distribution of elements within the Sun is fully homogeneous. The mixture of chemical elements is set to what is inferred to be the mixture in the local interstellar medium from which the Sun was formed. This is then evolved to the current age of the solar system, which is inferred from meteorite dating using radioactive elements. An example of a current reference model for the sound speed as a function of radius in the Sun is shown in Fig. 2.1. This model is evolved from zero-age up to the current age of the Sun, using the most recent physics.

Sometimes one starts solar and stellar evolution sequences in the final stages of contraction from an interstellar cloud. The reason for this is that some nuclear processes take place already before the star is completely in hydrostatic equilibrium. One can argue that when hydrostatic equilibrium is first attained in the Sun and stars the internal distribution of chemical elements is not exactly homogeneous. Although the departures from homogeneity are minor it is important to check whether starting with a homogeneous zero-age model introduces a bias in the modelling.

2.4 Numerical Solution Techniques

As is clear from the previous two sections, constructing a model of a star requires solving a sequence of boundary value problems. There are numerous textbooks presenting numerical methods for boundary value problems (e.g. [Stoer & Bulirsch (2002)], [Press *et al.* (1992)]). Details of such methods and implementations can be found there. Here just a brief overview of useful methods is given.

In standard shooting techniques for solving boundary value problems the differential equations are integrated from one boundary to the other. The mismatch between the boundary conditions at the other end and the values the numerical solution provides, is used to adjust initial conditions. One iterates until satisfactory matching of boundary conditions on all boundaries is achieved. Here, the coupled set of differential equations to solve are non-linear, which puts this into the class of *stiff* problems. Stiff boundary value problems have the property that the differential equations allow for solutions that have exponentially increasing behaviour as well as solutions with exponentially decreasing behaviour. The boundary conditions exclude one of these two. However, the discretisation required for numerical integration always introduces some admixture of the solution that is to be excluded. Because of the exponential behaviour the wrong solution will tend to dominate near one of the boundaries of the integration domain, if a standard solver for differential equations is used.

There are several ways to deal with this problem. One traditional way of eliminating the undesirable solution of the differential equations from the boundary value problem is to integrate from both boundaries inward to a matching point. The mismatch between the solution at this matching point is then used to adjust parameters on the boundaries, and the process is iterated. If convergence is still slow one can introduce more intermediate matching points, subdividing the integration range into successively smaller parts. When taken to the limit one obtains what is known as a relaxation method for solving boundary value problems. This is commonly referred to as the Henyey method for constructing stellar models (see also [Kippenhahn & Weigert (1994)]).

In relaxation methods every point of a discretised mesh is a matching point. This transforms the problem from solving the coupled differential equations into solving a large number of coupled non-linear equations. This is done with a multidimensional variant of a Newton-Raphson method, which requires repeated inversion of a matrix. The number of rows/columns in the matrix to factorise is equal to the number of mesh-points times the number of differential equations, which for solar models can amount to several thousand. The problem remains tractable because the matrix has a block banded structure, for which efficient factorisation algorithms exist. However, the convergence of such methods is only quadratic if the starting trial solution is reasonably close to the correct numerical solution. It is therefore not unusual to use a shooting method to provide a first trial solution for a relaxation method.

Another option is to use a particular class of solvers designed for stiff differential equations. The integrator of these *implicit* methods steps through the domain of integration as with standard solvers. However, at each step the dependent variables are used on both sides of the point for which a solution is being constructed. Since some of these have not yet been determined the solver requires at each step the inversion of a matrix. This matrix is much smaller than for relaxation methods however. Although more computationally intensive than standard methods, these have the advantage of being more robust and have better convergence properties than relaxation methods.

All of the above methods can be used for solar and stellar evolution calculations. However, if the eigenmodes and eigenfrequencies of the stellar models are needed, the numerical solution of the stellar structure equations needs to satisfy stringent requirements of accuracy. In order to see why this is the case, in the following two examples, it is illustrated how discretisation error can very substantially alter the inferred oscillation spectrum.

2.4.1 *Discretisation of a continuous potential*

As is shown in the subsequent section, the wave equation that needs to be solved to determine the global resonant modes of oscillation of the Sun and stars has a form similar to the following:

$$\frac{d^2\xi}{dx^2} + \left[\omega^2 - V(x)\right]\xi = 0 \qquad (2.15)$$

on the interval $[0, 1]$. The 'potential' V depends on the sound speed and possibly also other quantities. At this stage it is not necessary to consider exactly how V is determined. Instead a functional form is specified for which an analytical solution can be found, to facilitate comparison of frequencies between the exact and discretised versions of the potential. The potential V is set to:

$$V(x) = \frac{-1}{\cos^4\left(\omega(x - \frac{1}{2})\right)} \qquad (2.16)$$

with $0 < \omega < \pi$ so that the potential is finite on the domain. The boundary conditions that ξ must satisfy are:

$$\xi(0) = \xi(1) = 0 \ . \qquad (2.17)$$

By a change of independent variable to:

$$y = \tan\left(\omega(x - \frac{1}{2})\right) \qquad (2.18)$$

equation (2.15) is transformed to:

$$\frac{d^2}{dy^2}(1 + y^2)^{1/2}\xi + \frac{1}{\omega^2}(1 + y^2)^{1/2}\xi = 0 \qquad (2.19)$$

and therefore there are two types of solutions to (2.15) with (2.16):

$$\xi = \begin{cases} \cos\left(\omega(x - \frac{1}{2})\right)\sin\left(\frac{1}{\omega}\tan\left(\omega(x - \frac{1}{2})\right)\right) \\ \cos\left(\omega(x - \frac{1}{2})\right)\cos\left(\frac{1}{\omega}\tan\left(\omega(x - \frac{1}{2})\right)\right) \end{cases} . \qquad (2.20)$$

The boundary conditions imply that the frequency ω must satisfy

$$\frac{2}{\omega_n}\tan\left(\frac{\omega_n}{2}\right) = n\pi \qquad n = 1, 2, \ldots \qquad (2.21)$$

where the even and odd values of n correspond to the sin and cos types of solutions respectively. One property of note is that π is an accumulation point of the frequencies ω_n: for any δ the interval $(\pi - \delta, \pi]$ contains an infinite number of frequencies ω_n. This is similar to the oscillation spectrum for g-modes in the Sun, where the accumulation is to 0 from above.

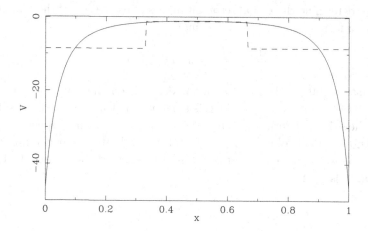

Fig. 2.2 The potential for $\omega = 0.75\pi$ and also a three-zone discretisation of it.

Now consider a crude discretisation of the potential V, which is obtained by replacing V with a piecewise constant potential over three sub-domains $[0, \frac{1}{3}]$, $[\frac{1}{3}, \frac{2}{3}]$ and $[\frac{2}{3}, 1]$. Its values are obtained by integrating V over these three sub-domains, and dividing by $\int dx = 1/3$ so that the integral of the piecewise constant version of V over the same sub-interval is identical to the integral of V over that sub-domain.

$$V(x) = \begin{cases} V_1 & \text{for } 0 \leq x < \tfrac{1}{3} \text{ and } \tfrac{2}{3} < x \leq 1 \\ V_2 & \text{for } \tfrac{1}{3} < x < \tfrac{2}{3} \end{cases} \quad (2.22)$$

where:

$$V_1 = \frac{-1}{\omega}\left[3\tan\left(\frac{\omega}{2}\right) - 3\tan\left(\frac{\omega}{6}\right) + \tan^3\left(\frac{\omega}{2}\right) - \tan^3\left(\frac{\omega}{6}\right)\right]$$
$$V_2 = \frac{-2}{\omega}\left[3\tan\left(\frac{\omega}{6}\right) + \tan^3\left(\frac{\omega}{6}\right)\right]. \quad (2.23)$$

The boundary conditions are the same as above. On each of the three sub-domains $[0, \frac{1}{3}]$, $[\frac{1}{3}, \frac{2}{3}]$ and $[\frac{2}{3}, 1]$ it is easy to see that the solution to this equation is:

$$\xi = A\sin(\omega_1 x) \quad \text{for} \quad 0 \le x < \frac{1}{3}$$
$$\xi = B\sin(\omega_2 x + \phi) \quad \text{for} \quad \frac{1}{3} < x < \frac{2}{3} \quad (2.24)$$
$$\xi = C\sin(\omega_1(x-1)) \quad \text{for} \quad \frac{2}{3} < x \le 1$$

with arbitrary constants A, B and C and where the frequencies $\omega_{1,2}$ are given by:

$$\omega_1^2 = \omega^2 - V_1$$
$$\omega_2^2 = \omega^2 - V_2 \ . \quad (2.25)$$

A solution for the entire domain $[0,1]$ requires that ξ be continuous and differentiable everywhere, and therefore also at the points $x = 1/3$ and $x = 2/3$. This means that:

$$A\sin(\omega_1/3) = B\sin(\omega_2/3 + \phi)$$
$$A\omega_1 \cos(\omega_1/3) = B\omega_2 \cos(\omega_2/3 + \phi)$$
$$B\sin(2\omega_2/3 + \phi) = C\sin(-\omega_1/3)$$
$$B\omega_2 \cos(2\omega_2/3 + \phi) = C\omega_1 \cos(-\omega_1/3) \ . \quad (2.26)$$

The first two of these equations can be used to eliminate the constants A and B between them, and similarly the next two can be used to eliminate B and C between them:

$$\omega_1 \cos(\omega_1/3)\sin(\omega_2/3 + \phi) = \omega_2 \sin(\omega_1/3)\cos(\omega_2/3 + \phi)$$
$$\omega_2 \cos(2\omega_2/3 + \phi)\sin(-\omega_1/3) = \omega_1 \sin(2\omega_2/3 + \phi)\cos(-\omega_1/3) \ . \quad (2.27)$$

Combining these equations yields a condition for the phase ϕ:

$$\phi = -\frac{\omega_2}{2} + k\frac{\pi}{2} \quad k = 0 \text{ or } 1 \quad (2.28)$$

and substituting this expression for ϕ results in:

$$\omega_1 \cos\left(\frac{\omega_1}{3}\right)\sin\left(\frac{\omega_2}{6} \pm k\frac{\pi}{2}\right) + \omega_2 \sin\left(\frac{\omega_1}{3}\right)\cos\left(\frac{\omega_2}{6} \pm k\frac{\pi}{2}\right) = 0 \ . \quad (2.29)$$

Using the definition (2.25) of $\omega_{1,2}$ in terms of ω, solutions of Eq. (2.29) give the oscillation spectrum for the discretised version of V, which makes it possible to compare frequencies. In the table below the third column shows the solutions for $n = 1,..10$ of (2.21) divided by π, the spectrum of the exact potential. The first and second column show the solutions of (2.29), for $k = 0$ and $k = 1$ respectively, also divided by π, which is the spectrum of the discretised potential. The results for $k = 0$ should correspond to the even values of n and $k = 1$ solutions should correspond to odd values of n.

$\omega_{k=0}/\pi$	$\omega_{k=1}/\pi$	ω_n/π
0.8612529	0.7538201	0.8512140
0.9257208	0.9202619	0.9309870
0.9466245	0.9452022	0.9550488
0.9572980	0.9567158	0.9666668
0.9638974	0.9635994	0.9735121
0.9684382	0.9682637	0.9780246
0.9717829	0.9716712	0.9812233
0.9743656	0.9742894	0.9836091
0.9764301	0.9763756	0.9854570
0.9781247	0.9780842	0.9869304

The frequencies are all calculated numerically, using a standard Newton-Raphson scheme, to 7 digits accuracy. There are clearly differences at the 1 % level or larger. Furthermore, because of the increasingly close spacing, the frequencies of the discretised model very quickly correspond better to those of different (lower) order numbers n for the exact potential. In all the table shows very little correspondence between the columns, apart from the qualitative property that all accumulate towards unity.

Evidently one would not use such a coarse discretisation in practice. This choice of potential diverges at 0 and 1 in the limit as ω approaches π from below, which makes this case particularly difficult to discretise. Although the potential relevant for helioseismology does not diverge in the domain of integration, it does have very narrow spikes which present essentially the same problem. Whatever the discretisation, discretisation error does propagate into errors in computed frequencies. Exactly the same reasons that make oscillation frequencies such sensitive measures of solar interior properties, also make the frequencies sensitive to numerical errors. Generally, discretisation meshes that are adequate for solar evolution calculations must be refined if one subsequently uses the models for helioseismic inference. For a wave equation that is cast in the form of Eq. (2.15), one useful recipe for the mesh width Δx would be to set:

$$\Delta x = \frac{1}{\eta \left|\frac{1}{V}\frac{\partial V}{\partial x}\right| + \frac{1}{\Delta_{max} - \Delta_{min}}} + \Delta_{min} \qquad (2.30)$$

which would guarantee a finer mesh where V changes rapidly, and a coarser mesh over ranges where V is constant. The small parameter $1/(\Delta_{max} - \Delta_{min})$ is introduced to set a maximum to the mesh size, at locations where the derivative of V becomes very small. The Δ_{min} term is introduced to prevent a mesh size of 0 near locations where the derivative of V is large: i.e. V is (nearly) discontinuous. One way of implementing this recipe is to introduce a mesh variable z as independent variable so that:

$$\frac{\partial}{\partial x} \rightarrow \left(\frac{dx}{dz}\right)^{-1} \frac{\partial}{\partial z}$$

$$\frac{dx}{dz} \equiv \frac{1}{\eta \left|\frac{1}{V}\frac{\partial V}{\partial x}\right| + \frac{1}{\Delta_{max} - \Delta_{min}}} + \Delta_{min} \qquad (2.31)$$

and then to solve the differential equation for x together with the coupled differential equations for ξ and $\partial \xi/\partial z$. The mesh parameters Δ_{min}, Δ_{max}, and η have to be adjusted to suit the behaviour of V and also to ensure that the eigenfunction is resolved. The latter implies that for a given choice of Δ_{max} there is an upper limit to the order n of the eigenmode which can be resolved, because generally the wavelength λ_n decreases with increasing n.

2.4.2 Discretisation in the presence of a discontinuity

It is at least in principle possible for the Sun or a star to exhibit a sound speed profile that is very nearly discontinuous at one or more points, for instance due to discontinuities in the chemical composition. Such discontinuities are of great interest to detect, and fortunately they have a characteristic signal. Here also, care needs to be taken with the model mesh when computing frequencies. Consider for simplicity an 'exact' potential with a single step at $x = \frac{1}{2}$, and a discretised version in which the discontinuity is located between two mesh points. The latter will normally occur in solar models unless an adaptive mesh is employed, which tracks and adjusts to possible discontinuities. The exact potential is:

$$V_e(x) = \begin{cases} V_1 & \text{for } 0 \leq x < \frac{1}{2} \\ V_2 & \text{for } \frac{1}{2} < x < 1 \end{cases}. \qquad (2.32)$$

This potential is shown together with a discretisation of it. A three-zone

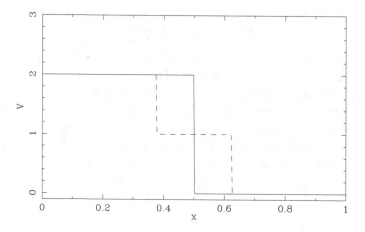

Fig. 2.3 The step potential and the three-zone discretisation of it.

discretisation, in which the middle zone is centered on the discontinuity and has an adjustable thickness $2\delta < \frac{1}{3}$, is:

$$V_d(x) = \begin{cases} V_1 & \text{for } 0 \leq x < \frac{1}{2} - \delta \\ \frac{V_1+V_2}{2} & \text{for } \frac{1}{2} - \delta \leq x < \frac{1}{2} + \delta \\ V_2 & \text{for } \frac{1}{2} + \delta < x < 1 \end{cases} . \qquad (2.33)$$

In both cases the boundary conditions are $\xi(0) = \xi(1) = 0$. Using the same process as described in the previous example, the oscillation spectrum can be determined for each of these potentials. The frequencies of the exact potential satisfy:

$$\omega_1^2 \equiv \omega^2 - V_1$$
$$\omega_2^2 \equiv \omega^2 - V_2$$
$$0 = \omega_1 \sin\left(\frac{\omega_2}{2}\right) \cos\left(\frac{\omega_1}{2}\right) + \omega_2 \sin\left(\frac{\omega_1}{2}\right) \cos\left(\frac{\omega_2}{2}\right) . \qquad (2.34)$$

The solution for the three-zone discretisation uses the same definitions for $\omega_{1,2}$ and in addition an ω_3:

$$\omega_3^2 \equiv \omega^2 - \frac{V_1+V_2}{2} = \frac{\omega_1^2+\omega_2^2}{2} . \qquad (2.35)$$

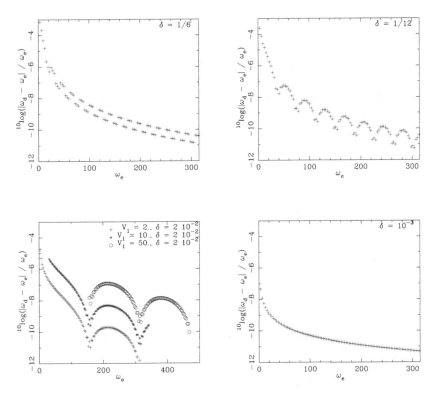

Fig. 2.4 The relative difference between frequencies calculated for the single and double step potentials, as a function of frequency for four values of δ. Top left panel: $\delta = \frac{1}{6}$, top right: $\delta = \frac{1}{12}$, bottom left $\delta = 2\,10^{-2}$, for three potential values V_1, bottom right $\delta = 10^{-3}$.

Requiring a continuous and differentiable solution imposes the conditions:

$$\begin{aligned}
&\omega_1 \cos\left(\omega_1\left(\frac{1}{2}-\delta\right)\right) \sin\left(\omega_3\left(\frac{1}{2}-\delta\right)+\phi\right) \\
&\quad = \omega_3 \sin\left(\omega_1\left(\frac{1}{2}-\delta\right)\right) \cos\left(\omega_3\left(\frac{1}{2}-\delta\right)+\phi\right) \\
&\omega_2 \cos\left(-\omega_2\left(\frac{1}{2}-\delta\right)\right) \sin\left(\omega_3\left(\frac{1}{2}+\delta\right)+\phi\right) \\
&\quad = \omega_3 \sin\left(-\omega_2\left(\frac{1}{2}-\delta\right)\right) \cos\left(\omega_3\left(\frac{1}{2}+\delta\right)+\phi\right)
\end{aligned} \tag{2.36}$$

which can be rewritten as:

$$\begin{pmatrix} a_{11} & a_{12} \\ a_{21} & a_{22} \end{pmatrix} \begin{pmatrix} \cos\phi \\ \sin\phi \end{pmatrix} = 0 \qquad (2.37)$$

in which the matrix elements are:

$$a_{11} = \omega_1 \cos\left(\omega_1\left(\frac{1}{2}-\delta\right)\right) \sin\left(\omega_3\left(\frac{1}{2}-\delta\right)\right)$$
$$- \omega_3 \sin\left(\omega_1\left(\frac{1}{2}-\delta\right)\right) \cos\left(\omega_3\left(\frac{1}{2}-\delta\right)\right)$$

$$a_{12} = \omega_1 \cos\left(\omega_1\left(\frac{1}{2}-\delta\right)\right) \cos\left(\omega_3\left(\frac{1}{2}-\delta\right)\right)$$
$$+ \omega_3 \sin\left(\omega_1\left(\frac{1}{2}-\delta\right)\right) \sin\left(\omega_3\left(\frac{1}{2}-\delta\right)\right)$$

$$a_{21} = \omega_2 \cos\left(\omega_2\left(\frac{1}{2}-\delta\right)\right) \sin\left(\omega_3\left(\frac{1}{2}+\delta\right)\right)$$
$$+ \omega_3 \sin\left(\omega_2\left(\frac{1}{2}-\delta\right)\right) \cos\left(\omega_3\left(\frac{1}{2}+\delta\right)\right)$$

$$a_{22} = \omega_2 \cos\left(\omega_2\left(\frac{1}{2}-\delta\right)\right) \cos\left(\omega_3\left(\frac{1}{2}+\delta\right)\right)$$
$$- \omega_3 \sin\left(\omega_2\left(\frac{1}{2}-\delta\right)\right) \sin\left(\omega_3\left(\frac{1}{2}+\delta\right)\right) . \qquad (2.38)$$

Hence a solution for ϕ can only be found if the determinant $a_{11}a_{22} - a_{12}a_{21} = 0$. To compare results of the exact and discretised version of the step potential, the values $V_1 = 2, V_2 = 0$ are chosen for all cases, and additionally cases $V_1 = 10, V_2 = 0$ and $V_1 = 50, V_2 = 0$ for a fixed width $\delta = 0.02$.

As in the previous example the frequencies ω_e and ω_d for the V_e and V_d potentials respectively are calculated numerically, using a standard Newton-Raphson scheme, to 7 digits accuracy. For each case the difference between the first 100 frequencies is calculated, and divided by ω_e. The result is plotted in Fig. 2.4 as a function of ω_e. There is clearly systematic behaviour which changes with δ. Overall the relative difference disappears asymptotically for high frequency as one should expect, and it also disappears as $\delta \downarrow 0$. However, as the size of the jump of the potential increases, so does the difference for a fixed width δ. This is particularly important because discontinuities, if they do arise in solar or stellar models, are significant features for which helioseismology and asteroseismology should provide diagnostics. One example of this is for instance in understanding convective overshoot. If the inferred location of a discontinuity is systematically displaced because of discretisation error in the models, this has profound consequences for our understanding of such phenomena.

2.5 The Resonances of Simple Systems

Sound waves can propagate in any physical object, or more precisely: in any medium in which the average distance between particles is much smaller than the wavelength of the waves. If a physical object has boundaries, whether it is tricycle, a piano or a star, it is possible to create sound waves inside it that form standing waves: resonances of the system. With tricycles and stars the frequencies of these standing sound waves are an accident of their shape, their size, and their temperature and composition. In the case of pianos there is of course a design involved in order to produce a combination of resonance frequencies that is pleasing to the human ear.

In general with each resonant frequency is associated a set of n node surfaces where the associated standing wave pattern has zero amplitude. It is therefore usual to number frequencies by the number of node surfaces in the body. In some cases frequencies can be *degenerate*: two (or more) different sets of node surfaces can have associated frequencies that are identical.

In all cases the set of resonances an object has is characteristic of it, which indicates that it might to be possible to completely reconstruct the dynamical properties of an object (size, shape, sound speed) from the observation of all of its frequencies. There are three practical problems with this. The number of resonance frequencies is usually infinite. Since frequencies have to be positive it is possible to identify a lowest possible frequency, which usually belongs to a resonant mode that has no node surfaces inside the body of the object. Since it is at least in principle possible to have an infinite number of node surfaces in a body, there is no upper limit to the value of resonant frequencies. In any case, since the number is infinite it is impossible to measure them all.

Not only is it impossible to measure all the frequencies, it is also impossible to measure any frequency with infinite precision. This is of course true for any measurement but in the case of reconstructing properties of an object through measurement of its frequencies there is a particular problem which in mathematical terms is known as 'ill-posedness'. The precision/resolution with which for instance the sound speed can be reconstructed can be extremely sensitive to even small errors in the data. The greater the detail that one wishes to resolve, the more cautious one has to be.

In order to reconstruct completely the properties of an object the set of standing wave patterns have to satisfy two mathematical properties: unicity and completeness. If a set of functions (i.e. the standing wave patterns) is complete it means that *any* kind of wave pattern in the object can be constructed from some linear combination of just the *standing* wave patterns. If a set of functions satisfies the unicity property it means that there is always just exactly one such linear combination for every possible wave pattern. Another way of expressing this is saying that the set of resonant wave patterns is a base-set of linearly independent functions. An example of a set of functions that satisfies these two criteria are sines and cosines of integer multiples of some length scale. The resonant wave patterns of a uniform rectangular membrane are sines and cosines, and the associated frequencies are integer multiples of some fundamental frequency from which the term 'harmonics' arises. There are not many objects for which there exists mathematical proof that the set of its standing wave patterns satisfies these criteria. Spheres inside a vacuum belong to the class of objects for which proofs exist that say that its resonant wave patterns ('eigenmodes') are unique and complete. Since the Sun and stars are approximately spherical and the gas density drops off quite sharply at some radius so that there is a reasonably clear boundary, it is usually assumed that the proofs hold (even though formally of course they do not because the density in reality never quite goes to zero).

None of these problems are such that they invalidate seismic inferences completely. However, they do imply that there are some fundamental limitations to the inferences that one can make from frequencies:

- Given a set of observed frequencies it is impossible to go beyond a certain spatial resolution *no matter how high the signal-to-noise of the data.*
- The frequencies of any observable set are not fully independent variables. The number of parameters of the Sun that one might want to determine, such as the number of positions at which the sound speed can be determined, is *always rather smaller than* the number of measured frequencies.

What makes a particular sound wave in a body a resonant wave is *constructive interference*. A wave front propagating out from any point inside the object and then being reflected at the surface and eventually coming back to its point of origin must have a phase that is an integer number times 2π at the time that it returns. If it does not, then the wave (partially) cancels itself out for every time it is reflected at the surface, eventually leading to full cancellation after many reflections. If the wave does match up with itself it will never destruct itself in this way. The amount by which the phase advances when a sound wave propagates from a point inside to the surface and back again is:

$$\phi = \omega \int \frac{\mathrm{d}s}{c_S} \tag{2.39}$$

where the integral is over any closed path from any point of origin to the surface and back. c_S is the sound speed. If the value of ϕ has to be $2\pi n$ then:

$$\omega_n \propto \left[\int \frac{\mathrm{d}s}{c_S} \right]^{-1} . \tag{2.40}$$

This is true for any object with resonances and therefore also for the Sun and stars. If one assumes that stars are spherical and that the sound speed depends only on radius r, the integration over any path can be replaced by a straightforward integration over r if a weight function $W_n(r)$ is introduced in the integration which in principle is different for every different mode:

$$\omega_n = \left[\int \frac{W_n(r)}{c_S(r)} \mathrm{d}r \right]^{-1} . \tag{2.41}$$

The weight function $W_n(r)$ expresses that different resonant sound waves traverse different regions of the star. For instance some waves have amplitudes that are only large very near the surface of the star, so for those the weight function is only large for $r \approx R$. Note that here a single index n is used for identifying the mode of pulsation. As already mentioned in section 1.4, three are used (n, l, and m) where this n refers to the number of nodes in the radial direction, l refers to the number of node lines on the surface, and $|m|$ is the number of nodes on the equator.

2.5.1 Resonant properties of some simple systems

Linear wave equations are all in essence versions of the harmonic oscillator. The acceleration that a bit of material (a piece of string, a section of a membrane, a parcel of gas) experiences is proportional to the displacement from an equilibrium position, and directed back towards the equilibrium position. For a particle with a mass m and a displacement from equilibrium z this implies:

$$m\frac{\partial^2 z}{\partial t^2} = -kz \tag{2.42}$$

where k is a spring constant, or an elasticity.

When looking at something other than point particles the mass m has to be replaced by the mass per unit length (or surface, or volume): the density ρ. The displacement from the equilibrium position must also be replaced. For instance in a string, if the entire string is displaced there is no force that pulls anything back. Also, if the string is tilted/rotated there is still no internal restoring force. There is only an elastic force if the string is bent/curves which means that the force is proportional to the second derivative of the displacement z with respect to the coordinate x along the string.

$$\rho\frac{\partial^2 z}{\partial t^2} = k\frac{\partial^2 z}{\partial x^2} \tag{2.43}$$

Usually this equation is divided by ρ and all terms are moved to the left-hand side. It is easy to see that k/ρ has the dimensions of the square of a velocity. This is the sound speed or its equivalent in the system. So the linear wave equation in one dimension will look like:

$$\frac{\partial^2 z}{\partial t^2} - c_S^2 \frac{\partial^2 z}{\partial x^2} = 0 \tag{2.44}$$

and in general it looks like:

$$\frac{\partial^2 \xi}{\partial t^2} - c_S^2 \nabla^2 \xi = 0 \tag{2.45}$$

where now ξ is the displacement to avoid confusion with the spatial coordinates. This equation applies to oscillations in stars as well. Equation (2.45) describes how waves propagate. However, in order to obtain the resonances of a given object it is necessary to add equations describing what happens to waves when they encounter the boundary of the object. Together these determine the resonant frequencies. The following three examples illustrate how the resonant frequencies for several simple systems, with a constant sound speed, can be obtained. This illustrates the procedure for obtaining oscillation frequencies of stellar models.

2.5.2 Example I: Linear waves in a rectangular membrane

The most appropriate two-dimensional form of equation (2.45) for a rectangular membrane is:

$$\frac{\partial^2 \xi}{\partial t^2} - c_S^2 \frac{\partial^2 \xi}{\partial x^2} - c_S^2 \frac{\partial^2 \xi}{\partial y^2} = 0 \ . \tag{2.46}$$

For boundary conditions one could for instance assume that the edges are clamped down, and so $\xi = 0$. Another possible condition is that they can oscillate freely without any restraining force in which case at the appropriate edges either $\partial \xi / \partial x = 0$ or $\partial \xi / \partial y = 0$. To see why this is so consider the difference in velocity ($\partial \xi / \partial t$) at two points, one of which is the end point. The closer these two points are, the smaller the velocity difference has to be so in the limit:

$$\frac{\partial}{\partial t} \frac{\partial \xi}{\partial x} = 0$$
$$\frac{\partial}{\partial t} \frac{\partial \xi}{\partial y} = 0 \tag{2.47}$$

which means that on the appropriate edges $\partial \xi / \partial x$ or $\partial \xi / \partial y$ must have a value that is constant in time:

$$\frac{\partial \xi}{\partial x} = f(x,y) \text{ for } x = 0, X$$
$$\frac{\partial \xi}{\partial y} = g(x,y) \text{ for } y = 0, Y \ . \tag{2.48}$$

For the solution to be continuous and differentiable everywhere it must have the form:

$$\xi = F(x,y) \tag{2.49}$$

and it has to satisfy:

$$\frac{\partial^2 F}{\partial x^2} + \frac{\partial^2 F}{\partial y^2} = 0 \ . \tag{2.50}$$

This describes a time-independent deformation of the membrane which is of no interest to the oscillation analysis. For any solution F one could always redefine $\xi_{new} = \xi_{old} - F$ which would still satisfy (2.46) and would (by definition) have the appropriate boundary conditions:

$$\frac{\partial \xi}{\partial x} = 0 \text{ for } x = 0, X$$
$$\frac{\partial \xi}{\partial y} = 0 \text{ for } y = 0, Y . \qquad (2.51)$$

The clamped boundary condition is often referred to as being of Dirichlet type, the free boundary condition is referred to as the Neumann type condition.

Since there are no mixed derivatives (e.g. $(\partial^2/\partial x \partial y)$) in (2.46) one may safely assume that the solution can be written as a separable function:

$$\xi = f_1(t) f_2(x) f_3(y) . \qquad (2.52)$$

Operators such as $(\partial^2/\partial t^2)$ suggest the use of functions such as $\cos \omega t$ and $\sin \omega t$ as possible solutions because when the operator is applied to these, the same function is returned, times a constant (the square of the frequency ω). The sin and cos functions are eigenfunctions of the operator. Using $A \cos + B \sin$, with arbitrary constants A and B, for each of the f_1, f_2, and f_3 in (2.52) and with the arguments ωt, $k_x x$, and $k_y y$ respectively, (2.46) becomes:

$$[-\omega^2 + c_S^2 (k_x^2 + k_y^2)] \xi = 0 . \qquad (2.53)$$

The boundary condition $\partial \xi / \partial x = 0$ for $x = 0, X$, together with the general form of $f_2(x) = A \cos k_x x + B \sin k_x x$ means that:

$$-A k_x \sin k_x x + B k_x \cos k_x x = 0 \text{ for } x = 0, X \qquad (2.54)$$

which means that $B = 0$ and that $k_x X = \pi(n+1)$ where n is any integer. Doing the same for the y-direction produces $k_y Y = \pi(l+1)$. Putting this back into equation (2.53) results in the values for the resonance frequencies:

$$\omega_{n,l}^2 = c_S^2 \pi^2 \left[\left(\frac{n+1}{X} \right)^2 + \left(\frac{l+1}{Y} \right)^2 \right] . \qquad (2.55)$$

The n and l are the number of node-lines in the two perpendicular directions across the membrane. If the membrane is square ($X = Y$) then there is a degeneracy in the frequencies: switching around the values for n and l produces the same frequency. Even if the membrane is not square but the ratio of the length of the sides is a rational number, there is some degeneracy. To see this, one can substitute $X = pZ$ and $Y = qZ$ where p and q are integers. For those values of n and l which satisfy $n = kp - 1$ and $l = k'q - 1$ (with both k and k' arbitrary integers) there is a degeneracy in the sense that the same frequency is obtained if k and k' are exchanged.

2.5.3 Example II: Linear waves in a circular membrane

In cylinder coordinates (r, ϕ) the 2-dimensional form of (2.45) is:

$$\frac{\partial^2 \xi}{\partial t^2} - c_S^2 \left[\frac{\partial^2 \xi}{\partial r^2} + \frac{1}{r} \frac{\partial \xi}{\partial r} + \frac{1}{r^2} \frac{\partial^2 \xi}{\partial \phi^2} \right] = 0 \ . \tag{2.56}$$

The 'boundary conditions' take a slightly different form in this case. For a proper resonance one should demand *periodicity* in the ϕ direction: going once full around the edge of the membrane one should end up with the same ξ when returning to $\phi = 0$. In the radial direction this time a clamped boundary condition is applied although there is no fundamental difficulty in using the same condition as in example I. Again one may assume that the solution can be written as a separable function because there are no mixed derivatives:

$$\xi = f_1(t) f_2(r) f_3(\phi) \ . \tag{2.57}$$

For the f_1 and f_3, one can use again $A \cos + B \sin$ as a general solution with as arguments ωt and $k_\phi \phi$ respectively. The periodicity condition means that k_ϕ has to be an integer: m. With these choices for f_1 and f_3 the following two expressions hold:

$$\begin{aligned} \frac{\partial^2 \xi}{\partial t^2} &= -\omega^2 \xi \\ \frac{\partial^2 \xi}{\partial \phi^2} &= -m^2 \xi \end{aligned} \tag{2.58}$$

and this can be substituted into equation (2.56). The functions f_1 and f_3 can be divided out at this stage and the result is:

$$-\omega^2 f_2 - c_S^2 \left[\frac{\partial^2 f_2}{\partial r^2} + \frac{1}{r} \frac{\partial f_2}{\partial r} - \frac{m^2}{r^2} f_2 \right] = 0 \ . \tag{2.59}$$

It is sometimes useful to remove the first derivative with respect to r from the equation (2.59) which can by done by substituting for $f_2 = r^{-1/2} g_m(r)$.

The differential equation for the function $g_m(r)$ is then:

$$\frac{d^2 g_m}{dr^2} + \left[\left(\frac{\omega}{c_S}\right)^2 - \frac{4m^2 - 1}{4r^2}\right] g_m = 0 . \qquad (2.60)$$

The index m on the function g is introduced because m appears as a parameter in the equation. At this point it is usual to introduce a dimensionless (phase-like) radius $x \equiv \omega r/c_S$, which reduces the equation (2.60) to:

$$\frac{d^2 g_m}{dx^2} + \left[1 - \frac{4m^2 - 1}{4x^2}\right] g_m = 0 . \qquad (2.61)$$

The solutions of this differential equation are $g_m(x) = \sqrt{x} J_m(x)$ where the $J_m(x)$ are known as Bessel functions. Their properties can be found in any mathematical text-book on special functions (eg. [Gradshteyn & Ryzhik (1994)]). (Note that removing the first derivative was not particularly useful here because $f_2 = J_m$ directly). For the nth resonant mode of the membrane the rim of the membrane with radius R has to coincide with the nth zero-point of the Bessel function (with n any positive integer). At very large values of the argument, Bessel functions resemble sine functions which means that their zero points are more and more nearly regularly spaced. The nth zero-point of Bessel function $J_m(x)$ is approximately at $x_0 \approx (2n + m - 1/2)\pi/2$ (cf. [Gradshteyn & Ryzhik (1994)]). Using this *asymptotic behaviour* it is straightforward to calculate the resonant frequencies approximately. If the zero-point coincides with the rim of the plate then:

$$\frac{\omega R}{c_S} = x_0 \quad \Longleftrightarrow \quad \nu_{n,m} \approx \frac{1}{4}(2n + m - \frac{1}{2})\frac{c_S}{R} \qquad (2.62)$$

where $\omega = 2\pi\nu$. This means that there is a (near)-degeneracy of mode frequencies so that $\nu_{n,m} \approx \nu_{n+1,m-2}$. Since this is an approximation which improves for increasing n and m one could say this is an *asymptotic approximation* of the frequencies.

2.5.4 Example III: Linear waves in a uniform sphere

In spherical coordinates (r, θ, ϕ) the 3-dimensional form of (2.45) is:

$$\frac{\partial^2 \xi}{\partial t^2} - c_S^2 \left[\frac{\partial^2 \xi}{\partial r^2} + \frac{2}{r}\frac{\partial \xi}{\partial r} + \frac{1}{r^2}\left(\frac{\partial^2 \xi}{\partial \theta^2} + \frac{\cos\theta}{\sin\theta}\frac{\partial \xi}{\partial \theta} + \frac{1}{\sin^2\theta}\frac{\partial^2 \xi}{\partial \phi^2}\right)\right] = 0 . \qquad (2.63)$$

In the ϕ direction and θ direction there are conditions on the resonant modes similar to the periodicity conditions of example II. The eigenmodes of spheres are usually written as a product of some function of just the radius r inside the sphere and another function of the two polar angles θ and ϕ. Evidently the places where this second function is zero are the node lines on the surface of any sphere of which the centre is coincident with the stellar centre. These functions are known as spherical harmonics with the notation $Y_l^m(\theta, \phi)$ where the *degree* l is the total number of node lines on the surface, and the absolute value of the *azimuthal order* m is the number of points on the equator ($\theta = \pi/2$) where $Y_l^m = 0$. One can prove that $|m| \leq l$. Some properties of spherical harmonics also can be found in appendix B, and further details can be found in any mathematical text-book on special functions (eg. [Gradshteyn & Ryzhik (1994)]). A few examples for low l values are shown in Fig. 2.5. For the purposes of the

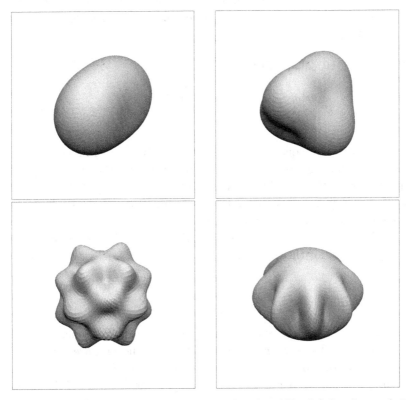

Fig. 2.5 Some examples of spherical harmonic functions. Top left $l = 2, m = 2$, top right $l = 3, m = 2$, bottom left $l = 8, m = 4$, bottom right $l = 8, m = 8$.

above equation the important property of the Y_l^m is that they satisfy the periodicity conditions, and that when substituting:

$$\xi = f(r) Y_l^m(\theta, \phi) \qquad (2.64)$$

the differential equation (2.63) reduces to:

$$\frac{\partial^2 f}{\partial r^2} + \frac{2}{r}\frac{\partial f}{\partial r} + \left[\left(\frac{\omega^2}{c_S^2}\right) - \frac{l(l+1)}{r^2}\right] f = 0 . \qquad (2.65)$$

As in example II it is useful to:

- Introduce a phase-like dimensionless radius $x \equiv \omega r/c_S$.
- Remove the first derivative term, this time by substituting $f(r) = r^{-1} g_{l+\frac{1}{2}}(r)$ (The reason to use the same symbol g as in example II but to give this an index $l + 1/2$ becomes clear when looking at the result).

The differential equation (2.65) is reduced to:

$$\frac{\partial^2 g_{l+\frac{1}{2}}}{\partial x^2} + \left[1 - \frac{l(l+1)}{x^2}\right] g_{l+\frac{1}{2}} = 0 . \qquad (2.66)$$

The solutions to this equation can be written using Bessel functions with half-integer indices $g_{l+\frac{1}{2}}(x) = \sqrt{x} J_{l+\frac{1}{2}}(x)$ (or $f_l(x) = J_{l+\frac{1}{2}}(x)/\sqrt{x}$). It can be seen that the two differential equations (2.61) and (2.66) are the same if in equation (2.61) the index m is substituted by $l + 1/2$. The properties of Bessel functions of integer and half-integer index are very similar (cf. [Gradshteyn & Ryzhik (1994)]). Therefore the same asymptotic relationship holds for the frequencies as (2.62):

$$\nu_{n,l} \approx \frac{1}{4}(2n + l)\frac{c_S}{R} . \qquad (2.67)$$

Note that apart from the (near)-degeneracy of mode frequencies $\nu_{n,l} \approx \nu_{n+1,l-2}$ there is also a degeneracy in that the frequencies are completely independent of m. The reason for this is that the velocity at which waves travel around a sphere parallel to the equator is independent of the directions ('East-West' or 'West-East'). If the sphere were to rotate (but remain spherical in shape) the frequencies for different m would be different because of a Doppler-shift, and the difference would be sensitive to the rate of rotation.

Finally it is useful to remark that if instead of a clamped boundary condition one were to take a free boundary condition, the resonant frequencies would shift slightly: instead of matching the zero points of the Bessel functions to the rim/edge one would have to match the minima or maxima (in between the zeros) to the rim. The position of these minima and maxima in fact obey much the same asymptotic behaviour as the zero points and the only effect on the frequencies is a shift:

$$\nu_{n,l} \approx \frac{1}{2}\left(2n + l + \frac{1}{2}\right)\frac{c_S}{R}. \tag{2.68}$$

Again this relationship is more accurate the higher l and n are.

2.6 Resonant Frequencies of the Sun and Stars

2.6.1 Derivation of the LAWE

In the Sun and in most stars the pulsation amplitude is not very large: the speed at which material moves in the compressions and dilations of the standing wave is much smaller than the sound speed. Therefore it is a good approximation to assume that the waves can be treated as small perturbations on an equilibrium state that is no different from what one would calculate in the absence of all waves. Another implication is that effects that are of higher order than linear in the small perturbations can be ignored: the frequencies are not affected by effects such as a wave front deforming into a shock wave or two waves belonging to different eigenmodes interacting with each other such as damping each other out or modifying each other's frequency.

The derivation of the equations for linear oscillations in the Sun or stars requires performing a perturbation analysis of the equations governing stellar structure. Since stars are self-gravitating bodies, there is a force from the buoyancy of material being moved around in an oscillation as well as the force from the pressure variations. Oscillations are then no longer pure sound waves but so-called gravito-acoustic waves. This leads to a system of ordinary differential equations which is a fourth-order system. The numerical codes that calculate linear adiabatic oscillations always solve this full set of equations. The reason to proceed in this way is that there are robust numerical techniques for solving linear systems of differential equations in order to get the resonant frequencies and the associated standing wave patterns. Also there are proofs that the set of all solutions of such a linear system, the *eigenmodes*, satisfies the requirements for completeness and unicity.

Starting point of the derivation of pulsation equations is a choice of coordinate system. There are two common choices. One is the Eulerian description in which all quantities describing fluid properties are functions of the independent variables time t and position $\mathbf{r}$. There is also the Lagrangian description in which the independent variables are time t and and the location $\mathbf{r}_0$ of a given fluid element at some reference time. This description therefore looks at variations that a given fluid element experiences, following the motion of that fluid element. The relationship between the Lagrangian perturbation δf of quantity f and the Eulerian perturbation f' is:

$$\delta f(\mathbf{r}) = f'(\mathbf{r}_0) + \boldsymbol{\delta r} \cdot \nabla f_0 . \tag{2.69}$$

For the purposes of this book it is most convenient to work within the Eulerian framework. Further details on the difference between these two descriptions and on the pulsation equations in Lagrangian form can be found in e.g. [Cox (1980)] or [Unno et. al (1989)].

The wave equation is derived from the equations governing conservation of mass, momentum and energy. Since the Sun is to a high degree spherically symmetric the most convenient choice of coordinate system is a spherical one, centered on the centre of mass of the Sun. Conservation of mass is expressed by:

$$\frac{\partial}{\partial t}\rho + \nabla \cdot (\rho \mathbf{v}) = 0 \tag{2.70}$$

in which v is the velocity of the plasma. Conservation of momentum is expressed by:

$$\frac{\partial}{\partial t}(\rho \mathbf{v}) + \nabla \cdot (\rho \mathbf{v}\mathbf{v}) = -\nabla p - \rho \nabla \Psi . \tag{2.71}$$

In principle the right-hand side of Eq. (2.71) could contain terms for other body forces and internal stresses, but for the global solar oscillations these are relatively unimportant and are omitted from the analysis. Under the same assumptions the equation of conservation of energy is:

$$\frac{\partial}{\partial t}\left(\frac{1}{2}\rho v^2 + \rho U\right) + \nabla \cdot \left[\rho \mathbf{v}\left(\frac{1}{2}v^2 + \rho U + p/\rho\right)\right] = -\rho \nabla \Psi \cdot \mathbf{v} + \rho \frac{\mathrm{d}q}{\mathrm{d}t} \tag{2.72}$$

in which U is the internal energy per unit mass, and $v = |\mathbf{v}|$. Eq. (2.72) can be reduced by making use of Eqs. (2.70) and (2.71) to become:

$$\frac{dq}{dt} = \frac{dU}{dt} - \frac{p}{\rho^2}\frac{d\rho}{dt} \quad (2.73)$$

which is a more familiar form of the first law of thermodynamics. This can be expressed in various other ways, by making use of thermodynamic relationships between p, ρ and T and the various derivatives:

$$\chi_T = \left(\frac{\partial \ln p}{\partial \ln T}\right)_\rho$$
$$\chi_\rho = \left(\frac{\partial \ln p}{\partial \ln \rho}\right)_T \quad (2.74)$$

which are both 1 for an ideal gas. For instance a change in temperature of plasma under heat exchange per unit mass dq with surroundings, can be expressed by making use of either one or the other of:

$$\frac{d\ln T}{dt} = (\Gamma_3 - 1)\frac{d\ln \rho}{dt} + \frac{1}{c_V T}\frac{dq}{dt}$$
$$\frac{d\ln T}{dt} = \frac{\Gamma_2 - 1}{\Gamma_2}\frac{d\ln p}{dt} + \frac{1}{c_P T}\frac{dq}{dt} \quad (2.75)$$

where c_V and c_P are the specific heats per unit mass at resp. constant volume and constant pressure. The adiabatic exponents are defined in Eq. (2.6) and (2.7). Either of these two equations can be used directly, or T can be eliminated between the two. The latter can then be rewritten using the identities:

$$\frac{c_P}{c_V} \equiv \gamma = \frac{\Gamma_1}{\chi_\rho}$$
$$\Gamma_1 = \chi_\rho + \chi_T(\Gamma_3 - 1) \quad (2.76)$$
$$\Gamma_3 - 1 = \frac{p\chi_T}{\rho c_V T}$$

in which case one obtains:

$$\frac{dq}{dt} = \frac{1}{\rho(\Gamma_3 - 1)}\left(\frac{dp}{dt} - \frac{\Gamma_1 p}{\rho}\frac{d\rho}{dt}\right). \quad (2.77)$$

In Eq. (2.72) and its alternative formulations Eq. (2.73) and Eq. (2.77) the material derivative of various quantities, such as q, with respect to time d/dt is used. This differs from the partial time derivative, and the two are related by:

$$\frac{d}{dt} \equiv \frac{\partial}{\partial t} + \mathbf{v} \cdot \nabla . \qquad (2.78)$$

Usually the oscillations in the Sun are assumed to be adiabatic, which according to the discussion of time scales in section 1.2.1 is likely to be a very good approximation everywhere, except very near the surface. For adiabatic behaviour $dq/dt = 0$ and therefore Eq. (2.77) reduces to:

$$\frac{dp}{dt} = \frac{\Gamma_1 p}{\rho} \frac{d\rho}{dt} . \qquad (2.79)$$

A detailed derivation of the equations describing linear adiabatic stellar oscillations can be found in e.g. [Unno et. al (1989)]. A particularly useful resource can be found at http://astro.phys.au.dk/~jcd/oscilnotes/ which is [Christensen-Dalsgaard (2003)]. The notation and derivation presented here are similar to the latter.

The linear adiabatic wave equation (LAWE) is obtained by writing out explicitly that every quantity f is the sum of its time-independent equilibrium value and a perturbation which is a function of space and time:

$$f(\mathbf{r}, t) = f_0(\mathbf{r}) + f'(\mathbf{r}, t) . \qquad (2.80)$$

Furthermore, since the equilibrium is hydrostatic: $\mathbf{v}_0 = 0$ and therefore:

$$\mathbf{v}'(\mathbf{r}, t) \equiv \mathbf{v}(\mathbf{r}, t) \equiv \frac{\partial}{\partial t} \delta \mathbf{r}$$
$$\frac{\partial^2}{\partial t^2} \delta \mathbf{r} \equiv \frac{\partial}{\partial t} \mathbf{v} . \qquad (2.81)$$

This is introduced in each of the equations (2.70), (2.71) and (2.72). Of the resulting terms in each of these equations, those terms containing only equilibrium quantities cancel out, by their definition of being solutions of the time-independent system. Terms that contain more than one primed quantity are presumed negligible, which is what the 'linear' in LAWE signifies. Following these steps, the equation of conservation of mass (2.70) is reduced to:

$$\frac{\partial}{\partial t} \rho' + \nabla \cdot (\rho_0 \mathbf{v}) = 0 . \qquad (2.82)$$

The equation of conservation of momentum (2.71) becomes:

$$\rho_0 \frac{\partial}{\partial t} \mathbf{v} = -\nabla p' - \rho_0 \nabla \Psi' - \rho' \nabla \Psi_0 . \tag{2.83}$$

The perturbation of the gravitational potential Ψ' is determined through Eq. (2.4):

$$\nabla^2 \Psi' = 4\pi G \rho' . \tag{2.84}$$

The equation of conservation of energy is most conveniently perturbed using it in the form of Eq. (2.77). At this point the 'adiabatic' part of LAWE is introduced, which is to say that any perturbation of q is set to 0. Using Eq. (2.78) for p and ρ yields:

$$\frac{d}{dt} p = \frac{\partial}{\partial t} p + \mathbf{v} \cdot \nabla p = \frac{\partial}{\partial t} p' + \mathbf{v} \cdot \nabla p_0 = \frac{\partial}{\partial t} [p' + \boldsymbol{\delta r} \cdot \nabla p_0]$$
$$\frac{d}{dt} \rho = \frac{\partial}{\partial t} \rho + \mathbf{v} \cdot \nabla \rho = \frac{\partial}{\partial t} \rho' + \mathbf{v} \cdot \nabla \rho_0 = \frac{\partial}{\partial t} [\rho' + \boldsymbol{\delta r} \cdot \nabla \rho_0] . \tag{2.85}$$

Substituting these in Eq. (2.77), eliminating higher order terms, and integrating with respect to time, then produces:

$$p' + \boldsymbol{\delta r} \cdot \nabla p_0 = \left(\frac{\Gamma_1 p}{\rho} \right)_0 (\rho' + \boldsymbol{\delta r} \cdot \nabla \rho_0) . \tag{2.86}$$

For an equilibrium state that is spherically symmetric it makes sense to separate radial and tangential components for all scalar and vector variables in all equations and also for the ∇ operator (see appendix B). In fact, the absence of mixed derivatives makes it possible to separate variables completely so that it is convenient for any perturbed quantity f' to substitute:

$$f'(r, \theta, \phi, t) = \tilde{f}(r) Y_l^m(\theta, \phi) e^{-i\omega t} \tag{2.87}$$

in which the $Y_l^m(\theta, \phi)$ are again the spherical harmonic functions already seen in section 2.5.4. After eliminating a factor ω, the equation of mass conservation Eq. (2.82) becomes:

$$\rho' = -\frac{1}{r^2} \frac{\partial}{\partial r} \left(\rho_0 r^2 \delta r_r \right) - \frac{\rho_0}{r} \nabla_h \cdot \boldsymbol{\delta r}_h . \tag{2.88}$$

The radial and tangential components of the equation of momentum conservation Eq. (2.83) become:

$$-\omega^2 \rho_0 \widetilde{\delta r}_r = -\frac{\partial}{\partial r}\widetilde{p} - \widetilde{\rho}\frac{\partial}{\partial r}\Psi_0 - \rho_0 \frac{\partial}{\partial r}\widetilde{\Psi}$$
$$-\omega^2 \rho_0 \boldsymbol{\delta r}_h = -\left[\widetilde{p} + \rho_0 \widetilde{\Psi}\right]\frac{1}{r}\nabla_h Y_l^m \ . \qquad (2.89)$$

The perturbed gravitational potential Eq. (2.84) becomes:

$$\frac{1}{r^2}\frac{\partial}{\partial r}\left(r^2 \frac{\partial \widetilde{\Psi}}{\partial r}\right) - \frac{l(l+1)}{r^2}\widetilde{\Psi} = 4\pi G \widetilde{\rho} \qquad (2.90)$$

and the adiabatic version of the equation of energy conservation Eq. (2.86) is reduced to:

$$\widetilde{p} + \widetilde{\delta r}_r \frac{\partial}{\partial r}p_0 = \left(\frac{\Gamma_1 p}{\rho}\right)_0 \left[\widetilde{\rho} + \widetilde{\delta r}_r \frac{\partial}{\partial r}\rho_0\right] \ . \qquad (2.91)$$

The second of Eqs. (2.89) is used in two ways. Firstly it specifies the vector $\boldsymbol{\delta r}_h$ in terms of the other perturbed variables:

$$\boldsymbol{\delta r}_h = \frac{1}{r\omega^2}\left(\frac{\widetilde{p}}{\rho_0} + \widetilde{\Psi}\right)\left(0, \frac{\partial Y_l^m}{\partial \theta}, \frac{1}{\sin\theta}\frac{\partial Y_l^m}{\partial \phi}\right)e^{-i\omega t} \ . \qquad (2.92)$$

Secondly, it can be used to eliminate the term $\nabla_h \cdot \boldsymbol{\delta r}_h$ from Eq. (2.88) which becomes:

$$\omega^2\left[\widetilde{\rho} + \frac{1}{r^2}\frac{\partial}{\partial r}\left(\rho_0 r^2 \widetilde{\delta r}_r\right)\right] = \left[\widetilde{p} + \rho_0 \widetilde{\Psi}\right]\frac{l(l+1)}{r^2} \ . \qquad (2.93)$$

Eq. (2.91) expresses a linear relationship between the perturbation in density and pressure and the radial displacement. This can be used to eliminate one of these three variables, eg. the density perturbation:

$$\widetilde{\rho} = \left(\frac{\rho}{\Gamma_1 p}\right)_0 \widetilde{p} + \rho_0 \widetilde{\delta r}_r \left(\frac{1}{\Gamma_{1,0}p_0}\frac{\partial p_0}{\partial r} - \frac{1}{\rho_0}\frac{\partial \rho_0}{\partial r}\right) \qquad (2.94)$$

which is substituted in the first of Eqs. (2.89):

$$\frac{\partial}{\partial r}\widetilde{p} = \left(\omega^2 - N^2\right)\rho_0 \widetilde{\delta r}_r + \frac{1}{\Gamma_{1,0}p_0}\frac{\partial p_0}{\partial r}\widetilde{p} - \rho_0 \frac{\partial}{\partial r}\widetilde{\Psi} \qquad (2.95)$$

in which the buoyancy frequency N is defined by:

$$N^2 \equiv \frac{\partial \Psi_0}{\partial r}\left(\frac{1}{\Gamma_{1,0}}\frac{\partial \ln p_0}{\partial r} - \frac{\partial \ln \rho_0}{\partial r}\right) \ . \qquad (2.96)$$

N is also referred to as the Brunt-Väisälä frequency. It is further convenient at this point to use the definition of the adiabatic sound speed c_s:

$$c_s^2 \equiv \left(\frac{\Gamma_1 p}{\rho}\right)_0 \qquad (2.97)$$

and to define a characteristic acoustic frequency:

$$S_l^2 \equiv \frac{l(l+1)c_s^2}{r^2}. \qquad (2.98)$$

Eq. (2.94) is also substituted in Eq. (2.90) and Eq. (2.93) which become respectively:

$$\frac{1}{r^2}\frac{\partial}{\partial r}\left(r^2\frac{\partial \widetilde{\Psi}}{\partial r}\right) = \frac{l(l+1)}{r^2}\widetilde{\Psi} + 4\pi G\left[\frac{1}{c_s^2}\widetilde{p} + \frac{\rho_0 N^2}{\partial \Psi_0/\partial r}\widetilde{\delta r}_r\right] \qquad (2.99)$$

$$\frac{\partial}{\partial r}\widetilde{\delta r}_r = -\left[\frac{2}{r} + \frac{1}{\Gamma_{1,0}p_0}\frac{\partial p_0}{\partial r}\right]\widetilde{\delta r}_r + \frac{1}{\rho_0 c_s^2}\left[\frac{S_l^2}{\omega^2} - 1\right]\widetilde{p} + \frac{l(l+1)}{\omega^2 r^2}\widetilde{\Psi}. \qquad (2.100)$$

To obtain a complete system of four first-order differential equations for the four variables $\widetilde{\delta r}_r$, $\widetilde{p}$, $\widetilde{\Psi}$, $\partial\widetilde{\Psi}/\partial r$, Eqs. (2.95), (2.99), and (2.100) are augmented with a trivial differential equation for $\widetilde{\Psi}$ in terms of $\partial\widetilde{\Psi}/\partial r$:

$$\frac{\partial}{\partial r}\widetilde{\Psi} = \frac{\partial \widetilde{\Psi}}{\partial r}. \qquad (2.101)$$

These equations describe the behaviour of linear adiabatic waves in the solar interior. Together with regularity conditions and boundary conditions at the centre and the surface, the solutions of these equations provide the eigenmodes and eigenfrequencies of the Sun.

In these equations two characteristic frequencies N and S_l appear. One can show that this gives rise to two distinctive classes of modes depending on the relative size of ω, N, and S_l. Throughout most of the solar interior $N < S_l$. If $\omega < N$ the oscillation behaviour is primarily controlled by bouyancy and the modes are referred to as g-modes. If on the other hand $S_l < \omega$ it is primarily pressure which controls the oscillation, and the modes are very similar to acoustic waves and referred to as p-modes. For $N < \omega < S_l$ one can show that there are no oscillatory solutions. It is further worth noting here that in layers that are stratified adiabatically, such as convection zones, $N^2 = 0$. A consequence of this is that g-modes do not propagate into the solar convection zone.

2.6.2 Boundary conditions

To fully specify the problem of oscillations in the Sun and stars, one needs the boundary conditions for the wave equations. At the centre of the Sun these boundary conditions are regularity conditions: physical quantities cannot diverge. If the pressure is assumed to disappear outside a certain radius, this surface is force-free.

The regularity conditions at $r = 0$ imply that for $l > 0$ the perturbations must vanish as:

$$\widetilde{\delta r}_r \propto r^{l-1}$$
$$\widetilde{p} \propto r^l$$
$$\widetilde{\Psi} \propto r^l \, . \qquad (2.102)$$

At the surface, one boundary condition is obtained by demanding that the perturbation of Ψ matches continuously to the solution of Poisson's equation which vanishes as $r \to \infty$. This leads to:

$$\left.\frac{\partial \widetilde{\Psi}}{\partial r}\right|_{r=R} + \frac{l+1}{r}\widetilde{\Psi}(R) = 0 \, . \qquad (2.103)$$

The other boundary condition depends on the treatment of the solar atmosphere (chromosphere and corona). If the boundary is presumed to be completely force-free, the Lagrangian pressure perturbation vanishes and therefore:

$$\widetilde{p}(R) + \left.\frac{\partial p_0}{\partial r}\right|_{r=R} \widetilde{\delta r}_r(R) = 0 \, . \qquad (2.104)$$

For a more realistic boundary some modifications can be necessary which are discussed in section 2.9.

In section 1.4 it is mentioned that near the surface for g-modes the motion is predominantly horizontal whereas for p-modes the motion is predominantly vertical. The above boundary conditions can be used to demonstrate that this is so. The ratio of horizontal to vertical amplitude is:

$$\left(\frac{|\widetilde{\delta \mathbf{r}_h}|}{|\widetilde{\delta r}_r|}\right)_R = \frac{1}{R\omega^2}\left(\frac{\widetilde{p}}{\rho_0} + \widetilde{\Psi}\right)_R \frac{(\partial p_0/\partial r)_R}{\widetilde{p}(R)} \qquad (2.105)$$

where Eqs. (2.92) and (2.104) are used to eliminate the displacements. The term $\widetilde{\Psi}$ is often quite small, so that the expression reduces to:

$$\frac{|\widetilde{\boldsymbol{\delta r}_h}|}{|\widetilde{\delta r_r}|} \approx \frac{GM}{R^3\omega^2} .\qquad(2.106)$$

For p-modes this ratio is very small because ω is large. For g-modes this ratio is large because ω is small.

2.7 Oscillations in the Cowling Approximation

Sometimes it is helpful to consider an approximate system of differential equations of lower order, because they are simpler to solve and there are several mathematical theorems applicable which guarantee certain properties of the solutions and the frequencies as regards unicity. One approximation that is very frequently used is known as the Cowling approximation. In this approximation the perturbations of the gravitational potential in the oscillation equations are set to 0. It can be shown that this approximation is better for larger n and/or larger l where l is the degree of the spherical harmonic and n corresponds to the number of radial nodes. In this approximation the order of the system is reduced from four to two. It is useful to define a vector of the dependent variables:

$$\mathbf{y} = \begin{pmatrix} \widetilde{\delta r}_r \\ \widetilde{p} \end{pmatrix} \qquad(2.107)$$

and to express the system of differential equations in matrix form:

$$\frac{\partial \mathbf{y}}{\partial r} = \mathbf{A} \cdot \mathbf{y} \qquad(2.108)$$

with:

$$\mathbf{A} \equiv \begin{pmatrix} -\dfrac{2}{r} + \dfrac{1}{\Gamma_{1,0} H_p} & \dfrac{1}{\rho c_s^2}\left(\dfrac{S_l^2}{\omega^2} - 1\right) \\ \rho_0\left(\omega^2 - N^2\right) & -\dfrac{1}{\Gamma_{1,0} H_p} \end{pmatrix} \qquad(2.109)$$

in which the pressure scale height H_p is defined as:

$$H_p \equiv -\left(\frac{1}{p_0}\frac{\partial p_0}{\partial r}\right)^{-1} . \qquad(2.110)$$

For the purposes of the JWKB analysis discussed in section 2.8 it is desirable to cast the differential equation in the form:

$$\frac{\partial^2 X}{\partial r^2} + K(r)X = 0 \qquad (2.111)$$

where X is some linear combination with factors v_1 and v_2 of the displacement $\widetilde{\delta r}_r$ and the pressure perturbation $\widetilde{p}$ respectively. In vector notation:

$$X \equiv \mathbf{v}^T \cdot \mathbf{y} \,. \qquad (2.112)$$

Introducing definition (2.112) in Eq. (2.111), and using the short-hand notation of a prime for taking partial derivatives with respect to r for all elements in a quantity (scalar, vector, or matrix) yields:

$$\left[\mathbf{v}^{T\prime\prime} + 2\mathbf{v}^{T\prime} \cdot \mathbf{A} + \mathbf{v}^T \cdot \mathbf{A}' + \mathbf{v}^T \cdot \mathbf{A} \cdot \mathbf{A} \right] \cdot \mathbf{y} + K(r)\mathbf{v}^T \cdot \mathbf{y} = 0 \qquad (2.113)$$

where the matrix $\mathbf{A}$ is defined in Eq. (2.109). The vector $\mathbf{y}$ can be eliminated so that the condition for $\mathbf{v}^T$ becomes:

$$\mathbf{v}^{T\prime\prime} + 2\mathbf{v}^{T\prime} \cdot \mathbf{A} + \mathbf{v}^T \cdot \mathbf{A}' + \mathbf{v}^T \cdot \mathbf{A} \cdot \mathbf{A} + \mathbf{v}^T K(r) = 0 \,. \qquad (2.114)$$

The same type of equation could even apply to the full system of four ODEs, but in either case in its general form this remains difficult to solve for $\mathbf{v}^T$, and for systems of more than two coupled first order ODEs there is no reason to assume that there would be a solution. A special case has been addressed however, in several variants. Consider a vector $\mathbf{v}$ which can be expressed as:

$$\mathbf{v} = \begin{pmatrix} \beta(r) & 0 \\ 0 & \frac{1}{\beta(r)} \end{pmatrix} \widetilde{\mathbf{v}} \qquad (2.115)$$

where the vector $\widetilde{\mathbf{v}}$ is either independent of r, or varies slowly with r ($O(r^{-1})$). In this case the first and second derivatives of $\mathbf{v}$ satisfy:

$$\mathbf{v}' = (\ln \beta)' \begin{pmatrix} 1 & 0 \\ 0 & -1 \end{pmatrix} \widetilde{\mathbf{v}}$$

$$\mathbf{v}'' = (\ln \beta)'' \begin{pmatrix} 1 & 0 \\ 0 & -1 \end{pmatrix} \widetilde{\mathbf{v}} + \left[(\ln \beta)' \right]^2 \widetilde{\mathbf{v}} \,. \qquad (2.116)$$

This is introduced in Eq. (2.114) which reduces to:

$$\mathbf{v}^T \left\{ \left[(\ln \beta')' \right]^2 \mathbf{I} + (\ln \beta)'' \begin{pmatrix} 1 & 0 \\ 0 & -1 \end{pmatrix} + 2 (\ln \beta)' \begin{pmatrix} 1 & 0 \\ 0 & -1 \end{pmatrix} \cdot \mathbf{A} \right.$$
$$\left. + \mathbf{A}' + \mathbf{A} \cdot \mathbf{A} + K(r)\mathbf{I} \right\} = 0 \,. \quad (2.117)$$

From this equation it is seen that the vector $\mathbf{v}^T$ sought is an eigenvector of a matrix. The matrix $\widetilde{\mathbf{A}}$ for which the eigenvector needs to be found is:

$$\widetilde{\mathbf{A}} \equiv (\ln \beta)'' \begin{pmatrix} 1 & 0 \\ 0 & -1 \end{pmatrix} + 2 (\ln \beta)' \begin{pmatrix} 1 & 0 \\ 0 & -1 \end{pmatrix} \cdot \mathbf{A} + \mathbf{A}' + \mathbf{A} \cdot \mathbf{A} \,. \quad (2.118)$$

The function $K(r)$ is then determined by the associated eigenvalue $\lambda(r)$ as:

$$K(r) = -\lambda(r) - \left[(\ln \beta)' \right]^2 \,. \quad (2.119)$$

If the four elements of $\widetilde{\mathbf{A}}$ are indicated by $\widetilde{a}_{ij}$ then the eigenvalues $\lambda_{1,2}$ are obtained by setting the determinant to 0 and solving the resulting quadratic equation:

$$\lambda_{1,2} = \frac{\widetilde{a}_{11} + \widetilde{a}_{22}}{2} \pm \sqrt{\left(\frac{\widetilde{a}_{11} - \widetilde{a}_{22}}{2} \right)^2 + \widetilde{a}_{12}\widetilde{a}_{21}} \,. \quad (2.120)$$

And the associated eigenvectors can be expressed as:

$$\mathbf{v} = \frac{1}{\sqrt{(\lambda_{1,2} - \widetilde{a}_{22})^2 + \widetilde{a}_{12}^2}} \begin{pmatrix} \lambda_{1,2} - \widetilde{a}_{22} \\ \widetilde{a}_{12} \end{pmatrix} \quad \text{or}$$
$$\mathbf{v} = \frac{1}{\sqrt{(\lambda_{1,2} - \widetilde{a}_{11})^2 + \widetilde{a}_{21}^2}} \begin{pmatrix} \widetilde{a}_{21} \\ \lambda_{1,2} - \widetilde{a}_{11} \end{pmatrix} \,. \quad (2.121)$$

Using Eqs. (2.109) and (2.118) the three factors that enter the expression (2.120) for λ can be evaluated:

$$\frac{\widetilde{a}_{11} + \widetilde{a}_{22}}{2} = (\ln \beta)' \left[-\frac{2}{r} + \frac{2}{\Gamma_{1,0} H_p} \right] + \frac{1}{r^2} + \frac{1}{c_s^2} \left(\frac{S_l^2}{\omega^2} - 1 \right) (\omega^2 - N^2)$$
$$+ \frac{1}{2} \left[-\frac{2}{r} + \frac{1}{\Gamma_{1,0} H_p} \right]^2 + \frac{1}{2\Gamma_{1,0}^2 H_p^2} \quad (2.122)$$

$$\frac{\widetilde{a}_{11} - \widetilde{a}_{22}}{2} = \frac{1}{2} \left\{ 2 (\ln \beta)'' - \frac{4}{r} (\ln \beta)' + \frac{2}{r^2} - \frac{1}{\Gamma_{1,0} H_p} \frac{\partial}{\partial r} \ln (\Gamma_{1,0} H_p) \right.$$
$$\left. + \left[-\frac{2}{r} + \frac{1}{\Gamma_{1,0} H_p} \right]^2 - \frac{1}{\Gamma_{1,0}^2 H_p^2} \right\} \quad (2.123)$$

$$\tilde{a}_{12}\tilde{a}_{21} = \frac{1}{c_s^2}\left(\frac{S_l^2}{\omega^2}-1\right)(\omega^2-N^2)$$
$$\times \left\{2(\ln\beta)' + \frac{1}{H_\rho} + \frac{\partial}{\partial r}\ln\left[\frac{1}{c_s^2}\left(\frac{S_l^2}{\omega^2}-1\right)\right] - \frac{2}{r}\right\}$$
$$\times \left\{-2(\ln\beta)' - \frac{1}{H_\rho} + \frac{\partial}{\partial r}\ln[\omega^2-N^2] - \frac{2}{r}\right\} \quad (2.124)$$

in which a density scale height H_ρ is used:

$$H_\rho \equiv -\left(\frac{1}{\rho_0}\frac{\partial\rho_0}{\partial r}\right)^{-1}. \quad (2.125)$$

The freedom of choice of the function β now allows for several cases which are straightforward to work through. Two possible choices of β are selected because each achieves that $\tilde{a}_{12}\tilde{a}_{21} = 0$:

$$\beta = \rho^{1/2} r c_s \left(\frac{S_l^2}{\omega^2}-1\right)^{-1/2} \quad \text{or}$$
$$\beta = \rho^{1/2} r \left(\omega^2-N^2\right)^{-1/2}. \quad (2.126)$$

With the former choice for β, one of the eigenvectors reduces to $\tilde{\mathbf{v}}^T = (1\ 0)$, and with the latter choice of β one of the eigenvectors is $\tilde{\mathbf{v}}^T = (0\ 1)$. Unfortunately in either case problems arise at the locations where resp. $\omega^2 = S_l'^2$ or $\omega^2 = N^2$. These locations correspond respectively to turning points for the p-modes and g-modes, where their behaviour changes from oscillatory to exponentially decaying (evanescent). A third possibility, originally presented by [Deubner & Gough (1984)] is to set $\beta = \rho^{1/2}$. Additionally it is assumed that the waves can be modelled as propagating in locally plane-parallel layers which amounts to ignoring terms containing factors of $1/r$ and terms that are derivatives of the local acceleration due to gravity $g = \partial\Psi_0/\partial r$. In this approximation the eigenvector reduces to $\tilde{\mathbf{v}}^T = (g\ -1)$. The expression for the associated eigenvalue is:

$$\lambda = \frac{S_l^2}{c_s^2}\left(\frac{N^2}{\omega^2}-1\right) + \frac{\omega^2}{c_s^2} - \frac{1}{2H_\rho}\frac{\partial}{\partial r}\ln H_\rho. \quad (2.127)$$

At this point it is useful to define one further characteristic frequency which is a cut-off frequency ω_c:

$$\omega_c^2 \equiv \frac{c_s^2}{4H_\rho^2}\left(1 - 2\frac{\partial H_\rho}{\partial r}\right). \quad (2.128)$$

From Eq. (2.119) the expression for the function $K(r)$ that enters Eq. (2.111), this becomes:

$$K(r) = \frac{1}{c_s^2}\left[S_l^2\left(\frac{N^2}{\omega^2} - 1\right) + \omega^2 - \omega_c^2\right]. \qquad (2.129)$$

With this choice of $\mathbf{v}$ the quantity $X = \mathbf{v}^T \cdot \mathbf{y}$ is related to the displacement vector $\boldsymbol{\delta r}$ as:

$$X \equiv c^2 \rho^{1/2} \nabla \cdot \boldsymbol{\delta r}. \qquad (2.130)$$

In any location in the Sun where ω_c rises above the frequency of a wave, the waves becomes exponentially decaying. ω_c rises rapidly near the photosphere of the Sun because the density decreases very rapidly (so H_ρ becomes very small) which means that most normally observed oscillation modes become exponentially decaying outside a certain radius. Only modes with very high frequencies remain oscillatory in character outside $r = R$, which means that these are not 'trapped' inside the star. The quantity N is the buoyancy frequency. It can be shown that $K(r)$ is positive, and solutions oscillatory in character, if either $\omega > \omega_+(r)$ or $\omega < \omega_-(r)$ where ω_- and ω_+ depend on the sound speed, the buoyancy frequency, and the cut-off frequency. Modes for which $\omega > \omega_+(r)$ are primarily pressure waves for which gravity is a minor effect which means they behave much like sound waves: these are known as *p-modes*. Modes for which $\omega < \omega_-(r)$ are primarily gravity waves for which pressure is a minor effect: these are known as *g-modes*. It is worth noting that for p-modes the frequency usually increases with increasing number of radial nodes n, whereas for g-modes the frequency usually decreases with increasing number of radial nodes. Since the frequencies of modes are always greater than zero it means that the frequencies of distinct g-modes accumulate towards 0: in the frequency interval between 0 and ϵ for any ϵ there is an infinite number of g-mode eigenfrequencies. This classification in p-modes and g-modes works fine for the Sun and for unevolved stars that do not have strong composition gradients. However for more evolved stars it is quite usual that an eigenfrequency satisfies $\omega > \omega_+(r)$ for a certain range of radii r and also satisfies $\omega < \omega_-(r)$ over a different range. In such cases the modes are referred to as *mixed modes*.

One can see that this equation is similar to the one for the homogeneous gas sphere (2.66). If gravity is ignored then $g = 0$ so $N^2 = 0$, and if also the density is uniform then $H \to \infty$ and so $\omega_c = 0$. Setting the sound speed to a constant value then makes the equations identical.

2.8 JWKB Analysis

In principle solutions to equation (2.111) could easily be obtained using numerical techniques for solving boundary value problems. However, if a numerical solution is required then it is takes essentially identical effort to solve the full system of four ODEs Eqs. (2.95), (2.99), (2.100), and (2.101), with the advantage of not being restricted by the Cowling approximation. The problem is less complicated than that of stellar structure calculations or stellar evolution, because the problem is linear. As illustrated before, the primary issue is to ensure that numerical errors in the frequencies, introduced by having obtained the stellar structure on a mesh with finite resolution, are well below the expected measurement errors in the frequencies, and unbiased.

The value of working with (2.111) is that there exists a widely used analytical approximation technique for solving it, which is the JWKB method. Many generic properties of oscillation frequencies for the Sun and stars can already be understood from this relatively straightforward analysis. Furthermore, one outcome of this type of analysis is an inverse problem for the sound speed which can be inverted analytically.

The abbreviation JWKB method arises through the initials of Jeffreys, Wentzel, Kramers and Brillouin, who can be said to have developed the method, either together or independently, although more mathematicians also seem to have developed the same idea. A formal treatment of the method can be found for instance in [Bender & Orszag (1978)]. Starting point is to assume that solutions X of (2.111) are either oscillatory or exponentially decaying in character, modulated by an envelope which is a slowly varying function of the independent variable. For the sake of clarity, Eq. (2.111) is restated in a slightly different form:

$$\varepsilon^2 \frac{\partial^2}{\partial r^2} X = Q(r) X \qquad (2.131)$$

where the small parameter ε is introduced to represent explicitly that $|Q(r)|$ is large. One condition for the method to be valid is therefore that $Q(r) \neq 0$. Next, $X(r)$ is expanded in a series of the form:

$$X(r) \propto \exp\left[\frac{1}{\delta} \sum_{n=0}^{\infty} \delta^n S_n(r)\right] \qquad (2.132)$$

where δ is a small parameter. Differentiating X twice yields:

$$X' \propto \left(\frac{1}{\delta}\sum_{n=0}^{\infty} \delta^n S'_n\right) \exp\left[\frac{1}{\delta}\sum_{n=0}^{\infty} \delta^n S_n\right] \quad (2.133)$$

$$X'' \propto \left[\frac{1}{\delta^2}\left(\sum_{n=0}^{\infty} \delta^n S_n\right)^2 + \frac{1}{\delta}\sum_{n=0}^{\infty} \delta^n S''_n\right] \exp\left[\frac{1}{\delta}\delta^n S_n\right] . \quad (2.134)$$

Introducing these into Eq. (2.131), and dividing out the exponential factors, yields:

$$\frac{\varepsilon^2}{\delta^2} S_0'^2 + \frac{\varepsilon^2}{\delta}(2S_0' S_1' + S_0'') + \cdots = Q(r) . \quad (2.135)$$

The first term on the left-hand side is also the largest, and this must match $Q(r)$. Setting the two small parameters equal $\varepsilon = \delta$ then produces a scheme to determine all the S_n iteratively, since at each order in ϵ separately, all terms must cancel:

$$S_0'^2 = Q(r)$$
$$2S_0' S_1' + S_0'' = 0$$
$$2S_0' S_n' + S_{n-1}'' + \sum_{j=1}^{n-1} S_j' S_{n-j}' = 0 \; \forall \, n \geq 2 . \quad (2.136)$$

The first of these equations is sometimes referred to as the *eikonal*, and the second as the transport equation. The solutions to these first two equations are:

$$S_0(r) = \pm \int^r dr' \sqrt{Q(r')} \quad (2.137)$$

$$S_1(r) = Q_0 - \frac{1}{4}\ln Q(r) \quad (2.138)$$

where Q_0 is a constant of integration. Expressions for higher-order terms can be found in [Bender & Orszag (1978)], but they are not necessary for the present purpose.

One problem that arises with the JWKB method as outlined above is that the $Q(r)$ in the case at hand does have locations r where $Q(r) = 0$, which are *turning points* of the solution. In the solar case, for oscillation frequencies ω that are sufficiently large, there are two turning points a, b in the interval $[0, 1]$. In the intervals $[0, a]$ and $(b, 1]$ Q is positive, and in (a, b)

Q is negative. Where Q is positive and large one would expect, by using Eq. (2.138) in Eq. (2.132), that X behaves as:

$$X_+(r) = C_+ \left[Q(r)\right]^{-1/4} \exp\left[-\frac{1}{\varepsilon}\int dr' \sqrt{Q(r')}\right] \qquad (2.139)$$

where Q is negative and large one would expect that X behaves as:

$$X_-(r) = C_- \left[-Q(r)\right]^{-1/4} \sin\left[\frac{1}{\varepsilon}\int dr' \sqrt{-Q(r')} + \phi\right] . \qquad (2.140)$$

The problem is now reduced to connecting these two types of solutions through the two turning points a, b, which then fixes the constants of integration $C_\pm$ and ϕ. In the case at hand one can approximate the function Q by a Taylor expansion near the turning points and the leading term is:

$$Q(r) \approx \begin{cases} \alpha_a(r-a) & |r-a| \ll 1 \\ \alpha_b(r-b) & |r-b| \ll 1 \end{cases} \qquad (2.141)$$

where $\alpha_{a,b} \neq 0$. In this case Eq. (2.131) reduces to:

$$\varepsilon^2 \frac{\partial^2}{\partial r^2} X = \alpha_q(r-q) X \qquad (2.142)$$

with either $q = a$ or $q = b$. Solutions of this equation are known as Airy functions Ai and Bi:

$$X(r) = C_A \mathrm{Ai}\left(\varepsilon^{-2/3} \alpha_q^{1/3}(r-q)\right) + C_B \mathrm{Bi}\left(\varepsilon^{-2/3} \alpha_q^{1/3}(r-q)\right) . \qquad (2.143)$$

For large positive values of the argument the leading asymptotic behaviour of Ai and Bi is:

$$\mathrm{Ai}\left(\varepsilon^{-2/3}\alpha_q^{1/3}(r-q)\right) \approx \frac{1}{2\sqrt{\pi}} \varepsilon^{1/6} \left(\alpha_q(r-q)^3\right)^{-1/12}$$
$$\times \exp\left[-\frac{2\left(\alpha_q(r-q)^3\right)^{1/2}}{3\varepsilon}\right]$$

$$\mathrm{Bi}\left(\varepsilon^{-2/3}\alpha_q^{1/3}(r-q)\right) \approx \frac{1}{\sqrt{\pi}} \varepsilon^{1/6} \left(\alpha_q(r-q)^3\right)^{-1/12}$$
$$\times \exp\left[\frac{2\left(\alpha_q(r-q)^3\right)^{1/2}}{3\varepsilon}\right] . \qquad (2.144)$$

For large negative values of the argument the leading asymptotic behaviour of Ai and Bi is:

$$\text{Ai}\left(\varepsilon^{-2/3}\alpha_q^{1/3}(r-q)\right) \approx \frac{1}{\sqrt{\pi}}\varepsilon^{1/6}\left(-\alpha_q(r-q)^3\right)^{-1/12}$$

$$\times \sin\left[\frac{2\left(-\alpha_q(r-q)^3\right)^{1/2}}{3\varepsilon} + \frac{\pi}{4}\right]$$

$$\text{Bi}\left(\varepsilon^{-2/3}\alpha_q^{1/3}(r-q)\right) \approx \frac{1}{\sqrt{\pi}}\varepsilon^{1/6}\left(-\alpha_q(r-q)^3\right)^{-1/12}$$

$$\times \cos\left[\frac{2\left(-\alpha_q(r-q)^3\right)^{1/2}}{3\varepsilon} + \frac{\pi}{4}\right]. \quad (2.145)$$

A proof of these asymptotic behaviours can be found in [Bender & Orszag (1978)]. These Airy functions can now be used to connect the oscillatory and exponential behaviours around the two turning points. In order to obtain a solution over the entire interval $[0,1]$ all solutions must match up. As is demonstrated in detail in [Bender & Orszag (1978)] the condition guaranteeing this matching everywhere in the domain is:

$$\int_a^b dr' \sqrt{-Q(r')}/\varepsilon = \left(n + \frac{1}{2}\right)\pi + O(\varepsilon). \quad (2.146)$$

As demonstrated by [Christensen-Dalsgaard & Pérez-Hernández (1992)], the JWKB analysis applied to Eq. (2.111) with K as in Eq. (2.129) reproduces well the behaviour of the frequencies known as the Duvall law ([Duvall (1982)]). The standard expression of this law is:

$$\frac{(n + \alpha(\omega))\pi}{\omega} = F\left(\frac{\omega}{\sqrt{l(l+1)}}\right)$$

$$F(w) = \int_{\ln r_t(w)}^{\ln R} d\ln r \left(1 - \frac{c_s^2}{w^2 r^2}\right)^{1/2} \frac{r}{c_s}. \quad (2.147)$$

To obtain this from Eq. (2.146) some additional manipulation is necessary. The first point to note is that the buoyancy term N^2 is absent, which is allowed as long as $\omega^2 \gg N^2$. The second problem lies in the upper integration limit, which is the upper turning point in Eq. (2.146) whereas in the Duvall law it is set to the solar radius. The reason for this is that R can be measured independently (see appendix C), whereas the former is not known a-priori for the Sun, although of course it can be calculated for any model. In [Christensen-Dalsgaard & Pérez-Hernández (1992)] it is shown that this difference can be accounted for by introducing the frequency-dependent phase function $\alpha(\omega)$ instead of the constant $1/2$ in Eq. (2.146).

In the same sense as for the frequencies of the isothermal gas sphere in section 2.5, it is possible to calculate asymptotic resonant frequencies for the differential equation (2.111) using expression (2.129) in Eq. (2.146) (with $\varepsilon K(r) = -Q(r)$). In a homogeneous isothermal gas sphere, $c_s = $ const. and $N = \omega_c = 0$. The upper turning point r_2 is set equal to R. The lower turning point $r_1 = \sqrt{l(l+1)}c_s/\omega$. In this limiting case the two expressions should produce the same result:

$$\int_{r_1}^{R} dr \, \frac{1}{c_s} \left[\omega^2 - S_l^2\right]^{1/2} \approx \left(n + \frac{1}{2}\right) \pi \, . \qquad (2.148)$$

The integral is re-written, using the integration variable $\cos u \equiv r_1/r$, and integration limit $\delta \equiv \mathrm{acos}(r_1/R)$:

$$\left(n+\frac{1}{2}\right)\pi \approx \omega \int_0^R dr \, \frac{1}{c_s} - \left\{\omega \int_0^{r_1} dr \, \frac{1}{c_s} - \omega \int_{r_1}^R dr \, \frac{1}{c_s} \left[1 - \left(1 - \frac{S_l^2}{\omega^2}\right)^{1/2}\right]\right\}$$

$$= \omega \int_0^R dr \, \frac{1}{c_s} - \left[\sqrt{l(l+1)} + \frac{\omega r_1}{c_s} \int_0^\delta du \, (1 - \sin u) \frac{\sin u}{\cos^2 u}\right]$$

$$\approx \omega \int_0^R dr \, \frac{1}{c_s} - \sqrt{l(l+1)} \left[1 + \left(\frac{\pi}{2} - 1\right)\right] \, . \qquad (2.149)$$

In the final approximation r_1/R is set to 0. This reproduces the lowest order term for the expression, if one replaces $l(l+1)$ with $(l+1/2)^2$. This difference arises in part due to inaccuracies in the core (i.e. the approximation that $\delta = 0$), and there is also some influence at higher order due to setting $\omega_c = 0$ which means that the upper turning point disappears and has to be replaced by R. A much more detailed analysis, for a sound speed that is not constant, can be found in [Tassoul (1980)]. From this the asymptotic approximation for the frequencies $\nu_{nl} = \omega_{nl}/2\pi$ of p-modes is shown to be:

$$\nu_{p,nl} \approx \left(n + \frac{l}{2} + \alpha_p(\omega)\right) \Delta\nu - (Al(l+1) - \beta(\omega))\frac{\Delta\nu^2}{\nu_{nl}} \qquad (2.150)$$

where:

$$\Delta \nu = \left[2 \int_0^R \frac{dr}{c} \right]^{-1}$$

$$A = \frac{1}{4\pi^2 \Delta \nu} \left[\frac{c(R)}{R} - \int_0^R \frac{dc}{dr} \frac{dr}{r} \right] . \quad (2.151)$$

The phases $\alpha(\omega)$ and $\beta(\omega)$ express various inaccuracies in the approximations that go into deriving this asymptotic expression. One of them is that in the Sun where the density does not completely vanish at the surface, the boundary condition is neither quite the free boundary condition nor the clamped one but something in between. In terms of the JWKB approximation the upper turning point is not the solar radius R but slightly below that and it is a function of ω.

The quantity $\Delta \nu$ is known as the large separation because the frequencies appear to be (nearly) regularly spaced with this separation:

$$\nu_{n+1,l} - \nu_{n,l} \approx \Delta \nu . \quad (2.152)$$

There is almost degeneracy of modes $\nu_{n+1,l} \approx \nu_{n,l+2}$ which gives rise to the definition of small separation:

$$\delta \nu_{nl} \equiv \nu_{nl} - \nu_{n-1,l+2} \approx (4l+6) \frac{\Delta \nu}{4\pi^2 \nu_{nl}} \left[\frac{c(R)}{R} - \int_0^R \frac{dc}{dr} \frac{dr}{r} \right] . \quad (2.153)$$

Using the same procedures a similar expression for g-modes can also be derived (cf. [Tassoul (1980)]) which to lowest order yields:

$$\omega_{g,nl} = \frac{\sqrt{l(l+1)}}{\pi [n + l/2 + \alpha_g]} \int_0^R dr \frac{N}{r} . \quad (2.154)$$

2.9 Surface Effects

In section 2.6.2 it is pointed out that the boundary conditions play an important role in determining the resonant frequencies and that it makes a difference whether or not it is taken into account that the Sun has an atmosphere. Different boundary conditions do not affect the overall frequency spacing, but do produce an overall shift of the frequencies which can be (weakly) frequency dependent. In the context of the JWKB analysis of the previous section 2.8 the treatment of the atmosphere governs the behaviour of ω_c and through that the location of the upper turning point.

As pointed out in the previous section, part of the modification of frequencies through 'surface effects' are a consequence of the approximations employed to produce a problem that can be addressed analytically. These are 'mathematical' surface effects in the sense that the frequencies of the true Sun will be reproduced better by a numerical solution of the full system of ODEs than the frequencies obtained from the asymptotic analysis presented in sections 2.7 and 2.8.

There are also more physical causes for discrepancies in the frequencies, that have their origin near the surface of the Sun. The reflection at the upper turning point just below the surface occurs through the rapid rise of ω_c, which in turn is due to the density scale height H_ρ becoming very small. The precise behaviour of H_ρ is crucial for locating the upper turning point. The problem lies in that this region of the Sun is extremely complicated to model properly. The medium is convective, with highly turbulent overturning motions with velocities approaching the speed of sound. It is no longer optically very thick, so that the energy transport is neither dominated by the convection nor by radiation. Furthermore, the energy density in magnetic fields becomes a significant fraction of the thermal energy density in localised regions: an influence of which the models take no account whatever. The precise behaviour of H_ρ close to the surface in fact can only be determined through either full 3-D radiative magneto-hydrodynamical simulations, or equivalently a sophisticated treatment in terms of Reynolds' stresses, as discussed in section 2.2. The approximate treatments currently in widespread use, introduce inaccuracies in H_ρ, and hence in ω_c. At an even more detailed level the stochastic nature of turbulent convection implies that the solar surface is 'corrugated': the upper turning point of waves is dependent on θ and ϕ and varies stochastically. At some level, this too modifies frequencies of modes.

In [Rosenthal et al. (1999)] the effect is studied of turbulent near-sonic convection on the frequencies by calculating the frequency difference between a standard solar model and a model which uses radiative hydro-dynamical simulations to provide an improved density profile of the solar atmosphere. There it is shown that the contributions of turbulent pressure on the density profile are indeed of the right sign and magnitude to substantially improve the agreement between model frequencies and measured oscillation frequencies for the Sun. It is shown by [Christensen-Dalsgaard & Pérez-Hernández (1992)] that one expects any surface effect to produce a frequency difference between Sun and model that is a function of ω only, rather than being a function of $\omega/(l+1/2)$. This is also true for the results of [Rosenthal et al. (1999)]. As is shown by [Christensen-Dalsgaard & Pérez-Hernández (1992)], this property makes it possible to separate out the effects on the frequencies due to imperfect modelling of the near-surface layers. Appropriate filtering, i.e. removing slowly varying behaviour with ω, suppresses the influence of the surface layers while leaving unaffected the signature in the frequencies of differences between the Sun and a model in interior layers. This is discussed in more detail in section 3.1.2.

2.10 Excitation and Damping of Modes

Up to this point the oscillations are treated as linear and adiabatic waves. As a result all oscillations can be described as purely sinusoidal in time, and there is no method for determining the amplitude of any mode. The amplitude of the modes is determined by a balance between the excitation and damping of the mode. In the Sun the mode amplitudes are observed to be a small fraction of the sound speed ($< 10^{-5}$) and therefore non-linear effects are expected to be small. The effect of gas compression and dilation on the propagation of radiation can work either as excitation or damping, depending on whether the gas radiates more effectively when compressed or becomes more opaque when compressed. Integrated over the entire solar interior, one can show that the total contribution of these effects for all oscillations is negative: the oscillations are damped. The reason that oscillations are nevertheless observed, over a substantial range in frequency, is that the convection in the Sun is a turbulent flow, with velocities that approach the sound speed. As originally described for turbulent flow in the absence of external forcing by Lighthill (cf. [Lighthill (1978)]), turbulent flows generate sound waves. The power radiated in the sound waves is proportional to M^8 where M is the Mach number of the turbulent flow $M \equiv v/c_s$. The reason for this high power is that the emission is related to the quadrupole moment of the fluctuations in unforced turbulent fluids (a discussion of this can also be found in section 75 of [Landau & Lifshitz (1987)]). The monopole term disappears because of the absence of source terms in the equation of conservation of mass. The dipole term disappears because of the assumption of the absence of external forces. Although the Lighthill mechanism for sound generation does not strictly apply, since there is forcing due to radiation in the solar atmosphere, one still expects a stochastic energy input into the oscillations. The source in this case can be expected to also have monopole and dipole contributions. It is the balance between the forcing and the non-adiabatic damping effects which produces a finite amplitude for the oscillations. Some special cases of generation of sound by forced turbulence are discussed in more detail in [Goldreich & Kumar (1988)], and are further developed in [Goldreich et al. (1994)]. The character of the source in such calculations is the most complicated problem and observations (cf. [Strous et al. (2000)]), appear not to confirm some of the predicted properties of the acoustic power of [Goldreich & Kumar (1988)]. For alternative views see e.g. [Musielak et al. (1994)] and [Gabriel (2000)]. The damping of sound waves in turbulent flows is equally a subject of research, because apart from radiative dissipation, mode coupling between acoustic and magnetic modes, and interaction with the velocity fields of rotation and meridional circulation also contribute (cf. [Shergelashvili & Poedts (2005)]).

The amplitude of the turbulent velocities increases rapidly towards the surface before dropping to 0 where the solar atmosphere becomes optically thin, and stable against convection. This means that the power input into oscillations is very strongly localised. In the simplest model for the excitation the excitation source is therefore described by a Dirac delta-function at some depth below the surface. Such a model is explored in [Roxburgh & Vorontsov (1995)] and [Roxburgh & Vorontsov (1997)]. Starting point is to introduce a stochastic forcing function in the wave equation in the Cowling approximation Eq. (2.111) with Eq. (2.129). [Roxburgh & Vorontsov (1995)] and [Roxburgh & Vorontsov (1997)] transform this equation to a slightly different form. They use an independent variable τ which similar to the acoustic depth. The standard definition of acoustic depth is:

$$\tau = \int_r^R dr \, \frac{1}{c_s} \, . \quad (2.155)$$

Changing the dependent variable in Eq. (2.111) to $\xi = c^{-1/2}X$ at the same time then produces a wave equation of the form:

$$\frac{\partial^2 \xi}{\partial \tau^2} + \left[\omega^2 - V(\tau)\right]\xi = S(\tau, \omega) \quad (2.156)$$

in which the function $S(\tau, \omega)$ introduced on the right-hand side is the (Fourier transform of) the forcing function. The treatment is in an adiabatic approximation, where damping is taken care of by 'leakage' of acoustic power through the barrier of the potential V outside $r = R$. The acoustic potential V written as a function of radius is:

$$V(r) = S_l^2 \left(1 - \frac{N^2}{\omega^2}\right) + \omega_c^2 - \frac{1}{2} c_s \frac{\partial^2 c_s}{\partial r^2} + \frac{1}{4}\left(\frac{\partial c_s}{\partial r}\right)^2 \qquad (2.157)$$

which is implicitly a function of τ as long as there is a one-to-one relationship between r and τ. The solution of this equation can be obtained by obtaining the solutions of the homogeneous system (i.e. $S = 0$), and to construct a Green's function from the Wronskian of these solutions (cf. [Bender & Orszag (1978)] or [Zwillinger (1989)]). For forcing functions that are more complicated than a single Dirac delta function, the solution for ξ can be obtained by convolving this Green's function solution with the forcing function. [Nigam et al. (1998)] follow essentially the same route. The forcing function in the time domain $s(\tau, t)$ is stochastic with 0 mean, but has a finite expectation value for the mean of the square. In the Fourier domain, the power as a function of frequency and of depth $|S(\tau, \omega)|^2$ is deterministic and given by specifying the physical mechanisms responsible for the forcing and damping. An alternative discussion can be found

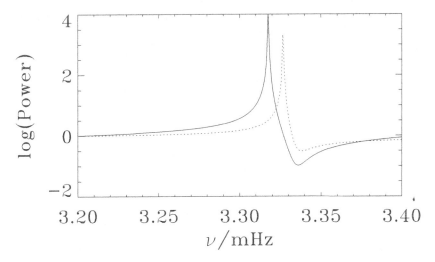

Fig. 2.6 The profile of acoustic power around a resonance for $l = 300$ in a polytropic atmosphere, for two different reflection conditions at the surface but with the same properties of the exciting source (from [Rosenthal (1998)] reproduced by permission of the AAS).

in [Rosenthal (1998)], which demonstrates the causes of asymmetry both through simplified toy models and also through a discussion improving on the treatment of [Roxburgh & Vorontsov (1995)] around the turning points of the modes. In [Rosenthal (1998)] it is clearly shown that the location of the minima in acoustic power as a function of frequency are a good indicator of the location and character of the exciting source. Fig. 2.6 illustrates that the maximum of acoustic power shifts when the boundary conditions change, but that the minimum remains at the same location, due to the fact that the properties of the exciting source have not been changed.

The difference between the treatments of [Roxburgh & Vorontsov (1997)] and [Nigam et al. (1998)] lies in the details of the forcing function and the correlation between the oscillations and the noise due to the convection. Subsequent work by [Georgobiani et al. (2003)] uses 3-D numerical hydrodynamical simulations to produce the forcing function, rather than a phenomenological description of the turbulent forcing. This work essentially confirms that the turbulent excitation mechanism does reproduce the observed asymmetries in the oscillation line profiles both in velocity and in intensity. The width and asymmetry of the lines therefore carry information concerning the excitation and damping of the modes. Models for the solar surface layers can thus be tested by comparing their predictions for the line widths and asymmetries with measurements obtained from the fitting procedure outlined in section 1.5.

Chapter 3

Global Helioseismology: Inverse Methods

Inverse methods rely on the relationship between the speed of propagation of waves throughout the interior of the Sun and the frequencies of the resonant oscillations. The frequencies are determined through solving a wave equation and are demonstrated to be measures of the propagation speed through an integral over the volume of the Sun. Inferring this propagation speed from oscillation frequencies therefore requires inverting an integral equation.

Inverse problems and their solutions are a distinct branch of mathematics. In fact the mathematical definition of inverse problems comprises a wider range of problems than is necessary to consider in the context of helio- and asteroseismology. In direct connection with measured data it is useful to consider in particular those methods that can be used in practice, for the type of inverse problems that occur. To focus the discussion, first it is useful to define the type of inverse problem of interest. In the most general form the problem to solve is the relationship between observables O_i for $i = 1, \ldots, N$, and the physical quantity or quantities of interest $Q_j(\mathbf{r})$ with $j = 1, \ldots, M$ wich are functions of location $\mathbf{r}$ within the Sun (or object).

$$O_i = \int d\mathbf{r}\, \mathcal{F}_i(Q_1, \ldots, Q_M, \mathbf{r}) \tag{3.1}$$

The function $\mathcal{F}$ of the Q can be non-linear. Although the formulation of Eq. (3.1) is general in terms of the number M of Q involved, the most common cases encountered in helioseismology are for $M = 1$ or 2. Furthermore, it is quite common to work with linearised problems, so that Eq. (3.1) reduces to:

$$O_i = \int d\mathbf{r}\, [K_{i1}(\mathbf{r})Q_1(\mathbf{r}) + K_{i2}(\mathbf{r})Q_2(\mathbf{r})] \, . \tag{3.2}$$

In the linear integral equation (3.2) the functions K_{i1} and K_{i2} are known as integration kernels or integration weights. The O_i are evidently weighted averages of the Q with the K as weighting functions. The K are also referred to as Fréchet derivatives, because they can be interpreted as a generalised partial derivative of the O_i with respect to the Q_j.

From Eqs. (3.1) and (3.2) one can see that the number of observables N is a finite number, and therefore so is the information content of any set of measurements, even in the absence of measurement errors. If the measurement errors are finite the information content is reduced. On the right hand side of Eqs. (3.1) and (3.2), each Q is a function on a domain. Since the functions Q describe physical quantities it is reasonable to assume they are at least piecewise continuous. They may well even be continuous everywhere on the domain or even differentiable, but the weaker condition suffices for this discussion. It is known that any piecewise continuous function on a bounded domain can be represented by an appropriate linear combination of orthogonal functions, but the number of such functions needed is at least in principle infinite. A more mathematical formulation of this is to state that the space of piecewise continuous functions is a Hilbert space which is infinite dimensional. Clearly then, the amount of information that would be needed to completely reconstruct the Q can never be obtained. It is this issue which makes all inverse problems *ill-posed*. Part of the Hilbert space is not constrained by a finite amount of measurements. Any element from this *null-space* can be added to a solution Q satisfying all the measurements and the result will satisfy all the measurements equally well. There is therefore no single unique solution.

Another property of ill-posed problems is that they are generally very sensitive to measurement errors. Even small changes in the measurements can produce solutions that look very different from one another. The reason for this is that although the data provide linearly independent constraints on the system, the associated functions under the integral sign are not orthonormal. Consider for example the kernels K of the linearised problem Eq. (3.2). Integrals of products of any two of such kernels are not usually equal to 0. A matrix with as its elements all these cross-product integrals is therefore filled, and symmetric. One can perform a singular value decomposition (SVD) of this matrix, which is essentially a projection: an orthonormal base set is created from a linear combination of the original kernels, which therefore spans the same sub-space of the Hilbert space that the original data kernels do. Suppressing for convenience the second index on the kernels K the procedure thus yields a matrix A with its SVD as:

$$A_{ij} = A_{ji} = \int d\mathbf{r}\, K_i(\mathbf{r}) K_j(\mathbf{r})$$
$$A = UDU^T \tag{3.3}$$

where the matrix U is orthonormal and the matrix D is diagonal, with the singular values as its elements. Normally one would arrange columns and rows of U in such a way that the singular values are arranged in D in descending order of magnitude, and usually the magnitude of the singular values decreases very rapidly. The linear combinations of data (and associated kernels) that go with these smallest of singular values are responsible for the sensitivity to data errors. If the data are altered within the errors in such a way that there is a change in the projection(s) associated with any of these small singular values a large change in the model is required to reproduce that change. A similar property also holds for the non-linear inverse problem. This means that although formally the N data span an N-dimensional sub-space of Hilbert space, operationally any finite signal-to-noise ratio reduces the dimension of this subspace to $N' < N$ because in some 'directions' in this Hilbert space an almost arbitrarily large change is allowed by the data with errors. The null-space of the data is therefore in some sense increased by the existence of measurement errors.

Any method for solving inverse problems must therefore contain some prescription which constrains the addition of elements from the null-space for the data at hand. Such prescriptions are referred to as regularisation. Regularisation can take different forms as is demonstrated in the following sections where various methods are discussed. To facilitate discussing the methods, the relationships between the observables, oscillations frequencies, and the quantities Q is presented first.

3.1 The Relationship between Frequencies and Sound Speed

From equations (2.111) and (2.129) it is clear that the sound speed comes into the wave equation and therefore one expects the frequencies to be sensitive to the sound speed. A precise way to formulate this is through what is known as the *variational principle* for oscillation frequencies.

The linear adiabatic wave equation with appropriate boundary conditions is known to be Hermitian or self-adjoint. That is to say, if the set of differential equations of the LAWE is expressed symbolically as a linear operator $\mathcal{L}$, with complex conjugate $\mathcal{L}^*$, acting on a function then Hermiticity means that:

$$\int \mathrm{d}V \; W \left\{ u^* \mathcal{L}(v) \right\} = \int \mathrm{d}V \; W \left\{ v \left[\mathcal{L}(u) \right]^* \right\} \qquad (3.4)$$

where V is a volume element and w an appropriate weighting function. In principle one can prove this by combining the set of four ODEs Eqs. (2.95), (2.99), (2.100), and (2.101) into a single fourth order system. This then defines the operator $\mathcal{L}$ and one works through the integrations. A more straightforward approach is to realise that the operator $\mathcal{L}$ is entirely real. One can carry out an integration by parts one or more times for all terms, so that one obtains an integral containing $\mathcal{L}(u^*)$ and since $\mathcal{L}$ is real this is identical to $[\mathcal{L}(u)]^*$. The above condition is therefore satisfied, but only as long as the boundary conditions of section 2.6.2 are satisfied, because they are used to eliminate the surface terms from the steps of integration by parts. From the discussion in section 2.9 one might worry that the true Sun does not exactly satisfy these conditions and that therefore Hermiticity is violated. However the departures can be demonstrated to be a small effect.

A consequence of the Hermiticity is that the spectrum of eigenfrequencies is discrete and that an inner product can be defined in such a way that the eigenfunctions form a base set for the Hilbert space of solutions: they are orthogonal and complete. Orthogonality means that the inner product between two eigenfunctions is only non-zero if the eigenfunctions are identical. Completeness means that all of Hilbert space is spanned by the eigenfunctions. Note that for this to be true one should also take into account toroidal modes, which all have frequency 0 in a spherically symmetric star, and for which all scalar perturbed variables are $= 0$ as well (cf. [Aizenman & Smeyers (1977)], [Unno et. al (1989)]). If one includes these then completeness is guaranteed as long as the boundary conditions are homogeneous (cf. [Courant & Hilbert (1953)]), i.e. expressible as:

$$\mathbf{B} \cdot \begin{pmatrix} \widetilde{\delta r_r} \\ \widetilde{p} \\ \widetilde{\Psi} \\ \frac{\partial \widetilde{\Psi}}{\partial r} \end{pmatrix} \qquad (3.5)$$

where $\mathbf{B}$ is a 4×2 matrix of coefficients. The zero-boundary conditions discussed in section 2.6.2, and also the more realistic boundary conditions considered thus far, can be formulated in this way. Any linear oscillation of the Sun and stars can therefore be decomposed uniquely into a linear combination of eigenmodes. The inner product has the form:

$$\langle \xi_a \cdot \xi_b \rangle \equiv \int dV \, W(\mathbf{r}) \xi_a \cdot \xi_b^* \qquad (3.6)$$

in which the ξ are eigenfunctions, and W is a weighting function. The $*$ superscript refers to taking the complex conjugate of the eigenfunction and the integral is over the volume of the Sun or star. Since there are several possible choices of variable with which to express the eigenfunction, there are various associated weighting functions W as well. The differential equation for the oscillations written can be written as an operator $\mathcal{L}_l$ working on the amplitude functions ξ of the oscillations so that:

$$\mathcal{L}_l(\xi_{n,l}) = \omega_{n,l}^2 \xi_{n,l} \,. \qquad (3.7)$$

The operator is given the subscript l because the spherical harmonic degree l, describing the azimuthal part of the oscillations, occurs explicitly in the operator. Taking the inner product with $\xi_{n,l}$ on left- and right-hand side then leads to an expression for the angular frequencies:

$$\omega_{n,l}^2 = \frac{\langle \xi_{n,l} \cdot \mathcal{L}_l(\xi_{n,l}) \rangle}{\langle \xi_{n,l} \cdot \xi_{n,l} \rangle} \,. \qquad (3.8)$$

It is demonstrated in [Chandrasekhar (1964)] that this frequency has a stationarity property in that $\delta\omega^2 = 0$ for arbitrary changes $\delta\xi$ around the eigenfunction ξ associated with ω. It is this property which is referred to as the variational principle.

An explicit expression for the frequencies is obtained following [Unno et. al (1989)]. For the eigenfunction ξ, the radial displacement vector is used so that the radial component is $\xi_r \equiv \widetilde{\delta r_r} Y_l^m(\theta, \phi) e^{-i\omega t}$ and the horizontal components $\xi_h \equiv \widetilde{\delta r_h} \left(0, \frac{\partial Y_l^m}{\partial \theta}, \frac{1}{\sin\theta}\frac{\partial Y_l^m}{\partial \phi}\right) e^{-i\omega t}$ (see Eq. 2.89). The appropriate weighting function is ρ_0 so that the inner product of ξ with itself is:

$$\int_0^R dr \int_{-1}^1 d\cos\theta \int_0^{2\pi} d\phi \, \rho_0 r^2 \xi \cdot \xi^* = \int_0^R dr \, \rho_0 r^2 \left|\widetilde{\delta r_r}\right|^2$$
$$+ \int_0^R dr \, \rho_0 r^2 \left|\widetilde{\delta r_h}\right|^2 \int_0^1 d\cos\theta \int_0^{2\pi} d\phi \left[\left|\frac{\partial Y_l^m}{\partial \theta}\right|^2 + \left|\frac{1}{\sin\theta}\frac{\partial Y_l^m}{\partial \phi}\right|^2\right]$$
$$= \int_0^R dr \, \rho_0 r^2 \left[\left|\widetilde{\delta r_r}\right|^2 + l(l+1)\left|\widetilde{\delta r_h}\right|^2\right]$$
$$\equiv E_{nl} \qquad (3.9)$$

where the reduction of the integrals over the angles is achieved by integration by parts. Using this definition of E_{nl} it is seen that:

$$\int dV\, \omega^2 \rho_0 r^2 \xi_{nl} \cdot \xi_{nl}^* = \omega^2 E_{nl}\,. \tag{3.10}$$

The next step is to use the oscillation equations to find a different expression for the same quantity that does not contain ω. By using Eq. (2.83), and also using that $\mathbf{v} = i\omega\xi$, one can see that:

$$\begin{aligned}\omega^2 \rho_0 r^2 \xi \cdot \xi^* &= \xi^* \cdot (\nabla p' + \rho_0 \nabla \Psi' + \rho' \nabla \Psi_0) \\ &= \nabla \cdot (p'\xi^* + \rho_0 \Psi' \xi^*) - p' \nabla \cdot \xi^* - \Psi' \nabla \cdot (\rho_0 \xi^*) \\ &\quad + \rho' \xi_r^* \frac{\partial \Psi_0}{\partial r}\,. \end{aligned} \tag{3.11}$$

From Eq. (2.82) it follows that $\rho' = -\nabla \cdot (\rho_0 \xi)$, and substituting this yields:

$$\begin{aligned}\omega^2 \rho_0 r^2 \xi \cdot \xi^* &= \nabla \cdot (p'\xi^* + \rho_0 \Psi' \xi^*) + p' \frac{\rho'^*}{\rho_0} + p' \frac{1}{\rho_0} \xi^* \cdot \nabla \rho_0 - \Psi' \rho'^* \\ &\quad + \rho' \xi_r^* \frac{\partial \Psi_0}{\partial r}\,. \end{aligned} \tag{3.12}$$

Eq. (2.94) can now be used to eliminate the ρ' terms. Note that although Eq. (2.94) refers to the quantities $\tilde{\rho}$, $\tilde{p}$, and $\tilde{\delta r}_r$, that are just the parts of the perturbation that are functions of r, the same relationship also holds for the primed quantities which is easily seen by multiplying with the factor $Y_l^m(\theta,\phi)e^{-i\omega t}$. The result is:

$$\begin{aligned}\omega^2 \rho_0 r^2 \xi \cdot \xi^* &= \nabla \cdot (p'\xi^* + \rho_0 \Psi' \xi^*) + \frac{1}{\rho_0 c_s^2} p'p'^* \\ &\quad + p' \xi_r^* \left(\frac{N^2}{\partial \Psi_0/\partial r} + \frac{1}{c_s^2}\frac{\partial \Psi_0}{\partial r} + \frac{\partial \ln \rho_0}{\partial r} \right) \\ &\quad + \rho_0 N^2 \xi_r \xi_r^* - \Psi' \rho'^* \\ &= \nabla \cdot (p'\xi^* + \rho_0 \Psi' \xi^*) + \frac{1}{\rho_0 c_s^2} p'p'^* \\ &\quad + \rho_0 N^2 \xi_r \xi_r^* - \Psi' \rho'^* \\ &= \nabla \cdot \left(p'\xi^* + \rho_0 \Psi' \xi^* + \frac{1}{4\pi G} \Psi' \nabla \Psi'^* \right) \\ &\quad + \frac{1}{\rho_0 c_s^2} p'p'^* + \rho_0 N^2 \xi_r \xi_r^* - \frac{1}{4\pi G} \nabla \Psi' \cdot \nabla \Psi'^* \end{aligned} \tag{3.13}$$

where the second equality follows from the definition (2.96), and the third equality follows after using Eq. (2.84). Integration over the volume of all terms and carrying through the integration of the divergence term (see appendix A) yields:

$$\omega^2 E_{nl} = \int_0^R dr \int_0^1 d\cos\theta \int_0^{2\pi} d\phi \left[\frac{p'p'^*}{\rho_0 c_s^2} + \rho_0 N^2 \xi_r \xi_r^* - \frac{1}{4\pi G} \nabla \Psi' \cdot \nabla \Psi'^* \right]$$
$$+ \left[p' \xi_r^* + \rho_0 \Psi' \xi_r^* + \frac{1}{4\pi G} \Psi' \nabla_r \Psi'^* \right]_{r=R}. \quad (3.14)$$

In this expression the terms under the integral are clearly symmetric in the eigenfunction which confirms the Hermiticity condition Eq. (3.4). The boundary conditions discussed in section 2.6.2 reduce the surface term to:

$$\left[p' \xi_r^* + \rho_0 \Psi' \xi_r^* + \frac{1}{4\pi G} \Psi' \nabla_r \Psi'^* \right]_{r=R} = \rho_0(R) \xi_r^*(R) \left[\Psi(R) + \Psi'(R) \right]$$
$$- \frac{l+1}{4\pi G} R |\Psi'(R)|^2. \quad (3.15)$$

The final term is also symmetric in the perturbation. The other two terms are proportional to $\rho_0(R)/R^2$ which is very small compared to the terms under the integral sign in Eq. (3.14), even if the density does not disappear. This therefore introduces only small corrections to the frequencies. Eqs. (2.100) and (2.92) are used to replace the $\tilde{p}$ terms, so that the integral is written only in terms of ξ_r and ξ_h and their derivatives. Perturbations of the gravitational potential are eliminated making use of Eqs. (2.84) and (2.82). The expression becomes:

$$\omega^2 E_{nl} = \int_0^R dr\, r^2 \left(\Gamma_{1,0} p_0 D_1^2 + 2 \frac{\partial p_0}{\partial r} D_1 \widetilde{\delta r}_r + \frac{\partial \ln \rho_0}{\partial r} \frac{\partial p_0}{\partial r} \widetilde{\delta r}_r^2 \right)$$
$$- \frac{8\pi G}{2l+1} \int_0^R dr \left[r^{-l+1} D_2(r) \int_0^r dr'\, r'^{l+2} D_2(r') \right]$$
$$+ \frac{8\pi G}{2l+1} \rho_0(R) \widetilde{\delta r}_r(R) R^{-l+1} \int_0^R dr\, D_2(r) r^{l+2}$$
$$+ \tilde{p}(R) \widetilde{\delta r}_r(R) R^2 - \frac{4\pi G}{2l+1} \rho_0^2(R) \widetilde{\delta r}_r^2(R) R^3 \quad (3.16)$$

in which use is made of the quantities D_1 and D_2 which are the amplitudes of the divergence of $\boldsymbol{\delta r}$ and $\rho_0 \boldsymbol{\delta r}$ respectively:

$$D_1 \equiv \frac{1}{r^2} \frac{\partial}{\partial r} (r^2 \widetilde{\delta r}_r) - \frac{l(l+1)}{r} \widetilde{\delta r}_h$$
$$D_2 \equiv \frac{1}{r^2} \frac{\partial}{\partial r} (r^2 \rho_0 \widetilde{\delta r}_r) - \frac{l(l+1)}{r} \rho_0 \widetilde{\delta r}_h. \quad (3.17)$$

As a consequence of the variational principle, Eq. (3.14) or Eq. (3.16) provide a means for calculating frequencies with a high accuracy even if the numerical accuracy with which the eigenfunctions are known is limited.

Even if the model available for the Sun or star is not correct but a reasonably good approximation, one can still use the eigenfunctions $\xi_{n,l}$ of the (slightly wrong) model to relate the difference in frequency between the star and the model to a difference $\delta\mathcal{L}_l$ between the star and the model. To see this, it is necessary to do a first-order perturbation analysis of Eq. (3.8):

$$\omega_{n,l}^2 + \delta\omega_{n,l}^2 = \frac{\langle (\xi_{n,l} + \delta\xi) \cdot (\mathcal{L}_l + \delta\mathcal{L}_l)(\xi_{n,l} + \delta\xi) \rangle}{\langle (\xi_{n,l} + \delta\xi) \cdot (\xi_{n,l} + \delta\xi) \rangle} . \qquad (3.18)$$

Since for any fixed l the $\xi_{n,l}$ are a basis-set there is a unique set of coefficients a_{nl} so that $\delta\xi = \sum_{n'} a_{n'l}\xi_{n',l}$ and therefore:

$$\delta\xi = a_{nl}\xi_{n,l} + \sum_{n' \neq n} a_{n',l}\xi_{n',l} . \qquad (3.19)$$

Therefore, since the $\xi_{n,l}$ are all orthogonal, it is true for any n, l pair that:

$$\langle \xi_{n,l} \cdot \delta\xi \rangle = a_{nl} \langle \xi_{n,l} \cdot \xi_{n,l} \cdot \rangle \qquad (3.20)$$

The denominator of (3.18) then reduces to:

$$\langle (\xi_{n,l} + \delta\xi) \cdot (\xi_{n,l} + \delta\xi) \rangle \approx (1 + 2a_{nl})\langle \xi_{n,l} \cdot \xi_{n,l} \rangle \qquad (3.21)$$

where second order terms have been discarded. The first order terms in the numerator of (3.18) are:

$$\langle \delta\xi \cdot \mathcal{L}_l \xi_{n,l} \rangle = \omega_{n,l}^2 \langle \delta\xi \cdot \xi_{n,l} \rangle = \omega_{n,l}^2 a_{nl} \langle \xi_{n,l} \cdot \xi_{n,l} \rangle$$

$$\langle \xi_{n,l} \cdot \mathcal{L}_l \delta\xi \rangle = \left\langle \xi_{n,l} \cdot \left(\sum_{n'} a_{n',l}\omega_{n',l}^2 \xi_{n',l} \right) \right\rangle = \omega_{n,l}^2 a_{nl} \langle \xi_{n,l} \cdot \xi_{n,l} \rangle$$

$$\text{and} \quad \langle \xi_{n,l} \cdot \delta\mathcal{L}_l \xi_{n,l} \rangle . \qquad (3.22)$$

Putting all of these terms back into (3.18), and eliminating the zero-order terms on both sides as well as an additional second order term, the result is:

$$\delta\omega_{n,l}^2 = \frac{\langle \xi_{n,l} \cdot \delta\mathcal{L}_l(\xi_{n,l}) \rangle}{\langle \xi_{n,l} \cdot \xi_{n,l} \rangle} . \qquad (3.23)$$

What is further often done is to divide left- and right hand sides by $\omega_{n,l}^2$ and therefore work with relative frequency differences:

$$\frac{\delta\omega_{n,l}^2}{\omega_{n,l}^2} = 2\frac{\delta\omega_{n,l}}{\omega_{n,l}} = \frac{\langle\xi_{n,l}\cdot\delta\mathcal{L}_l(\xi_{n,l})\rangle}{\omega_{n,l}^2\langle\xi_{n,l}\cdot\xi_{n,l}\rangle}. \qquad (3.24)$$

The denominator is merely a normalisation factor which can be calculated exactly from the model. The unknown lies in $\delta\mathcal{L}_l$, and since that occurs inside an inner product Eq. (3.24) is explicitly a linear integral relation between relative frequency differences between the true Sun and a model and the difference in quantities entering the wave equation between Sun and model and therefore the operator $\mathcal{L}_l$.

One can see from Eq. (3.16) that two state quantities enter the operator: the operator is determined completely by for instance the sound speed c and the density ρ of the equilibrium model, but any other independent combination of state variables would do as well. In the linear approximation the $\delta\mathcal{L}_l$ operator must therefore be expressible in terms of these two state variables.

$$\frac{\delta\omega_{n,l}}{\omega_{n,l}} = \int_0^R \left[K_{nl}^{c,\rho}(r)\frac{\delta c}{c}(r) + K_{nl}^{\rho,c}(r)\frac{\delta\rho}{\rho}(r) \right] dr \qquad (3.25)$$

where the weighting functions or 'kernels' K can be calculated exactly because they only depend on the calculated model and the oscillation mode functions for that model. These kernels K have superscripts c, ρ or ρ, c because the the stellar oscillations are affected by both sound speed and density. Differences in sound speed and differences in density between a star and a model affect the frequencies in separate ways, so one is the kernel for sound speed differences when there are no density differences (between star and model), and the other is the kernel for density differences when there are no sound speed differences. These two together fully determine the dynamical state of the gas. Only two are necessary because only two forces are considered: gas pressure and gravity. If for instance a magnetic field were to be introduced in Eq. (2.71) the order of the system of differential equations which is the LAWE would increase by two: the magnetic field is a vector quantity with three components but since it obeys $\nabla\cdot\mathbf{B} = 0$ only two are independent. Equation (3.25) would then also gain at least one term expressing the effect of changes in total magnetic field strength $\delta|B|/|B|$. The second additional term could then be for instance the ratio between poloidal and toroidal field strengths if the magnetic field is decomposed in this manner (cf. [Chandrasekhar (1961)] and appendix

B). Formulations of the dependence of frequencies on magnetic fields can be found in ([Dziembowski & Goode (1989)], [Gough & Thompson (1990)] and [Goode & Thompson (1992)]), although explicit calculations are only carried out for axisymmetric configurations of the magnetic field.

In the absence of such additional forces it is possible, making use of an EOS, to transform an integral equation for the pair c^2, ρ into any other linearly independent pair (cf. [Basu & Christensen-Dalsgaard (1997)], see also [Lin & Däppen (2005)]). Once the kernels for any one pair of variables are known, the others are straightforward to obtain making use of linear transformations where the transformation coefficients are derivatives among the various thermodynamic quantities. An EOS is needed to specify these derivatives between thermodynamic variables, and therefore uncertainties in the EOS must be taken into account when carrying out the inverse problem. This implies that changes in the composition of the material must also be present as terms in such transformed versions of this integral equation. A consequence of this is that uncertainties in the results of structure inversions do not just depend on the propagated data errors but also on the limitations to our knowledge in other areas of physics. For example it is possible to formulate an inverse problem to determine variations in the gravitational constant G from helioseismology, but the effect from variations in G is much smaller than other uncertainties in the models so that no useful constraint can at present be obtained (cf. [Christensen-Dalsgaard et al. (2005)]).

3.1.1 *Derivation of the variational structure kernels*

The variational structure kernels can be determined by perturbing all equilibrium quantities in Eq. (3.16) as is presented in [Gough & Thompson (1991)]. Note that in this section the perturbations are not functions of time. The differences indicated by δ are differences between a model and the true Sun, or between two models, at a given radius r. Carrying out the perturbation of the terms on the left-hand side yields the frequency perturbation and also a term in $\delta\rho/\rho$. This arises because ρ is the weighting function in the inner product, which contributes to frequency changes. On the right hand-side of Eq. (3.16) there is a double integral with terms δD_2. Since the perturbation of these terms require a bit more manipulation it is useful to consider these first:

$$\delta\left\{\int_0^R dr\left[r^{-l+1}D_2(r)\int_0^r dr'\,r'^{l+2}D_2(r')\right]\right\}$$

$$= \int_0^R dr\left[r^{-l+1}\delta D_2(r)\int_0^r dr'\,r'^{l+2}D_2(r')\right]$$

$$+ \int_0^R dr\left[r^{-l+1}D_2(r)\int_0^r dr'\,r'^{l+2}\delta D_2(r')\right]$$

$$= \left[\int_0^R dr\,r^{-l+1}D_2(r)\right]\left[\int_0^R dr\,r^{l+2}\delta D_2(r)\right]$$

$$+ \int_0^R dr\left[\delta D_2(r)\left(r^{-l+1}\int_0^r dr'\,r'^{l+2}D_2(r')\right.\right.$$

$$\left.\left.- r^{l+2}\int_0^r dr'\,r'^{-l+1}D_2(r')\right)\right]$$

$$= \int_0^R dr\,\delta D_2(r)\left[r^{-l+1}\int_0^r dr'\,r'^{l+2}D_2(r')\right]$$

$$+ \int_0^R dr\,\delta D_2(r)\left[r^{l+2}\int_r^R dr'\,r'^{-l+1}D_2(r')\right] \qquad (3.26)$$

which is obtained after an integration by parts of the outer integral, and then rearranging terms, by adjusting integration limits. The expression for δD_2 can be found using its definition Eq. (3.17):

$$\delta D_2 = \rho_0 D_1 \frac{\delta\rho}{\rho} + \widetilde{\delta r}_r \frac{\partial \delta\rho}{\partial r}. \qquad (3.27)$$

The first term in δD_2 is proportional to the relative density perturbation as required. The second term requires a further integration by parts:

$$\int_0^R dr\,\frac{\partial\delta\rho}{\partial r}\left[\widetilde{\delta r}_r r^{-l+1}\int_0^r dr'\,r'^{l+2}D_2(r')\right]$$

$$= -\int_0^R dr\,\delta\rho\left[r^3 D_2\widetilde{\delta r}_r + (-l+1)r^{-l}\widetilde{\delta r}_r\int_0^r dr'\,r'^{l+2}D_2(r')\right.$$

$$\left.+ \frac{\partial\widetilde{\delta r}_r}{\partial r}r^{-l+1}\int_0^r dr'\,r'^{l+2}D_2(r')\right] \qquad (3.28)$$

and

$$\int_0^R dr\, \frac{\partial \delta\rho}{\partial r} \left[\widetilde{\delta r}_r r^{l+2} \int_r^R dr'\, r'^{-l+1} D_2(r')\right]$$

$$= -\int_0^R dr\, \delta\rho \left[-r^3 D_2 \widetilde{\delta r}_r + (l+2)r^{l+1}\widetilde{\delta r}_r \int_r^R dr'\, r'^{-l+1} D_2(r')\right.$$

$$\left. + \frac{\partial \widetilde{\delta r}_r}{\partial r} r^{l+2} \int_r^R dr'\, r'^{-l+1} D_2(r')\right] \qquad (3.29)$$

in which it is assumed that there is no perturbation in density at the surface $\delta\rho(R) = 0$. In Eqs. (3.28) and (3.29) it is convenient to make the substitution:

$$\frac{\partial \widetilde{\delta r}_r}{\partial r} = D_1 - \frac{2}{r}\widetilde{\delta r}_r + \frac{l(l+1)}{r}\widetilde{\delta r}_h \,. \qquad (3.30)$$

Substituting these results back into Eq. (3.26), the D_1 terms and the $r^3 D_2$ cancel. Eq. (3.26) reduces to:

$$\delta\left\{\int_0^R dr\, \left[r^{-l+1} D_2(r) \int_0^r dr'\, r'^{l+2} D_2(r')\right]\right\}$$

$$= \int_0^R dr\, \frac{\delta\rho}{\rho}(l+1)\rho_0 r^{-l} \left[\widetilde{\delta r}_r - l\widetilde{\delta r}_h\right] \int_0^r dr'\, r'^{l+2} D_2(r')$$

$$- \int_0^R dr\, \frac{\delta\rho}{\rho} l\rho_0 r^{l+1} \left[\widetilde{\delta r}_r + (l+1)\widetilde{\delta r}_h\right] \int_r^R dr'\, r'^{-l+1} D_2(r') \,. \qquad (3.31)$$

All terms are written such that they appear as a ratio with the unperturbed value to express explicitly that the differences must remain small for the method to be valid. Collecting all perturbed terms, and making use of Eq. (3.31), the result of perturbing Eq. (3.16) is:

$$2\omega^2 E_{nl} \frac{\delta\omega}{\omega} + \omega^2 \int_0^R dr\, \rho_0 r^2 \frac{\delta\rho}{\rho} \left[\widetilde{\delta r}_r^2 + l(l+1)\widetilde{\delta r}_h^2\right] = \qquad (3.32)$$

$$\int_0^R dr\, \rho_0 r^2 \left[2D_1^2 c_s^2 \frac{\delta c_s}{c_s} + D_1^2 c_s^2 \frac{\delta \rho}{\rho}\right] + \int_0^R dr\, r^2 \frac{\partial \delta p}{\partial r}\left[2D_1 \widetilde{\delta r}_r + \widetilde{\delta r}_r^2 \frac{\partial \ln \rho_0}{\partial r}\right]$$

$$- \int_0^R dr\, \frac{\delta \rho}{\rho}\frac{\partial}{\partial r}\left(r^2 \widetilde{\delta r}_r^2 \frac{\partial p_0}{\partial r}\right)$$

$$- \frac{8\pi G}{2l+1}\int_0^R dr\, \frac{\delta\rho}{\rho}(l+1)\rho_0 r^{-l}\left[\widetilde{\delta r}_r - l\widetilde{\delta r}_h\right]\int_0^r dr'\, r'^{l+2} D_2(r')$$

$$+ \frac{8\pi G}{2l+1}\int_0^R dr\, \frac{\delta\rho}{\rho}l\rho_0 r^{l+1}\left[\widetilde{\delta r}_r + (l+1)\widetilde{\delta r}_h\right]\int_r^R dr'\, r'^{-l+1} D_2(r')$$

$$- \frac{8\pi G}{2l+1}\rho_0(R)\widetilde{\delta r}_r(R)R^{-l+1}\int_0^R dr\, \frac{\delta\rho}{\rho}l\rho_0 r^{l+1}\left[\widetilde{\delta r}_r + (l+1)\widetilde{\delta r}_h\right].$$

Note that in Eq. (3.16) in the first term on the right-hand side the substitution $\Gamma_{1,0}p_0 = \rho_0 c_s^2$ is made to yield the first term on the right-hand side of Eq. (3.32). The third term on the right-hand side requires perturbing $\partial \ln \rho/\partial r$, which is treated by performing an integration by parts. There are assumed to be no perturbations at the surface, so that the final two terms in Eq. (3.16) disappear. The second and third terms in Eq. (3.32) can be reduced further, making use of the fact that hydrostatic equilibrium must continue to be satisfied in the perturbation so that:

$$\frac{\partial \delta p}{\partial r} = -\frac{Gm}{r^2}\delta\rho - \frac{G\rho_0}{r^2}\int_0^r dr'\, 4\pi r'^2 \delta\rho \qquad (3.33)$$

in which the mass m interior to r is defined by:

$$m(r) = \int_0^r dr'\, 4\pi r'^2 \rho_0\,. \qquad (3.34)$$

Introducing this into the second term on the right-hand side in Eq. (3.32), and performing an integration by parts, yields:

$$\int_0^R dr\, r^2 \frac{\partial \delta p}{\partial r}\left[2D_1 \widetilde{\delta r}_r + \widetilde{\delta r}_r^2 \frac{\partial \ln \rho_0}{\partial r}\right]$$

$$= -\int_0^R dr\, \frac{\delta\rho}{\rho}Gm\rho_0 \widetilde{\delta r}_r\left[2D_1 + \widetilde{\delta r}_r \frac{\partial \ln \rho_0}{\partial r}\right]$$

$$- \int_0^R dr\, \frac{\delta\rho}{\rho}4\pi G\rho_0 r^2 \int_r^R dr'\, \rho_0 \widetilde{\delta r}_r \left[2D_1 + \widetilde{\delta r}_r \frac{\partial \ln \rho_0}{\partial r}\right] \qquad (3.35)$$

where the surface term from the integration by parts is accounted for by changing the integration limits in the second term. The third term in Eq. (3.32) is reduced to:

$$\int_0^R dr \, \frac{\delta\rho}{\rho} \frac{\partial}{\partial r} \left(r^2 \widetilde{\delta r}_r^2 \frac{\partial p_0}{\partial r} \right)$$
$$= -\int_0^R dr \, \frac{\delta\rho}{\rho} \left[2Gm\rho_0 \widetilde{\delta r}_r \frac{\partial \widetilde{\delta r}_r}{\partial r} + Gm \widetilde{\delta r}_r^2 \frac{\partial \rho_0}{\partial r} + 4\pi G r^2 \rho_0^2 \widetilde{\delta r}_r^2 \right] . \quad (3.36)$$

At this point all terms are reduced to a form in which are linear in either $\delta c_s/c_s$ or $\delta\rho/\rho$. This means that by rearranging the terms, the frequency perturbation can be cast into the form of Eq. (3.25). The expression for the sound speed kernel, with fixed ρ, is then:

$$K_{nl}^{c_s,\rho}(r) = \frac{\rho_0 r^2}{\omega^2 E_{nl}} c_s^2 D_1^2 . \quad (3.37)$$

The expression for the density kernel, with fixed c_s, is obtained by collecting all terms from Eqs. (3.32), (3.35), and (3.36), some of which cancel out:

$$K_{nl}^{\rho,c_s}(r) = \frac{\rho_0 r^2}{\omega^2 E_{nl}} \left\{ -\frac{1}{2}\omega^2 \left[\widetilde{\delta r}_r^2 + l(l+1)\widetilde{\delta r}_h^2 \right] + \frac{1}{2} c_s^2 D_1^2 \right.$$
$$+ \frac{2Gm}{r^3} \widetilde{\delta r}_r \left[\widetilde{\delta r}_r \left(\frac{\pi\rho_0 r^3}{m} - 1 \right) + \frac{1}{2} l(l+1) \widetilde{\delta r}_h \right]$$
$$- 4\pi G \int_r^R dr' \, \rho_0 \widetilde{\delta r}_r \left[D_1 + \frac{1}{2} \widetilde{\delta r}_r \frac{\partial \ln \rho_0}{\partial r} \right]$$
$$- \frac{4\pi G}{2l+1} \left[(l+1) r^{-l-2} \left(\widetilde{\delta r}_r - l \widetilde{\delta r}_h \right) \int_0^r dr' \, r'^{l+2} D_2(r') \right.$$
$$- l r^{l-1} \left(\widetilde{\delta r}_r + (l+1) \widetilde{\delta r}_h \right) \int_r^R dr' \, r'^{-l+1} D_2(r')$$
$$\left. \left. \times \rho_0(R) \widetilde{\delta r}_r(R) \left(\frac{r}{R} \right)^{l-1} l \left(\widetilde{\delta r}_r + (l+1) \widetilde{\delta r}_h \right) \right] \right\} . \quad (3.38)$$

An example of these kernels for one mode of a standard solar model is shown in Fig. 3.1. In order to derive the kernels the perturbation of the frequencies has been manipulated in such a way as to obtain terms that are linear in the pair of variables $(\delta c_s/c_s, \delta\rho/\rho)$. By making use of the definition

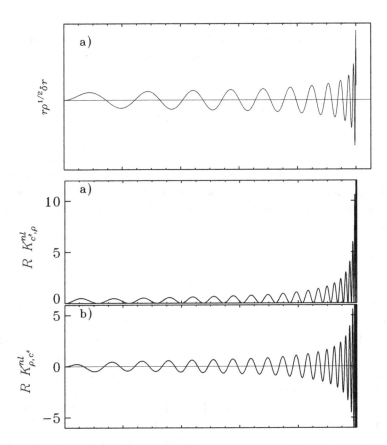

Fig. 3.1 Upper panel: the radial displacement eigenfunction $\widetilde{\delta r}_r$ scaled with a factor $r\rho^{1/2}$ for a mode with $(n,l) = (23,0)$ plotted as a function of radius from 0 to $R_\odot$, for a standard solar model. Middle and lower panels are the variational kernels for $K_{nl}^{c_s,\rho}$ and K_{nl}^{ρ,c_s} for the same model and $(n,l) = (21,0)$. (figure from [Christensen-Dalsgaard (2003)])

of c_s one can see that:

$$2\frac{\delta c_s}{c_s} = \frac{\delta \Gamma_1}{\Gamma_1} + \frac{\delta p}{p} - \frac{\delta \rho}{\rho} \ . \tag{3.39}$$

As pointed out by [Gough & Thompson (1991)], this together with Eq. (3.33) can be used the transform the integral for $\delta\omega$ to one with a pair of kernels for $(\delta\Gamma/\Gamma, \delta\rho/\rho)$, or any other pair that might be convenient.

In addition to the relationships between the frequencies and the relative difference in two dynamical variables, it is usual to add one additional constraint with the same form as the above relationships:

$$0 = \int_0^R \mathrm{d}r\, 4\pi r^2 \rho_0 \frac{\delta\rho}{\rho} \qquad (3.40)$$

which expresses that the difference in total mass between the Sun and the model must be equal to 0. The mass of the Sun is well constrained by measurements of the motion of planetary bodies in the solar system (cf. appendix C).

3.1.2 *Filtering out surface effects*

The kernels derived in the previous section allow frequency differences between two models, or between the Sun and a model, to be interpreted in terms of differences in sound speed and/or density between the two. This is also true for the frequencies calculated in the Cowling approximation from section 2.7 and 2.8. [Christensen-Dalsgaard & Pérez-Hernández (1992)] perturb the frequencies from the Duvall law in the same way as is done above, which results in an integral relationship between the frequency perturbation and the sound speed difference. This is a differential form of the Duvall law, and the integral is an Abel integral equation, discussed in more detail in the next section. Importantly however, in the analysis in [Christensen-Dalsgaard & Pérez-Hernández (1992)] the effect of surface changes is also investigated, by looking in more detail at the phase function α. An expression for $\alpha(\omega)$ is derived by taking at the difference between the Duvall law and the asymptotic relation for the frequencies from the Cowling approximation with $\omega^2 \gg N^2$. The appendix of [Christensen-Dalsgaard & Pérez-Hernández (1992)] shows that:

$$\frac{\pi(n+\epsilon)}{\omega} = F\left(\frac{\omega}{L}\right) - \frac{I_1 + I_2 + I_3}{\omega} \qquad (3.41)$$

in which the function $F(w)$ is the Duvall law (see Eq. (2.147)) and the I are three integrals:

$$I_1 = \omega \int_{r_2}^{R} dr \frac{1}{c_s} \left[1 - \frac{S_l^2}{\omega^2}\right]^{1/2}$$

$$I_2 = \omega \int_{r_1}^{r_2} dr \frac{1}{c_s} \left[\left(1 - \frac{S_l^2}{\omega^2}\right)^{1/2} - \left(1 - \frac{S_l^2 + \omega_c^2}{\omega^2}\right)^{1/2}\right]$$

$$I_3 = \omega \int_{r_t}^{r_1} dr \frac{1}{c_s} \left[1 - \frac{S_l^2}{\omega^2}\right]^{1/2}. \tag{3.42}$$

Given the integration limits for I_1 the term S_l^2/ω^2 can be neglected, so that the remainder is clearly a function of ω only (rather than depending on ω/L). I_3 does depend on ω/L but is smaller than the other two terms by a factor ω_c^2/ω^2 and is therefore negligible. With some additional manipulation (see [Christensen-Dalsgaard & Pérez-Hernández (1992)]) I_2 can also be shown to be a function of ω only. The result is that the sum of the three integrals can be defined to be the phase function α which is a function of ω only. This expression for the phase function can also be perturbed to investigate in detail where in the Sun the 'surface corrections' arise. This is done in several ways to demonstrate that the boundary conditions by themselves, or equivalently the precise location R taken for that boundary, has very little influence on the frequencies. The Hermiticity condition even if not mathematically satisfied, is satisfied to a sufficiently high degree that the effect on the frequencies is well below what can currently be measured. However, the acoustic cut-off frequency, which peaks slightly below the surface, does have an influence on the frequencies. In the same manner as shown in the previous section, it is possible to express the frequency differences in terms of an integral equation over a pair of perturbed variables, of which one is the acoustic cut-off frequency ω_c. These kernels have a large amplitude primarily very near the surface and the effect this type of perturbation has is that the frequency difference due to it is a function of ω but not of l.

The property of near-surface differences to produce an effect on the frequencies that is nearly independent of l can also be deduced directly from the expressions (3.37) and (3.38), together with the boundary conditions for the oscillations. From the boundary conditions (2.106) it is clear that the ratio of horizontal to vertical displacement $\widetilde{\delta r}_h/\widetilde{\delta r}_r$ becomes very small near the surface for p-modes. All terms containing factors of l or $l+1$ in Eqs. (3.37) and (3.38) can either be written such that they occur as a product with the square of that ratio or they occur in a ratio with other terms in l and the apparent dependence cancels out. Therefore near the surface these terms have very little influence on the frequency changes. From this one deduces that the frequency differences for near-surface changes remain essentially independent of l until l is so large that it compensates for the small ratio $\widetilde{\delta r}_h/\widetilde{\delta r}_r$. In the Sun one expects about equal magnitude of the terms for $l \approx 1000$.

As a consequence of this one can separate out near-surface changes formally by writing:

$$\frac{\delta\omega}{\omega} = \mathcal{H}_1\left(\frac{\omega}{\sqrt{l(l+1)}}\right) + E_{nl}^{-1}\mathcal{H}_2(\omega) \ . \qquad (3.43)$$

By using for instance spline functions in ω to represent $\mathcal{H}_2$, and subtracting this contribution from the frequency differences, one can focus on the deeper lying differences between models or between Sun and model. More details on the filtering procedure in the Cowling approximation can be found in [Pérez Hernández & Christensen-Dalsgaard (1994)] and for the general case with the variational kernels from the previous section in [Basu et al. (1996)]. In the latter paper it is shown that the filtering procedure can be approached either by transforming the data appropriately and subsequently applying linear inverse methods, or by adding additional linear constraints to those linear inverse methods. A more detailed discussion is therefore deferred to section 3.6.

3.2 The Abel Transform

An approach to the problem of determining c through measured frequencies, that does not rely on kernels which are calculated using a reference model, follows directly from the JWKB analysis of section 2.8. To understand how this works it is necessary to consider a particular class of integral equations known as Abel integral equations. Generalised Abel integral equations have the form:

$$F(u) = \int_{\xi^*}^{u} d\xi \, \frac{f(\xi)}{(u-\xi)^\lambda} \ . \qquad (3.44)$$

The special case $\lambda = 1/2$ is sometimes referred to as simply the Abel integral equation. These type of integrals can be inverted analytically (cf. [Courant & Hilbert (1953)]) and the expression is:

$$f(\xi) = \frac{\sin(\pi\lambda)}{\pi} \frac{d}{d\xi} \int_{u^*}^{\xi} du \, \frac{F(u)}{(\xi - u)^{1-\lambda}}$$

$$= \frac{\sin(\pi\lambda)}{\pi} \left[\frac{F(u^*)}{(\xi - u^*)^{1-\lambda}} + \int_{u^*}^{\xi} du \, \frac{dF'(u)/du}{(\xi - u)^{1-\lambda}} \right]. \quad (3.45)$$

For what follows an alternative form of these equations is more convenient, which is obtained by making the variable changes:

$$\begin{aligned} \zeta &\equiv (\xi^* + u^*) - \xi \\ v &\equiv (\xi^* + u^*) - u \end{aligned} \quad (3.46)$$

and defining the functions:

$$\begin{aligned} f^*(\zeta) &\equiv f((\xi^* + u^*) - \zeta) = f(\xi) \\ F^*(v) &\equiv F((\xi^* + u^*) - v) = F(u) \end{aligned} \quad (3.47)$$

so that Eq. (3.44) can also be written as:

$$F^*(v) = \int_v^{u^*} d\zeta \, \frac{f^*(\zeta)}{(\zeta - v)^\lambda} \quad (3.48)$$

and the inverse of this equation is obtained by making the same substitutions in Eq. (3.45)

$$f^*(\zeta) = -\frac{\sin(\pi\lambda)}{\pi} \frac{d}{d\zeta} \int_\zeta^{\xi^*} dv \, \frac{F^*(v)}{(v - \zeta)^{1-\lambda}}. \quad (3.49)$$

Comparing Eq. (3.44) with (3.48) it is clear that these are in fact identical but for the integration limits and the denominator under the integral sign. The corresponding inverse relations Eqs. (3.45) and (3.49) differ in the same manner, and in an additional minus sign. It was pointed out by [Gough (1984)] that the Duvall law can be cast into this second form of an Abel integral. Starting from:

$$F(w) = \int_{\ln r_t}^{\ln R} d\ln r \, \left(1 - \frac{c^2}{r^2 w^2}\right)^{1/2} \frac{r}{c} \qquad (3.50)$$

one takes the derivative with respect to w:

$$\frac{dF}{dw} = \int_{\ln r_t}^{\ln R} d\ln r \, \left(1 - \frac{c^2}{r^2 w^2}\right)^{-1/2} \frac{c}{rw^3} . \qquad (3.51)$$

Now one identifies $\xi = r^2/c^2$ and $u = 1/w^2$ so that this becomes:

$$\frac{-2}{w^3} \frac{dF}{du} = \int_{\ln r_t}^{\ln R} d\ln r \, (\xi - u)^{-1/2} \frac{1}{w^3} . \qquad (3.52)$$

Now one multiplies with w^3 and changes the integration variable from $\ln r$ to ξ:

$$-2 \frac{dF}{du} = \int_u^{R^2/c(R)^2} d\xi \, (\xi - u)^{-1/2} f(\xi) \qquad (3.53)$$

in which f is:

$$f(\xi) \equiv \frac{d\ln r}{d\xi} . \qquad (3.54)$$

Making use of Eqs. (3.48) and (3.49) then results in:

$$\frac{d\ln r}{d\xi} = \frac{2}{\pi} \frac{d}{d\xi} \int_\xi^{R^2/c(R)^2} du \, (u - \xi)^{-1/2} \frac{dF}{du} . \qquad (3.55)$$

Clearly one can integrate with respect to ξ on left- and right-hand side. Restoring the original variables and switching around the integration limits then produces:

$$\ln\left(\frac{r}{R}\right) = -\frac{2}{\pi} \int_{c(R)/R}^{c(r)/r} dw \, (w^{-2} - r^2/c(r)^2)^{-1/2} \frac{dF}{dw} . \qquad (3.56)$$

With this equation one determines r as a function of $c(r)/r$, given $F(w)$, which is implicitly a determination of $c(r)$ as a function of r.

This approach to performing the inversion for solar structure, in particular to obtain the sound speed, has advantages as well as disadvantages. An advantage is that no reference model is necessary in a quantitative sense. There is therefore no need to run an evolution code to obtain a solar model, nor a need for calculating the spectrum of resonant frequencies of such a model. Implicitly some model-based assumption is used, in that Eq. (3.50) assumes that an asymptotic representation of the frequencies is sufficiently accurate, i.e. one can use the Cowling approximation and JWKB analysis. Furthermore one assumes that the frequencies are large enough that the dispersion relation for the frequencies can be simplified to exclude effects of the buoyancy frequency. It turns out that this is possible in the Sun, as models suggested, but can not necessarily be generalised to other stars. However, the freedom from using a reference model in itself makes this method quite powerful.

The problem with using this equation lies in the function F, the derivative of which is under the integral sign. The function $F(w)$ is not known: from the measurements of ω and the mode identification in terms of n and l one obtains the values of F, ie. $(n + \alpha)\pi/\omega$, for values of its argument $w \equiv \omega/L$. n and L are discrete numbers and the frequencies ω are members of a discrete spectrum, and are measured with a finite precision. It is therefore fundamentally impossible to know F for all values of its argument, and even where it is known there is an uncertainty both in the argument and in the value. Strictly the derivative dF/dw is therefore not defined. This problem is dealt with by fitting a function or set of functions of w to the data $(n + \alpha)\pi/\omega$. This function will of course have a derivative. It is generally the case that a function that fits better to the data needs to have a higher order. If one uses polynomials in w this means adding higher powers of w. If one uses an orthogonal set of base functions in w to fit to the data, then this means adding functions that vary more over smaller changes of the argument. When taking the derivative of this fitting function, a variation over small changes of the argument is amplified. One can see from Eq. (3.56) that increasing variations in dF/dw produce increasing variations of the sound speed over small scales. Since some of the variations in F are produced by measurement errors, eventually it is effectively measurement error that is amplified to produce unrealistic small-scale variations in c. This is a generic property of inverse problems and a direct consequence of their *ill-posed*ness. One has to deal with this in practice by imposing limits on the number of components to fit F with. In terms of a reduced χ^2 of the fit, one should use no more than the smallest number of components that produces a reduced χ^2 measure between 1 and a few times that. What value of χ^2 exactly one must aim for depends on the character of the errors. In principle it is possible to asses this by doing simulations. One calculates frequencies given a sound speed profile, perturbs these frequencies with a random component satisfying the statistical properties of the measurement errors, and then carries out the fitting procedure to recover c. Ideally one repeats this several times for different realisations of the random noise representing measurement errors, which amounts to doing a *Monte-Carlo analysis*. One can then judge at what stage increasing the number of fitting components does not improve the agreement between the original and recovered values of c.

3.3 The Relationship between Frequencies and Rotation

Up to this point the Sun has been modelled as a static system. In fact the Sun rotates slowly and this has an effect on the frequencies. The ratio of the integrated kinetic energy in rotation and the total gravitational potential energy of the Sun is $\sim 10^{-6}$. This suggests that the rotation can be regarded as a small perturbation and one can therefore calculate its effect on the frequencies using the variational principle in the same way as is done in section 3.1. The perturbation analysis of section 2.6.1 requires some modifications to accommodate the equilibrium rotational velocity $\mathbf{v}_0$. This implies that the relationship between the time derivative of the displacement and the Eulerian velocity perturbation is modified because the displacement is relative to the moving fluid:

$$\frac{\partial \boldsymbol{\delta r}}{\partial t} = v' + \boldsymbol{\delta r} \cdot \nabla \mathbf{v}_0 - \mathbf{v}_0 \cdot \nabla \boldsymbol{\delta r} \tag{3.57}$$

in which v' is the Eulerian velocity perturbation associated with the oscillations. The perturbed equation of conservation of mass becomes:

$$\begin{aligned} 0 &= \frac{\partial \rho'}{\partial t} + \nabla \cdot (\rho' \mathbf{v}_0 + \rho_0 \mathbf{v}') \\ &= \frac{\partial}{\partial t} [\rho' + \nabla \cdot (\rho_0 \boldsymbol{\delta r})] + \nabla \cdot (\rho' \mathbf{v}_0 + \rho_0 \mathbf{v}_0 \cdot \nabla \boldsymbol{\delta r} - \rho_0 \boldsymbol{\delta r} \cdot \nabla \mathbf{v}_0) \ . \end{aligned} \tag{3.58}$$

The final two terms can be reduced as follows:

$$\begin{aligned}
\nabla \cdot &[\rho_0 \mathbf{v}_0 \cdot \nabla \boldsymbol{\delta r} - \rho_0 \boldsymbol{\delta r} \cdot \nabla \mathbf{v}_0] \\
&= \nabla \cdot [\rho_0 \mathbf{v}_0 \cdot \nabla \boldsymbol{\delta r} - \boldsymbol{\delta r} \cdot \nabla (\rho_0 \mathbf{v}_0) + \mathbf{v}_0 \boldsymbol{\delta r} \cdot \nabla \rho_0] \\
&= \nabla \cdot [\rho_0 \mathbf{v}_0 \cdot \nabla \boldsymbol{\delta r} - \boldsymbol{\delta r} \cdot \nabla (\rho_0 \mathbf{v}_0) + \mathbf{v}_0 (\nabla \cdot (\rho_0 \boldsymbol{\delta r}) - \rho_0 \nabla \cdot \boldsymbol{\delta r})] \\
&= \nabla \cdot [\mathbf{v}_0 \nabla \cdot (\rho_0 \boldsymbol{\delta r}) + \rho_0 \mathbf{v}_0 \cdot \nabla \boldsymbol{\delta r} - \boldsymbol{\delta r} \cdot \nabla (\rho_0 \mathbf{v}_0) - \mathbf{v}_0 \rho_0 \nabla \cdot \boldsymbol{\delta r}] \\
&= \nabla \cdot [\mathbf{v}_0 \nabla \cdot (\rho_0 \boldsymbol{\delta r}) + \nabla \times (\boldsymbol{\delta r} \times \rho_0 \mathbf{v}_0) - \boldsymbol{\delta r} \nabla \cdot (\rho_0 \mathbf{v}_0)] \\
&= \nabla \cdot [\mathbf{v}_0 \nabla \cdot (\rho_0 \boldsymbol{\delta r})] \ . \quad (3.59)
\end{aligned}$$

In the first and second equality the first of Eqs. (A.3) is used. The third equality is a re-arranging of terms. The fourth equality uses the third of Eqs. (A.4). The final equality relies on the second of Eqs. (A.2) and on $\nabla \cdot (\rho_0 \mathbf{v}_0) = 0$. This result means that Eq. (3.58) reduces to:

$$\begin{aligned}
0 &= \frac{\partial}{\partial t}[\rho' + \nabla \cdot (\rho_0 \boldsymbol{\delta r})] + \nabla \cdot [\mathbf{v}_0 (\rho' + \nabla \cdot (\rho_0 \boldsymbol{\delta r}))] \\
&= \rho_0 \frac{\partial}{\partial t}\left[\frac{1}{\rho_0}(\rho' + \nabla \cdot (\rho_0 \boldsymbol{\delta r}))\right] + \nabla \cdot \left[\rho_0 \mathbf{v}_0 \frac{1}{\rho_0}(\rho' + \nabla \cdot (\rho_0 \boldsymbol{\delta r}))\right] \\
&= \rho_0 \frac{\partial}{\partial t}\left[\frac{1}{\rho_0}(\rho' + \nabla \cdot (\rho_0 \boldsymbol{\delta r}))\right] + \rho_0 \mathbf{v}_0 \cdot \nabla \left[\frac{1}{\rho_0}(\rho' + \nabla \cdot (\rho_0 \boldsymbol{\delta r}))\right] \\
&= \rho_0 \frac{d}{dt}\left[\frac{1}{\rho_0}(\rho' + \nabla \cdot (\rho_0 \boldsymbol{\delta r}))\right] \quad (3.60)
\end{aligned}$$

where in the first equality use is made of the fact that ρ_0 is time-independent and in the second equality again of $\nabla \cdot (\rho_0 \mathbf{v}_0) = 0$. The third equality is the definition of the material time derivative. This implies that the perturbation quantity $\rho' + \nabla \cdot (\rho_0 \boldsymbol{\delta r})$ is time independent, and the only constant value that makes any physical sense is 0. This means that Eq. (2.88) is satisfied also when the equilibrium state is not static. The oscillations remain adiabatic and so the perturbed equation of conservation of energy is not modified either. The equation of motion Eq. (2.83) does need modification on its left-hand side. Starting point is to combine the left-hand terms of (2.71) with (2.70) and then to carry out a first-order perturbation, reducing terms by making use of Eqs. (3.57) and (3.59), and of $\rho' = -\nabla \cdot (\rho_0 \boldsymbol{\delta r})$:

$$\left[\rho_0 \frac{\partial \mathbf{v}}{\partial t} + \rho_0 \mathbf{v} \cdot \nabla \mathbf{v} \right]'$$

$$= \rho_0 \frac{\partial \mathbf{v}'}{\partial t} + \rho' \mathbf{v}_0 \cdot \nabla \mathbf{v}_0 + \rho_0 \mathbf{v}' \cdot \nabla \mathbf{v}_0 + \rho_0 \mathbf{v}_0 \cdot \nabla \mathbf{v}'$$

$$= \rho_0 \frac{\partial^2 \boldsymbol{\delta r}}{\partial t^2} + \rho_0 \mathbf{v}_0 \cdot \nabla \frac{\partial \boldsymbol{\delta r}}{\partial t} - \rho_0 \left(\frac{\partial \boldsymbol{\delta r}}{\partial t} \right) \cdot \nabla \mathbf{v}_0 + \rho' \mathbf{v}_0 \cdot \nabla \mathbf{v}_0$$

$$+ \rho_0 \left(\frac{\partial \boldsymbol{\delta r}}{\partial t} \right) \cdot \nabla \mathbf{v}_0 - \rho' \mathbf{v}_0 \cdot \nabla \mathbf{v}_0 + \nabla \times (\boldsymbol{\delta r} \times \rho_0 \mathbf{v}_0) \cdot \nabla \mathbf{v}_0$$

$$+ \rho_0 \mathbf{v}_0 \cdot \nabla \frac{\partial \boldsymbol{\delta r}}{\partial t} + \rho_0 \mathbf{v}_0 \cdot \nabla (\mathbf{v}_0 \cdot \nabla \boldsymbol{\delta r}) - \rho_0 \mathbf{v}_0 \cdot \nabla (\boldsymbol{\delta r} \cdot \nabla \mathbf{v}_0)$$

$$= \rho_0 \frac{\partial^2 \boldsymbol{\delta r}}{\partial t^2} + 2\rho_0 \mathbf{v}_0 \cdot \nabla \frac{\partial \boldsymbol{\delta r}}{\partial t} . \tag{3.61}$$

In the final equality terms of second order and higher in $\mathbf{v}_0$ are set to zero. This assumption implies that the velocity $\mathbf{v}_0$ is assumed to be small compared to the phase speed of the waves. Since the equilibrium structure is time-independent the perturbations are still separable with time dependence $e^{-i\omega t}$ so that the equivalent of Eq. (2.89) is:

$$-\omega^2 \rho_0 \widetilde{\delta r}_r - 2i\omega \rho_0 (\mathbf{v}_0 \cdot \nabla \boldsymbol{\delta r})_r = -\frac{\partial}{\partial r}\widetilde{p} - \widetilde{\rho}\frac{\partial}{\partial r}\Psi_0 - \rho_0 \frac{\partial}{\partial r}\widetilde{\Psi}$$

$$-\omega^2 \rho_0 \boldsymbol{\delta r}_h - 2i\omega \rho_0 (\mathbf{v}_0 \cdot \nabla \boldsymbol{\delta r})_h = -\left[\widetilde{p} + \rho_0 \widetilde{\Psi}\right] \frac{1}{r} \nabla_h Y_l^m . \tag{3.62}$$

This expression shows explicitly that the terms on the right hand side taken together form the operator on the eigenfunctions of oscillation as it was before without the term $\mathbf{v}_0$. The second term on the left-hand side can therefore be identified with the perturbation of that operator. In [Lynden-Bell & Ostriker (1967)] one can find a comprehensive treatment of the stability of rotating systems and proof of the Hermiticity of the resulting oscillation equations. The implication of this is that it is possible to use the variational principle in the same manner as is done in section 3.1.1, and produce an expression for the change in frequencies due to this perturbation. The explicit form of Eq. (3.23) with this perturbation of the operator becomes:

$$E_{nl}\delta\omega = -i \int_0^R dr \int_{-1}^1 d\cos\theta \int_0^{2\pi} d\phi \, \rho_0 r^2 \boldsymbol{\delta r}^* \cdot (\mathbf{v}_0 \cdot \nabla \boldsymbol{\delta r}) . \tag{3.63}$$

For pure rotation around a unique axis, and with a spherical coordinate system aligned appropriately, the velocity $\mathbf{v}_0$ only has a ϕ component so that (cf. appendix B):

$$\mathbf{v}_0 = \Omega(r,\theta) r \sin\theta \widehat{e}_\phi$$

$$\mathbf{v}_0 \cdot \nabla \boldsymbol{\delta r} = \Omega \frac{\partial}{\partial \phi} \boldsymbol{\delta r} + \Omega \left[-\delta r_\phi \sin\theta \widehat{e}_r - \delta r_\phi \cos\theta \widehat{e}_\theta \right.$$
$$\left. + (\delta r_r \sin\theta + \delta r_\theta \cos\theta) \widehat{e}_\phi \right]$$
$$= im\Omega \boldsymbol{\delta r} - \Omega \left[\widetilde{\delta r}_h \frac{\partial Y_l^m}{\partial \phi} \widehat{e}_r + \widetilde{\delta r}_h \cot\theta \frac{\partial Y_l^m}{\partial \phi} \widehat{e}_\theta \right.$$
$$\left. - \left(\widetilde{\delta r}_r \sin\theta Y_l^m + \widetilde{\delta r}_h \cos\theta \frac{\partial Y_l^m}{\partial \theta} \right) \widehat{e}_\phi \right] . \quad (3.64)$$

Substituting Eq. (3.64) in (3.63) and making use of the properties of the spherical harmonics Y_l^m then yields an expression for the frequency perturbations:

$$E_{nl}\delta\omega = m \int_0^R \mathrm{d}r \int_{-1}^1 \mathrm{d}\cos\theta\, \rho_0 r^2 \Omega(r,\theta) \left\{ \widetilde{\delta r}_r^2 P_l^{m\,2} - 2 P_l^{m\,2} \widetilde{\delta r}_r \widetilde{\delta r}_h \right.$$
$$\left. + \widetilde{\delta r}_h^2 \left[\left(\frac{\mathrm{d}P_l^m}{\mathrm{d}\theta} \right)^2 + \frac{m^2}{\sin^2\theta} P_l^{m\,2} \right] - 2 P_l^m \widetilde{\delta r}_h^2 \cot\theta \frac{\mathrm{d}P_l^m}{\mathrm{d}\theta} \right\} \quad (3.65)$$

which is, by construction, a linear integral equation in Ω. One can see that the frequency perturbation is proportional to m, although note that there is a further m-dependence under the integral sign. Modes with the same n and l but different m are degenerate in spherically symmetric configurations. The rotation breaks this spherical symmetry and produces a regular comb of frequencies centered on the frequency for $m=0$ for each n,l pair. Usually an explicitly symmetrised version of this equation is used where the data is the rotational splitting D_{nlm}:

$$D_{nlm} = \frac{\omega_{nl\,m} - \omega_{nl\,-m}}{2m} . \quad (3.66)$$

The above relationship can be used directly for inversions for rotation from the rotational splitting of the oscillation frequencies. The number of measured splittings currently available is large ($\sim 10^5$), which allows for a mapping of Ω as a function of r and θ at high spatial resolution. When using a regularised least-squares algorithm for the inversion (cf. section 3.7.1) it is not unusual to attempt to determine of the order of 10000 parameters.

Carrying out the inversion then involves factorising matrices which are $N \times M$ in size where is the number of data and M the number of parameters to fit for. In optimally localised averages (see section 3.5.2) one also needs to factorise a matrix which in this case is $N \times N$ in size. To carry out the inversions therefore becomes a very considerable computational task, and this is one reason that alternative ways of dealing with this data volume have been explored.

Another reason for representing the splitting data in different forms is that the width of the oscillation resonances at higher frequencies is of very similar magnitude to the rotational splitting and therefore it becomes increasingly difficult to isolate individual frequencies. Instead of relying on individual frequencies in multiplets one can fit orthogonal polynomials in m/l to the frequencies, and use the fitting coefficients as data. This is effectively a linear transformation. If one uses the same number of polynomials as there are m values in a multiplet $(2l+1)$, there is no loss of information. With decreasing signal-to-noise it makes sense to retain fewer coefficients for the polynomials and discard those corresponding to polynomials of higher degree. An appropriate choice of polynomials proposed in [Ritzwoller & Lavely (1991)] is now widely used, and such data is usually referred to as *a-coefficients*. Since the transformation between the frequency splittings and the a-coefficients is linear, one can formulate the linear inverse problem directly in terms of the integral relationships between the a-coefficients and the rotation, with appropriate kernels. Explicit expressions for the polynomials and the kernels can be found in [Pijpers (1997)].

Other than discarding high order a-coefficients to cut down the data volume one can pursue several paths to deal with the large data volume. One path is to rely on fast iterative techniques for factorising large ill-conditioned matrices such as [Larsen & Hansen (1997)]. Another path is to make use of the particular structure of the kernels. As pointed out in [Sekii (1993)] the rotational kernels are dominated by a term that is separable in r and θ, and furthermore separates out the dependence on m within each multiplet from the dependence on the multiplets, i.e.:

$$K_\Omega^{nlm}(r,\theta) \approx F_\Omega^{nl}(r) G_\Omega^{lm}(\theta) \qquad (3.67)$$

which means that the full 2D inversion can be split up into the product of two 1D inversion steps on subsets of data, with considerable savings in terms of computing effort. Taking this idea one step further [Pijpers & Thompson (1996)] describes that the full rotational integral equation can be written as:

$$D_{nlm} = \int_0^R dr \int_{-1}^1 du \, [F_1^{nl}(r)G_1^{lm}(u) + F_2^{nl}(r)G_2^{lm}(u)] \, \Omega(r,u) \quad (3.68)$$

where $u \equiv \cos\theta$. The functions $F_{1,2}$ are defined by:

$$\begin{aligned} F_1^{nl} &\equiv \rho(r)r^2 \left[\widetilde{\delta r}_{r\,nl}^2(r) - 2\widetilde{\delta r}_{r\,nl}(r)\widetilde{\delta r}_{h\,nl}(r) + L^2\widetilde{\delta r}_{h\,nl}^2(r)\right]/E_{nl} \\ F_2^{nl} &\equiv \rho(r)r^2 \left[\widetilde{\delta r}_{h\,nl}^2(r)\right]/E_{nl} \end{aligned} \quad (3.69)$$

and the functions $G_{1,2}$ by:

$$\begin{aligned} G_1^{lm} &\equiv [P_l^m(u)]^2 \\ G_2^{lm} &\equiv \frac{1}{2}(1-u^2)\frac{d^2 G_1^{lm}}{du^2} \end{aligned} \quad (3.70)$$

where the normalisation of the Legendre polynomials is such that:

$$\int_{-1}^1 du \, G_1^{lm}(u) = 1 = -\int_{-1}^1 du \, G_2^{lm}(u) \, . \quad (3.71)$$

This separation of the rotational kernel into a sum of two separable functions is not unique but it is convenient as shown in [Pijpers & Thompson (1996)] and [Pijpers (1997)]. A similar separation is also possible for the kernels where the data used are the a-coefficients. This form allows a generalisation of the original method with essentially the same computational advantages that the original approximate treatment has. These methods are referred to as $\mathbb{R}^1 \otimes \mathbb{R}^1$ methods because of the separation of a single 2-dimensional inverse problem into a set of 1-dimensional inverse problems. There is an additional advantage in using a-coefficients when doing SOLA $\mathbb{R}^1 \otimes \mathbb{R}^1$ inversions (cf. section 3.5.2). By using the a-coefficients as data, the inversions to be carried out to obtain localised estimates of Ω in latitude, require matrix factorisations of matrices that are tri-diagonal. This is less computational effort than factorising filled matrices, as is necessary for the rotational splittings. Effectively the projection of the splittings onto the a-coefficient fitting functions achieves a pre-conditioning of the matrices to be factorised. Further details on the implementation of the method with explicit expressions for the matrix elements relevant for SOLA can be found in [Pijpers & Thompson (1996)] and [Pijpers (1997)].

3.4 Linearisation

The relationship between oscillation frequencies and sound speed, rotation or other variables, are all non-linear in their most general form. Nevertheless, the problem is normally approached in such a way as to linearise the equations, as is demonstrated in section 3.1. Although inversion of non-linear integrals is possible in some cases, as in the Abel integral equation, it is difficult to assess the influence of measurement errors on the solution. Since inverse problems are notoriously sensitive to data errors it is of interest to linearise problems where possible.

In all the data analysis steps discussed in this chapter, inverse problems occur. Clearly the seismic analysis of oscillation frequencies is not the only place where inverse methods are to be applied, and so it is useful to consider non-linear inverse problems in a more general way. The procedure is quite straightforward. Consider again Eq. (3.1), in the case of the true Sun (or star) for which the Q are unknown, and for a model for which the Q are known and the O can be calculated. For the true Sun (or object) all quantities are given a $*$ subscript, for the model a subscript m:

$$O_{i*} = \int d\mathbf{r}\, \mathcal{F}_i(Q_{1*},\ldots,Q_{M*},\mathbf{r})$$
$$O_{im} = \int d\mathbf{r}\, \mathcal{F}_i(Q_{1m},\ldots,Q_{Mm},\mathbf{r}) \,. \tag{3.72}$$

Subtracting these two yields:

$$\begin{aligned} O_{i*} - O_{im} &= \int d\mathbf{r}\, [\mathcal{F}_i(Q_{1*},\ldots,Q_{M*},\mathbf{r}) - \mathcal{F}_i(Q_{1m},\ldots,Q_{Mm},\mathbf{r})] \\ &= \int d\mathbf{r}\, \sum_j \left.\frac{\partial \mathcal{F}_i}{\partial Q_j}\right|_m (Q_{j*} - Q_{jm}) \\ &\quad + \mathcal{O}\left((Q_{j*} - Q_{jm})^2\right) \end{aligned} \tag{3.73}$$

where the subscript m on the derivative refers to the derivative being taken at the model values of variables. The linearisation is done by setting all terms of second order and higher to 0. Evidently this requires that the model is already close to the true Sun (or object), by some measure.

In order to show explicitly that all the O are weighted averages of the

Q it makes sense to normalise the $\mathcal{F}$ and define a normalised function $\widehat{\mathcal{F}}$:

$$\widehat{\mathcal{F}}_i \equiv \mathcal{F}_i / \int d\mathbf{r}\, \mathcal{F}_i$$
$$= \mathcal{F}_i / O_{im} \,. \tag{3.74}$$

By construction the integral of $\widehat{\mathcal{F}}$ is 1. Introducing this in Eq. (3.73), and dropping the higher order terms yields:

$$O_{i*} - O_{im} = O_{im} \int d\mathbf{r} \sum_j \left.\frac{\partial \widehat{\mathcal{F}}_i}{\partial Q_j}\right|_m (Q_{j*} - Q_{jm}) \,. \tag{3.75}$$

One now views the relative differences as observables in their own right, and the relative difference in each Q as the parameters to be determined:

$$\widehat{O}_i = \frac{O_{i*} - O_{im}}{O_{im}}$$
$$\widehat{Q}_j = \frac{Q_{j*} - Q_{jm}}{Q_{jm}} \,. \tag{3.76}$$

Further define a kernel K as:

$$K_{ij} = Q_j \frac{\partial \widehat{\mathcal{F}}_i}{\partial Q_j} \,. \tag{3.77}$$

Introducing these definitions into Eq. (3.75) an equation of the type (3.2) is obtained:

$$\widehat{O}_i = \int d\mathbf{r} \sum_j K_{ij}(\mathbf{r}) \widehat{Q}_j(\mathbf{r}) \,. \tag{3.78}$$

Comparing Eqs. (3.78), (3.25) and (3.23) or (3.24) one can see explicitly that the expressions from section 3.1 are the same, if one makes the identifications:

$$\widehat{O}_{nl} = \frac{\delta \omega_{nl}}{\omega_{nl}} \tag{3.79}$$

$$\frac{\xi_{nl} \delta \mathcal{L}_l(\xi_{nl}^*)}{2\omega_{n,l}^2 \langle \xi_{nl} \cdot \xi_{nl} \rangle} = K_{nl}^{c,\rho}(r) \frac{\delta c}{c}(r) + K_{nl}^{\rho,c}(r) \frac{\delta \rho}{\rho}(r)$$
$$= \sum_{j=1,2} K_{nl\,j}(\mathbf{r}) \widehat{Q}_j(\mathbf{r}) \,. \tag{3.80}$$

The single index i of the observable $\widehat{O}$ is replaced by the nl pair for the relative frequency differences since this is the usual labelling of oscillation data. For a finite number of measured oscillation frequencies, the two ways of labelling are entirely equivalent.

3.5 Linear Methods

In linear methods for inverse problems each datum is multiplied by a weighting factor and then summed over all available frequencies. The way that these linear weights are determined differs among methods, but since the methods are linear, there is always a linear transformation that produces one from the other. Two methods in particular are in widespread use, and merit separate discussion.

3.5.1 *Regularised least-squares*

Perhaps the most straightforward way to attempt solving the inverse problem (3.2), is to parameterise the Q. One example of that is to express the Q in terms of a base set of orthonormal functions $b_{jm}(\mathbf{r})$. The inverse problem then reduces to solving a system of linear equations in which the unknowns are the coefficients of the orthogonal functions. One reason for choosing orthogonal functions to represent the Q is that the technique is then *robust*: the value that the coefficient for a particular component takes is independent of the number of components used to represent Q. Clearly the number of orthogonal components used can never exceed the number of independent data available, otherwise their coefficients become underdetermined. This theoretical limit can in practice not be attained either for two reasons. The kernels K are not necessarily in all cases linearly independent, which means that the dimension of the space they span is smaller than the number of functions. The data also have a finite signal-to-noise ratio which reduces their information content further.

The orthogonal functions $b_{jm}(\mathbf{r})$ can be any convenient set. For instance it is quite usual to parameterise the $\widehat{Q}$ by discretising space using M mesh points or nodes, and representing $\widehat{Q}$ in terms of piecewise constant or linear functions between the nodes. For each of the $\widehat{Q}_j$ therefore there are sets of coefficients q_{jm} to be determined:

$$\widehat{Q}_j = \sum_m q_{jm} b_{jm}(\mathbf{r}) \ . \tag{3.81}$$

Introducing this expression into Eq. (3.78) produces:

$$\begin{aligned}\widehat{O}_i &= \int d\mathbf{r} \sum_j K_{ij}(\mathbf{r}) \sum_m q_{jm} b_m(\mathbf{r}) \\ &= \sum_j \sum_m q_{jm} \int d\mathbf{r}\, K_{ij}(\mathbf{r}) b_m(\mathbf{r}) \ . \end{aligned} \quad (3.82)$$

Since the functions K and b both are known, the integrals can be evaluated trivially. At this point there is little distinction between the index j, for the several physical quantities occurring in the integral, and the index m for the number of components used to represent each of these. The two can be merged into a single m', as long as one keeps in mind that the number of components is $J \times M$. The next step is to represent the data by an N-element vector $\widehat{\mathbf{O}}$, and to collect the coefficients in a vector $\mathbf{q}$. A matrix $\mathbf{A}$ is defined by:

$$\mathbf{A}_{im'} \equiv \int d\mathbf{r}\, K_{ij}(\mathbf{r}) b_m(\mathbf{r}) \ . \quad (3.83)$$

With these definitions the problem then reduces to solving a set of N linear equations in $J \times M$ unknowns, in matrix form:

$$\widehat{\mathbf{O}} = \mathbf{A} \cdot \mathbf{q} \ . \quad (3.84)$$

This matrix can be large, but with $N > J \times M$ this is a standard linear least-squares problem for which many appropriate techniques exist. The problem with it lies in the ill-posed nature of inverse problems, and the implied sensitivity to data errors. A solution of Eq. (3.84) as is would become very non-smooth or 'noisy' because small measurement errors would be magnified many times. The solutions therefore would generally not be considered as physically acceptable. The approach to this is to regularise the inverse problem by modifying the matrix A. Regularisation is a collective name for any technique which is used to limit solutions to inverse problems to a class that have some properties considered desirable. This therefore involves to some extent the prejudice of physicists of what is acceptable. A usual choice for regularising the matrix of Eq. (3.83) is to add a second matrix which has the property that it suppresses large variations on small scales in the solutions. The most wide-spread method is known as Tikhonov regularisation. In Tikhonov regularisation, a balance is found between obtaining the best fit to the data in the least-squares sense and obtaining a solution which has a small overall mean square of the gradient C:

$$C^2 \propto \int dr \left(\frac{\partial Q}{\partial r}\right)^2 \qquad (3.85)$$

where Q is the parameter which is being inverted for (eg. $\delta c_s/c_s$ or Ω). This type of regularisation suppresses large gradients in the inferred Q which, although generally desirable, can also have adverse effects in suppressing real gradients such as occur in the rotation rate near the bottom of the convection layer in the Sun (the tachocline). Minimisation of C^2 leads to a set of linear equations for the coefficients: a smoothing matrix $\mathbf{D}$ which can take various forms, depending on the choice of the b functions. The amount of regularisation needed to obtain acceptable solutions, depends on the data errors. Therefore a trade-off parameter μ is used for tuning the method to the signal-to-noise available. With regularisation A is replaced by $\mathbf{A}'$:

$$\mathbf{A}' = \mathbf{A} + \mu \mathbf{D} . \qquad (3.86)$$

This regularised least-squares method will produce acceptable solutions and is widely in use. The only potential issue with it lies in the character of the regularisation. For instance, from solar structure models it is known that at the boundary between radiative and convective zones, some quantities have large gradients or second derivatives. The regularisation suppresses this and would thus in part suppress real behaviour together with spurious behaviour induced by measurement errors. Careful choice of the location of the mesh points can aid in alleviating this problem.

3.5.2 Optimally localised averages

A different point of view one can take is not to try to reconstruct a function. Instead one tries to obtain averages of the quantity sought over small spatial regions, around selected target locations $\mathbf{r}$. This is achieved by applying appropriately chosen linear weights to the data. With a vector of data-weights $\mathbf{w}$ the weighted average localised around $\mathbf{r}$ is $\langle Q_j \rangle(\mathbf{r})$:

$$\langle Q_j \rangle(\mathbf{r}) = \mathbf{w} \cdot \hat{\mathbf{O}} . \qquad (3.87)$$

Clearly a mechanism must then be found to choose the correct weights to achieve this. Substituting Eq. (3.78) for the $\widehat{O}_i$ yields:

$$\langle Q_j \rangle(\mathbf{r}) = \int d\mathbf{r}' \sum_{j'} \sum_i w_i K_{ij'}(\mathbf{r}') \widehat{Q}_{j'}(\mathbf{r}')$$
$$= \int d\mathbf{r}' \sum_{j'} \mathcal{K}_{j'}(\mathbf{r}') \widehat{Q}_{j'}(\mathbf{r}') \quad (3.88)$$

where the averaging kernels $\mathcal{K}_j$ are defined by:

$$\mathcal{K}_j(\mathbf{r}) \equiv \sum_i w_i K_{ij}(\mathbf{r}) \ . \quad (3.89)$$

In order to obtain a perfectly localised 'average' of the Q_j it would be desirable to have the $\mathcal{K}$ satisfy:

$$\mathcal{K}_j(\mathbf{r}') = \delta_{jj'} \delta(\mathbf{r}' - \mathbf{r}) \quad (3.90)$$

in which the first δ is the Kronecker delta, and the second is the Dirac δ-function. However this would require a complete set (i.e. infinite number) of error free data. This is evidently not realisable, and so the strategy is to make $\mathcal{K}$ as sharply peaked as the data allows.

A method currently often referred to as optimally localised averages (OLA) or multiplicative optimally localised averages (MOLA) was originally proposed by [Backus & Gilbert (1968)] [Backus & Gilbert (1970)], and the method is therefore also referred to as Backus & Gilbert method. For simplicity the index j on the $\mathcal{K}$ is suppressed here. In this method the weights are determined by minimising for the w:

$$\int d\mathbf{r}' [\mathcal{K}(\mathbf{r}')]^2 |\mathbf{r}' - \mathbf{r}|^2 \ . \quad (3.91)$$

The role of the function $|\mathbf{r}' - \mathbf{r}|^2$ is to penalise large values of $\mathcal{K}$ everywhere, except near $\mathbf{r} = \mathbf{r}'$ where $\mathcal{K}$ can be large. This minimisation also reduces to solving a set of coupled linear equations:

$$\mathbf{A} \cdot \mathbf{w} = \mathbf{b} \quad (3.92)$$

in which the matrix $\mathbf{A}$ is symmetric, with elements:

$$A_{lm} = \int d\mathbf{r}' K_l(\mathbf{r}') K_m(\mathbf{r}') |\mathbf{r}' - \mathbf{r}|^2 \ . \quad (3.93)$$

Regularisation is necessary, which is now achieved by adding to **A** the matrix of measurement error variances and co-variances **E**, with a trade-off parameter μ:

$$\mathbf{A}' = \mathbf{A} + \mu \mathbf{E} \ . \qquad (3.94)$$

This choice of regularisation reflects the fact that the weighted averages tend to produce error estimates for the result that are very large, which renders the determination essentially useless. The smaller domain over which one attempts to determine a localised average, the worse this effect becomes. There is therefore a trade-off between spatial resolution and acceptable error estimates. Eq. (3.94) is not complete however, without also adding the constraint that the weights w must sum to 1. Without such a constraint it is easy to see that setting the w all to 0 would minimise (3.91) perfectly, and free of errors, but this does not produce an average. By using a Lagrange multiplier, one can see that this constraint merely adds a row and column of 1's to the matrix $\mathbf{A}'$ and an extra diagonal element = 0. The vector **b** has all elements = 0 except the final one, corresponding to the constraint that the sum of the w be 1. This element is = 1. The matrix **A**' also has the property that it is symmetric.

This size of the matrix $\mathbf{A}'$ is now $(N+1) \times (N+1)$ which means that it is larger than the matrix resulting from the regularised least-squares algorithm described in the previous section. This is inconvenient, but more problematic is that the matrix elements depend on the target location **r**. For every target location the matrix elements change and the matrix needs to be factorised again. For large data-sets the computational burden is sufficiently large that the method in this form is rarely used.

A variation, called subtractive optimally localised averages (SOLA), circumvents this problem. In SOLA the weights are determined by minimising for the w:

$$\sum_j \int \mathrm{d}\mathbf{r}' \left[\mathcal{K}_j(\mathbf{r}') - \mathcal{T}_j(\mathbf{r}') \right]^2 \qquad (3.95)$$

where $\mathcal{T}_j$ is a target function for parameter Q_j which has the required behaviour. Note that in this formulation it is easier to express that under the integral sign there may be a linear combination of physical variables Q_j that need to be determined in isolation. For localisation around **r** a useful choice is a Gaussian:

$$\mathcal{T}_j(\mathbf{r}') = \frac{1}{\Delta} e^{-|\mathbf{r}' - \mathbf{r}|^2/\Delta^2} \ . \qquad (3.96)$$

The target function for all other j would be set to zero: $T_{j'\neq j}(\mathbf{r}') = 0$. Although there is sufficient freedom, by varying Δ, to keep magnification of measurement errors under control, normally one would include the same regularisation as in MOLA. The matrix to factorise is the same size as in MOLA but the elements of the matrix $\mathbf{A}$ and the vector $\mathbf{b}$ are now composed of:

$$A_{lmj} = \int d\mathbf{r}' \, K_{lj}(\mathbf{r}') K_{mj}(\mathbf{r}')$$

$$b_{lj} = \int d\mathbf{r}' \, K_{lj}(\mathbf{r}') T_j \,. \tag{3.97}$$

For the case where there is more than one independent variable j, one sums over the j with trade-off weights μ_j. Furthermore, to the matrix $\mathbf{A}$ again the error (co-)variance matrix is added with weight μ_e and the matrix is augmented by a row and column of 1's and a zero diagonal element. $\mathbf{b}$ is also augmented by a corresponding element which is set to 1, and summing with trade-off weights is involved if there is more than one independent variable.

$$\begin{aligned} \mathbf{A}' &= \sum_j \mu_j \mathbf{A}_j + \mu_e \mathbf{E} \\ \mathbf{b}' &= \sum_j \mu_j \mathbf{b}_j \end{aligned} \tag{3.98}$$

The vector of weights $\mathbf{w}$ is now obtained by solving:

$$\mathbf{A}' \cdot \mathbf{w} = \mathbf{b}' \,. \tag{3.99}$$

From Eq. (3.97) it can be seen that the matrix no longer depends on the target location, which means that there is only one matrix to factorise, as in the regularised least-squares case. Another advantage over MOLA is that it is straightforward to obtain a localised average of each of the Q_j separately. Also, for some applications it is useful to have the option to choose a target function T that is not equal to a Gaussian. As pointed out in [Pijpers & Thompson (1994)] SOLA allows one to estimate directly the gradient in quantities. This is applied to data in [Charbonneau et al. (1999)]. It should be noted that in [Jeffrey (1988)] a method is proposed that is very similar in spirit to SOLA, but with a target function that is a Dirac δ function. This implies an unrealistically high resolution which results in undesirable properties of the inversions.

3.6 Choosing Regularisation Weighting

In all regularisation scenarios, a parameter appears which weights the relative importance of fitting the data and of remaining close to the 'prejudice' for what a physically reasonable solution is. There is some discussion whether there are objective ways for choosing this parameter. Two tools are discussed in the literature. One is the L-curve criterion the other is weighted cross validation.

When plotting the error variance in the estimates of $\delta Q/Q$ obtained in the inversion against the spatial resolution that is achieved, one tends to obtain a curve which has the axes as asymptotes and a point of maximum curvature at some distance from the origin. The regularisation parameter determines where on this curve the inversion lies, and the L-curve criterion is to choose that value of the regularisation parameter that locates the inversion on the point of maximum curvature of the L-curve (cf. [Gough & Thompson (1991)] and [Pijpers & Thompson (1994)] for examples).

Another method is weighted cross validation, proposed in [Wahba (1977)]. This method is most easily understood for linear methods, because in their discretised from these all reduce to factorising a matrix. For SOLA this matrix is symmetric and can therefore be decomposed in a singular value decomposition where:

$$A_{\text{SOLA}} = U \cdot D \cdot U^T \tag{3.100}$$

in which U is a unitary matrix: the rows of U are orthonormal vectors. D is a diagonal matrix with the singular values as the diagonal elements. This decomposition is possible because A_{SOLA} is a square and symmetric matrix. For the RLS method the matrix is not square but it is nevertheless possible to do singular value decomposition so that:

$$A_{\text{RLS}} = V \cdot D \cdot U^T \tag{3.101}$$

in which D again is diagonal with the singular values as diagonal elements, and U and V again have rows that are orthonormal. In the unregularised form of SOLA and RLS the diagonal elements very rapidly decrease in magnitude, which is another expression of the ill-posedness of inverse problems. What the regularisation does is to produce a different diagonal matrix in which the elements decrease in magnitude down to a certain level after which the spectrum of eigenvalues flattens out. By adjusting the regularisation parameter the level at which the eigenvalues flatten out is raised or lowered. In weighted cross-validation the λ is adjusted so that one achieves an overall minimum in the χ^2 measure defined as:

$$\chi^2(\lambda) = \frac{1}{n}\sum_i (d_j - K_{ji}f_i)^2 w_i$$

$$w_i = \left[\frac{1}{n}\sum_i d_i\right]^{-1} d_i \ . \qquad (3.102)$$

The weighted cross-validation and L-curve criteria produce similar but not identical solutions to the inverse problem. In both cases the balance between regularisation on the one hand and fitting the data on the other is acceptable. The precise regularisation to use therefore remains to some extent a matter of taste or expedience, where the L-curve criterion or the generalised cross validation provide guidance.

The SVD itself and some related matrix factorisation schemes are also valuable tools in investigating what various types of regularisation achieve, and therefore which one is optimal for any given application. A detailed discussion of the SVD and some variants can be found in [Christensen-Dalsgaard et al. (1993)]. One such variant is a truncated SVD, in which regularisation is achieved by ignoring the smallest singular values and the associated vectors of U and V when inverting the matrix A. One can think of this as setting $D_i^{-1} \to 0$ if $D_i < \epsilon$. Although effective, this does not allow a great deal of control over for instance the smoothing that it implies in RLS inversions. [Christensen-Dalsgaard et al. (1993)] proposes instead to apply a generalised SVD (GSVD) on the pair of matrices A and D of Eq. (3.86). The GSVD of a pair of matrices A (with $M \times N$ elements) and B (with $P \times N$ elements) has the following form:

$$A = U \cdot D_A \cdot W^{-1}$$
$$B = V \cdot D_B \cdot W^{-1} \ . \qquad (3.103)$$

D_A is a diagonal matrix of which all elements are non-negative. With appropriate normalisation the first $N - P$ of these elements can be set $= 1$. D_B is also diagonal with the first $N - P$ elements $= 0$ and the rest is non-negative. The final $N - P$ diagonal elements $d_{i\,A,B}$ of the two matrices satisfy:

$$d_{i\,A}^2 + d_{i\,B}^2 = 1 \ . \qquad (3.104)$$

The rows of both U and V are orthonormal, whereas the rows of W are merely linearly independent. The generalised singular values of this matrix pair (A, B) are:

$$\gamma_i = \begin{cases} 1 & i = 1, \ldots, N - P \\ d_{iA}/d_{iB} & i = N - P + 1, \ldots, N \end{cases}. \quad (3.105)$$

In the case that $B = I$ the GSVD reduces to a normal SVD.

Making use of the GSVD, the discrete inverse problem $A \cdot x = b$ can be transformed to:

$$U^T A \cdot x = U^T \cdot b. \quad (3.106)$$

In effect a unitary transformation is applied to the data and simultaneously to the associated kernels. [Christensen-Dalsgaard et al. (1993)] argues that the matrix W is expected to be reasonably well-conditioned and so regularisation is achieved as:

$$x = \tilde{D} U^T \cdot b \cdot W \quad (3.107)$$

in which the elements of the diagonal matrix are:

$$d_i = \frac{1}{d_{iA}} \times \begin{cases} 1 & i = 1, \ldots, N - P \\ \dfrac{\gamma_i^2}{\gamma_i^2 + \mu^2} & i = N - P + 1, \ldots, N \end{cases} \quad (3.108)$$

which can be interpreted as applying an optimal filter to the d_{iA}^{-1}. One can see that the GSVD of the matrix pair need be carried out only once. The regularisation weight μ appears only in the filter factors, and so it requires little computational effort to explore values of μ to determine which value is optimal. [Christensen-Dalsgaard et al. (1993)] also shows some examples of the kernels for rotation after the unitary transformation obtained through the GSVD, i.e. the kernels $\tilde{K}_i$ obtained by:

$$\tilde{K}_i(\mathbf{x}) \equiv \sum_j U_{ji} K_j(\mathbf{x}). \quad (3.109)$$

From [Christensen-Dalsgaard et al. (1993)] it becomes clear that the $\tilde{K}_i$ form something close to a wavelet basis, designed by the GSVD to be the most appropriate one for the problem given the original kernels and the smoothing requirements.

3.7 Non-Linear Methods

The linear inverse methods discussed in the previous section have found most widespread use. However, in some cases non-linear methods have been successfully employed and it is therefore useful to consider these as well. A comparison of a variety of inverse methods using non-linear techniques can be also found in [Jeffrey & Rosner (1986a)] and [Jeffrey & Rosner (1986b)] which include some applications to helioseismology although this is not the exclusive focus of these papers. Methods for inversion can be non-linear either because the problem itself is not linearised, or because the regularisation of the problem introduces non-linearity. An example of the former can be found in [Marchenkov et al. (2000)]. An example of the latter are [Corbard et al. (1998)] and [Corbard et al. (1999)]. In [Corbard et al. (1998)] several regularisations are described. One type of regularisation described in [Corbard et al. (1998)] is the *total variation minimisation* (TVM) in which the balance is between the data misfit and the total variation V_{tot} of Q:

$$V_{\text{tot}} \propto \int \mathrm{d}r \left| \frac{\partial Q}{\partial r} \right| . \qquad (3.110)$$

This method is less restrictive in allowing high gradients but [Corbard et al. (1998)] note that the method appears unstable in regions where the gradient in Q is small. In [Corbard et al. (1999)] a variation is also proposed which is a combination of this weighting function and Tikhonov regularisation (cf. section 3.5.1) so that it reduces to Tikhonov smoothing where the gradient is small and to TVM where the gradient is large. This method produces satisfactory results for inversions for the solar rotation rate.

3.7.1 *The maximum entropy method*

Probably the most well-known inverse methods which employs a non-linear regularisation function is the Maximum Entropy Method (MEM). Although it has been suggested as a potential alternative to more traditional inverse methods it has not yet seen much use within helioseismology. MEM has been used for time-series analysis in other fields cf. [Press et al. (1992)]) and therefore it could also find application to that aspect of helio- and asteroseismology. There is therefore sufficient reason to discuss the method here.

The starting point is Bayes' relation between conditional probabilities, which when applied to the case at hand concerns the probabilities of an object and the measurements:

$$P(\{\text{object}\}|\{\text{measurements}\})$$
$$P(\{\text{measurements}\}|\{\text{object}\}) \frac{P(\{\text{object}\})}{P(\{\text{measurements}\})} \quad . \quad (3.111)$$

For definiteness the individual measurements are d_j where $j = 1, \ldots, N$. The unknown {object}) is a function f of coordinates $\mathbf{x}$, which is parameterised so that:

$$f(\mathbf{x}) = \sum f_i b_i(\mathbf{x}) \quad (3.112)$$

where the b_i are a base set of functions (ideally orthogonal and normalised). The conditional probability $P(\{\text{object}\}|\{\text{measurements}\})$ is the probability for the parameters f_i to have certain values, given the set measurements {measurements}. It is this probability which must be maximised by the algorithm. $P(\{\text{measurements}\}|\{\text{object}\})$ is the probability of finding measurements {measurements} when given an object {object}. These probabilities can be calculated in a straightforward manner. One normally assumes that the measurement noise N arises due to uncorrelated Gaussian random processes with variance σ^2 yields and therefore that one can calculate $P(\{\text{measurements}\}|\{\text{object}\})$ from:

$$P(\{\text{measurements}\}|\{\text{object}\}) \propto \prod_j \exp\left(-\frac{N_j^2}{2\sigma_j^2}\right)$$
$$\propto \prod_j \exp\left[-\frac{(d_j - \sum_i K_{ji} f_i)^2}{2\sigma_j^2}\right] \quad . \quad (3.113)$$

The matrix elements K_{ji} are obtained by evaluating:

$$K_{ji} = \int d\mathbf{x}\, K_j(\mathbf{x}) b_i(\mathbf{x}) \quad (3.114)$$

where the K_j are the kernels associated with the measurements d_j. This equation generalises in a straightforward manner if there are two or more unknown functions $f^{(1)}, f^{(2)}, \ldots$ and the associated kernels $K_j^{(1)}, K_j^{(2)}, \ldots$ by adding a summation over the unknown functions. If the relationship between data and unknown is non-linear one also can generalise the expression but normally one would carry out the linearisation as demonstrated formally in section 3.4.

The a-priori probability of the measurements $P(\{\text{measurements}\})$ is independent of the unknown $\{\text{object}\}$ so in a maximisation of the likelihood of $\{\text{object}\}$ it is unimportant since its role is that of normalisation. The a priori distribution of objects $P(\{\text{object}\})$ is where the entropy is required: it reflects our prejudice of how likely any given object is.

$$P(\{\text{object}\}) \equiv P(f_i) \propto \exp S(f_i) \qquad (3.115)$$

Here $S(f_i)$ is the entropy of the object which is parameterised with the f_i. In other words we need to be able to assign a probability p_i for the ith parameter to have value f_i. The advantage of defining the entropy like this is that it is additive for independent systems. The probabilities p_i in the case of observing a number of photons which are distributed over a number of pixels/bins are the fraction in pixel i of the total number of photons i.e. $p_i = f_i / \sum_i f_i$. Since the number of photons is positive semi-definite it is trivial to construct a quantity that is positive semi-definite over its domain and normalised to unity, and therefore a valid probability distribution function. In astronomy inversion problems for which such a-priori information is applicable or which can be slightly reformulated so that this applies (such as the deconvolution of images) are quite common so this is not considered to be a severe restriction and can even be an advantage.

Introducing Eqs. (3.113) and (3.115) in (3.111) leads to:

$$\ln P(\{\text{object}\}|\{\text{measurements}\}) = S(f_i) - \frac{1}{2}\sum_j \left(\frac{d_j - \sum_i K_{ji}f_i}{\sigma_j}\right)^2 \qquad (3.116)$$

which is the function to be minimised for the f_i. It is clear that equation (3.116) is a least squares or maximum likelihood estimator which is regularised by the non-linear entropy function S. Because this function is non-linear the minimisation algorithm is not particularly simple and is usually implemented using some form of iterative minimisation (cf. [Press et al. (1992)] for algorithms). Because of the non-linearity, estimating the propagation of measurement errors to the final result also is not a trivial matter and usually is achieved with Monte Carlo simulations . This process can become very cumbersome computationally if the amount of data is very large. More discussion of various applications of MEM in astronomy can be found in the review [Narayan & Nityananda (1986)] and some discussion and an implementation can also be found in [Press et al. (1992)].

For application in helio- or asteroseismology it is not obvious what choice(s) of entropy are suitable. In the case of inversion for Ω one option would be to assume that there is zero probability for regions of the star to be counter-rotating, and that the probability is low for there to be large departures of Ω from some mean value:

$$P(\Omega) = \begin{cases} 0 & \Omega \leq 0 \\ \frac{n^n}{\Omega_m(n-1)!} \left(\frac{\Omega}{\Omega_m}\right)^n \exp\left(-n\frac{\Omega}{\Omega_m}\right) & \Omega > 0 \end{cases} \quad (3.117)$$

where the numerical factor normalises P and n is a parameter which makes the probability distribution function more peaked around Ω_m as it increases. With the P the entropy $\ln P$ becomes:

$$S(\Omega) = n \left[\frac{\Omega}{\Omega_m} - \ln\frac{\Omega}{\Omega_m}\right] + \text{const.} \quad (3.118)$$

The entropy is normally added to the maximum likelihood of Eq. (3.113) with a regularisation weighting factor which is a free parameter, and setting this is equivalent to setting n. The constant, which is the normalisation of the probability distribution function, is irrelevant for the minimisation, and therefore omitted from the algorithm.

If one for instance were to parameterise Ω with a tiling of tophat functions b_i, which are non-zero over sub-domains of $\mathbf{x}$, then the entropy in discretised form is:

$$S(\Omega) = n \sum_i \left[\frac{\Omega_i}{\Omega_m} - \ln\frac{\Omega_i}{\Omega_m}\right] . \quad (3.119)$$

An appropriate choice for the mean value Ω_m could be:

$$\Omega_m \equiv \left[\frac{4}{3}\int_0^R dr \rho_0 r^4\right]^{-1} \int_0^R dr \int_{-1}^1 d\cos\theta\, \rho_0 r^4 (1 - \cos^2\theta)\Omega(r,\theta) \quad (3.120)$$

which is that mean rotation rate which produces the same total angular momentum of the Sun if the Sun were to be rotating uniformly. This can be determined accurately using only the a_1 a-coefficient for all multiplets (cf. [Pijpers (1998)] and using the expressions of [Pijpers (1997)]). The total angular momentum changes slowly: over evolutionary timescales, because of (small) angular momentum loss through the solar wind. Ω_m will therefore be essentially constant over epochs of observation and a good reference value to use. The entropy defined in this way would not necessarily interfere with large gradients in the rotation rate but would weight against solutions with large excursions. In order to properly account for having used the a_1 coefficients to determine Ω_m one would have to remove one such a_1 coefficient from the data set before proceeding with the inversions. Evidently other forms of entropy are possible and for structure inversions one might instead opt for a null hypothesis that the probability decreases monotonically with the absolute values or the squares of of the $\delta Q/Q$.

3.8 Optimal Mask Design as an Inverse Problem

An example of a case in which an inverse problem in helioseismology does not arise from attempts to interpret the frequencies, is the problem of designing optimal masks for isolating modes. This problem is discussed in section 1.4 but it is worthwhile revisiting in the light of the inverse methods presented in this chapter.

The starting point is that one has available a number of measurements which are the average of the flux over a certain subsection of the solar surface. For convenience each of these regions are referred to as 'pixels' but one must remember that these pixels are not necessarily square nor arranged in a rectangle. The different subsections as shown in the LOI detector arrangement (see Fig. 1.4) are also pixels. The data is therefore:

$$I_i = \int_{D_i} d\cos\theta \, d\phi \, I(\mu)\mu^q \tag{3.121}$$

which uses the same definitions as Eq. (1.31): μ is the cosine of the angle between line of sight and the local normal to the surface, $q = 1$ or 2 depending on whether the observation is done in intensity or velocity and D_i is the domain or shape of the region on the solar surface which in projection corresponds to the pixel: the 'footprint' of the pixel.

For the purposes of the discussion here it is more convenient to express this footprint as a response function so that the integral is one over the entire visible part of the solar surface but under the integral sign there is an integration kernel K_i which is non-zero only on the domain D_i:

$$I_i = \int_0^\pi d\cos\theta \int_{-\pi}^\pi d\phi \, K_i(\theta,\phi)I(\mu)\mu^q \tag{3.122}$$

in which the coordinate system is chosen such that $\theta = 0, \pi$ corresponds to the solar rotational north- and south pole respectively, and $(\theta, \phi) = (\pi/2, 0)$ corresponds to the centre of the visible solar disc. The task to design an optimal filter for a specific mode can then be interpreted as the need to find linear weights a_i such that:

$$\sum_i a_i(l_0, m_0) I_i = \int_0^\pi d\cos\theta \int_{-\pi}^\pi d\phi \sum_i a_i(l_0, m_0) K_i(\theta, \phi) I(\mu) \mu^q$$
$$\approx I_{l_0}^{m_0} \delta_{l\, l_0} \delta_{m\, m_0} \,. \qquad (3.123)$$

I is the sum of a temporal mean value $I_0(\mu)$ and all the modes with their intensity/velocity patterns:

$$I(\mu) = I_0(\mu) \left[1 + \sum_{l\,m} I_l^m Z_l^m(\theta, \phi) \right] \,. \qquad (3.124)$$

The temporal mean $I_0(\mu)$ and the amplitudes $I_l^m I_0(\mu)$ depend on the cosine of the angle between the local normal to the surface and the line-of-sight μ because of limb-darkening. With the chosen coordinate system $\mu \equiv \sin\theta \cos\phi$. As mentioned in section 1.4, a 'localisation' in (l, m) space is not likely to be perfect if one cannot observe the entire solar surface, and can only observe with finite resolution. One can use the techniques from this chapter however, to minimise such cross-talk.

Taking the SOLA method as an example, integrals over a continuous variable are replaced by summations over the discrete variables l and m and one would set the weights by minimising for the a_i:

$$\sum_{l\,m} \left\{ \sum_i a_i(l_0, m_0) S_i(l, m) - \mathcal{T}(l - l_0, m - m_0) \right\}^2$$
$$+ \mu_e \sum a_i a_j E_{ij} \qquad (3.125)$$

where the S_i are:

$$S_i(l, m) \equiv \int_0^\pi d\cos\theta \int_{-\pi/2}^{\pi/2} d\phi\, K_i(\theta, \phi) I_0(\sin\theta \cos\phi) Z_l^m(\theta, \phi)(\sin\theta \cos\phi)^q$$
$$(3.126)$$

which is identical to expression (1.31), with the appropriate substitution for μ, and rewriting the integral over a domain D_i as an integral over the visible part of the solar disc with a response kernel K_i. E is the (co)variance matrix of the errors in the measured flux S_i in each pixel. $\mathcal{T}$ is the target function for localisation in l and m. The elements of the matrix $\mathbf{A}'$ that requires inverting in the SOLA algorithm are in this case:

$$A'_{ij} = \sum_{lm} S_i(l,m) S_j(l,m) + \mu_e E_{ij} \qquad (3.127)$$

and the elements of the vector $\mathbf{b}'$ satisfy:

$$b_i = \sum_{lm} S_i(l,m) \mathcal{T}(l - l_0, m - m_0) \ . \qquad (3.128)$$

Although in principle this involves a summation over an infinite range in l (and hence m), in practice there is a finite spatial resolution for any detector which means that the S_i will become small very rapidly for $l > l_{\max}$, so that one can cut off the summation at a finite l. The workload of the calculation of the integrals in the expression for S_i appears considerable, but can be reduced somewhat by making use of the symmetry properties of the I_0 and Z_l^m. Also, for a given detector these need be calculated only once.

From this discussion it is clear that the design of an optimal filter is a discrete inverse problem, and a solution can therefore be obtained in a variety of ways, of which the SOLA method is only one. A regularised least-squares method is described in [Toutain & Kosovichev (2000)]. A final point worth noting, is that for helioseismology the spatial filter will be needed in particular to be able to distinguish modes with (nearly) identical temporal frequencies. From the asymptotic character of the frequencies (cf. Eqs. (2.150) & (2.154)), one can see that in particular spatial cross-talk between modes for which $|l - l_0| = $ even needs to be suppressed, whereas cross-talk for which $|l - l_0| = $ odd is less problematic. One can therefore split up the inversion in two parts: one for the even values of l and one for the odd l, with appropriate summations over only the even l or only the odd l in Eqs. (3.127) and (3.128). This relaxes the restrictions on $\mathcal{T}$ which in turn can reduce sensitivity of the filter to measurement errors.

Chapter 4

Local Helioseismology

Local helioseismology refers to those techniques which do not rely on the variability signal to arise from global resonant modes of the Sun. The surface motions of the Sun are produced by stochastic excitation through convective overturning motion. At frequencies near the global resonant modes energy is retained longer than away from them, with the result that the global resonant modes are superposed on a broad background of 'solar noise'. By cross-correlating this mixture of periodic and stochastic fluctuations at distinct locations or patches on the Sun it is possible to infer the time scale of propagation of signals between these patches. This reflects conditions in the regions around and between the points of cross-correlation which allows in principle for tomographic mapping of structures in these regions.

4.1 The Data: Reduction and Analysis

The data necessary for local area helioseismology is in fact the same type of data that is also used for doing global helioseismology for intermediate and high degree modes. What is necessary is data at a reasonably high spatial resolution over a fairly large field of view. The larger the number of resolved elements within the field of view the more detail one can hope to resolve. Either intensity or velocity measurements could be used, but in practice workers in the field have relied on the high-resolution velocity maps or 'Dopplergrams' produced by the MDI instrument on-board SoHO, or the GONG network and others. Contrary to the strategy for determining global mode frequencies, in local area helioseismology relatively short time strings are used. The minimum required length of the time string is twice the typical travel time for signals from the surface down to the depth to which one aims to attempt tomography. Some 8 hours suffices, but time strings of up to a few days in length are used. This corresponds to time series of from ~ 500 to ~ 5000 Doppler images of 1k × 1k pixels.

The data volume that needs to be dealt with makes the use of fast Fourier transforms (FFTs) almost unavoidable, and therefore it is even more important to avoid the problems of missing data than it is for global helioseismology. In this chapter it is therefore implicitly assumed that exactly regularly sampled data is available. In practice for the total length of the time strings involved this can realistically be achieved both for MDI and for the ground-based networks. A further difference between global and local helioseismology lies in what is done with the data. In global helioseismology the signal is decomposed in terms of spherical harmonics. For local helioseismology the global modes are of little relevance and instead the data is Fourier transformed in the two spatial directions in conjunction with the Fourier transform in time, so that the entire data-cube is in the Fourier domain:

$$v(x,\ y,\ t) \to V(k_x,\ k_y,\ \omega)\ . \tag{4.1}$$

After this, different routes can be taken. For ring diagram analysis (cf. section 4.3.1), little further processing is necessary, since the technique makes use of fitting circles to maxima of acoustic power at fixed frequency. For time-distance analysis techniques (cf. section 4.3.2) the signal between different surface locations is cross-correlated and filtered, and usually also averaged in some way over small regions in space in order to improve the signal-to-noise ratio. In the sections that follow these steps are therefore discussed in more detail.

4.1.1 Tracking, cross-correlation and filtering

The techniques and purposes of local helioseismology bear great resemblance to those of exploration seismology on Earth, which are used to locate and map resources such as oil or water in subsurface layers of porous rock. In exploration seismology an impulsive or periodic source is used to send sound waves into the ground. Geophones are placed at a range of distances to this source. Placed at increasing distances from the source the geophones will pick up waves reflected at increasing depth. The arrival times of the reflected signals are then a measure of the sound speed in the subsurface layers.

In the Sun there is evidently a source of waves as discussed in section

2.10. This source is stochastic and distributed however, and therefore does not allow for a straightforward measurement of arrival times. Instead one has to cross-correlate the time series of Doppler measurements between different locations on the surface, and for different delay times between the measurements. The signal at each location is a response to the source, after the signal has passed between the source and the two surface locations chosen. The expectation is then that the cross-correlation function will show a large signal at a delay time corresponding to the travel time of signals between the two surface locations through the interior. The first attempt to use this principle is reported in [Duvall et al. (1993b)].

The cross-correlation is most easily carried out in the Fourier domain since the Fourier transform of the cross-correlation is the product of the Fourier transform of the two time series as expressed in chapter 1 (Eq. 1.28). The cross-correlation $\Gamma(\theta_1, \phi_1, \theta_2, \phi_2, t)$ between the signals $S_{1,2}$ at locations (in polar coordinates on the solar surface) (θ_1, ϕ_1) and (θ_2, ϕ_2), as a function of delay time t between the time series is:

$$\Gamma(\theta_1, \phi_1, \theta_2, \phi_2, t) = \int dt' S(\theta_1, \phi_1, t') S(\theta_2, \phi_2, t+t') . \qquad (4.2)$$

Since in practice the extent of the fields in θ and ϕ studied with time-distance helioseismology is relatively small, they are usually treated as being flat. A simple de-projection then suffices, to correct for the angle between the normal to the field at its centre and the line-of-sight. Instead of polar coordinates θ and ϕ simple Cartesian coordinates x, y can then be used so that Eq. (4.2) becomes:

$$\Gamma(x_1, y_1, x_2, y_2, t) = \int dt' S(x_1, y_1, t') S(x_2, y_2, t+t') . \qquad (4.3)$$

A cross-correlation of the raw data does not show any identifiable structure however, for the short time series considered here. Structure does emerge without filtering for longer time series. The reason for this is that the signals $S_{1,2}$ are a mixture of waves with wave vectors k of which the ratio between the vertical component at the surface, $k_z(R_\odot)$, and the horizontal component k_h vary over a wide range. This collection of waves sample a large region of the solar atmosphere between the surface points and the arrival at location 2 of a disturbance at location 1 is therefore dispersed over a large range in time. In order to obtain a reasonably clear signal, for time series of short durations, it is necessary to apply a filter. k_z is not an appropriate variable to filter in, but k_h and ω are direct Fourier transforms of the observables. Thus the way to filter the data is to do so in in (k_h, ω) space. By doing this one achieves that the signals selected are sensitive to conditions over a smaller region in space below and between the measurement locations.

To see what purpose the filtering serves it is useful to consider a toy model in which gravity is absent and spherical effects are ignored. The near surface regions of the Sun are then represented by a half-space of gas. The temperature is assumed to increase linearly with depth so that the sound speed c_s satisfies:

$$c_s^2 = \begin{cases} 0 & \text{for } z > 0 \\ c_s(0)^2 \left[1 - \frac{z}{D}\right] & \text{for } z < 0 \end{cases} \tag{4.4}$$

in which D is constant. A plane sound wave with frequency ω propagating through the half-space at an angle θ_0 to the normal has a wave vector k with a vertical and horizontal components k_z and k_h respectively. The wave vector k has to satisfy the dispersion relation for sound waves at every depth so that:

$$|k| = \frac{\omega}{c_s(z)} . \tag{4.5}$$

The horizontal part of the wave vector k_h must remain constant if there is no horizontal variation in the sound speed, so that:

$$\sqrt{k_h^2 + k_z(z)^2} = \frac{\omega}{c_s(z)} \tag{4.6}$$

which can be rearranged so that:

$$k_z(z) = -\sqrt{\frac{\omega^2}{c_s(z)^2} - k_h^2} . \tag{4.7}$$

Since $c_s(z)$ increases with depth, k_z must decrease in absolute value and at some depth $z_{\min}$ $k_z = 0$:

$$z_{\min} = D \left(1 - \frac{\omega^2}{k_h^2 c_s(0)^2}\right) . \tag{4.8}$$

The wave is refracted by the medium and eventually propagates outward again. The point at which acoustic power is focussed again at the surface, and the time it takes to travel, can be calculated approximately by using the ray approximation according to which the waves travel along a path described by the differential:

$$\frac{dh}{dz} = \frac{k_h}{k_z}$$
$$= \left[\frac{\omega^2}{k_h^2 c_s(z)^2} - 1\right]^{-1/2}. \tag{4.9}$$

It is convenient to define a variable θ:

$$\sin\theta \equiv \frac{k_h c_s(z)}{\omega} = \frac{k_h c_s(0)}{\omega}\left(1 - \frac{z}{D}\right)^{1/2} \tag{4.10}$$

and by using this in Eq. (4.9) it becomes:

$$dh = \frac{2\omega^2 D}{k_h^2 c_s(0)^2}\sin^2\theta\, d\theta \tag{4.11}$$

which can be integrated to yield:

$$\begin{aligned} h &= \frac{2\omega^2 D}{k_h^2 c_s(0)^2}\int_{\theta_0}^{\pi/2} d\theta\, [1 - \cos(2\theta)] \\ &= \frac{\omega^2 D}{k_h^2 c_s(0)^2}[\pi - 2\theta_0 + \sin(2\theta_0)] \\ &= D\frac{\pi - 2\theta_0 + \sin(2\theta_0)}{\sin^2\theta_0}. \end{aligned} \tag{4.12}$$

The time that it takes for a sound signal to travel can be found through integration of:

$$dt = \frac{ds}{c_s(z)} \tag{4.13}$$

where $ds^2 = dh^2 + dz^2$. Using again Eq. (4.9), and making use of a variable v:

$$\cos v \equiv -1 + 2\frac{k_h^2 c_s(z)^2}{\omega^2} = -1 + 2\frac{k_h^2 c_s(0)^2}{\omega^2}\left[1 - \frac{z}{D}\right] \tag{4.14}$$

one obtains that:

$$dt = \frac{\omega D}{k_h c_s(0)}dv \tag{4.15}$$

which can be integrated to yield:

$$t = 2D\frac{\omega}{k_h c_s(0)^2}(\pi - 2\theta_0)$$
$$= 2\frac{D}{c_s(0)}\frac{\pi - 2\theta_0}{\sin\theta_0}. \quad (4.16)$$

In Fig. 4.1 the travel time and distance are plotted for a range of angles θ_0, with $c_s(0) = 7$ km/s and $D = 800$ km which corresponds roughly to the properties of near-surface layers of the Sun. The solid curve corresponds

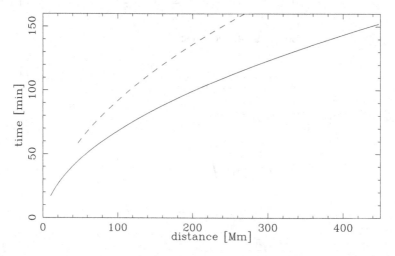

Fig. 4.1 Ray travel time versus horizontal distance travelled, for a toy model with $c_s(0) = 7$ km/s and $D = 800$ km. Solid line corresponds to signals travelling directly between two points on the surface. Dashed line corresponds to signals that are reflected once at an intermediate point

to signals travelling directly between two points a distance h apart. The dashed line corresponds to signals that are reflected at the surface once, at an intermediate point at half that distance. The former is sometimes referred to as 'first bounce' and the latter as 'second bounce'. It would be possible to have 'third bounce' signals etc., but in practice there is attenuation of the signal strength so that the subsequent bounces become increasingly hard to detect.

From this simplified scenario of the behaviour of waves in the solar atmosphere one can deduce that by selecting signals over a limited range in θ_0 one can concentrate on a limited range of rays, or 'ray bundle'. This ray bundle samples a restricted section of space, with a shape similar to that of a banana, between two surface locations. Selecting a limited range of θ_0 can be seen to correspond to a limited range in horizontal phase speed ω/k_h. For a cross-correlation of the signal at two locations a given distance h apart one would therefore tune a horizontal phase speed filter to select a ray bundle with end points that are that same distance apart. Evidently this filtering requires that one already knows with good accuracy the horizontally and temporally averaged sound-speed in the Sun as a function of depth. Also, if the fluctuations of the true sound speed around this average become too large, this simple picture breaks down, because rays and ray bundles become too much deformed. Fortunately global seismology can establish the near-surface sound speed profile well, and the average appears to be quite stable over time scales of months or years.

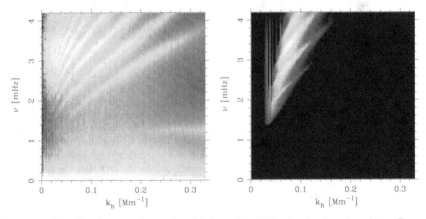

Fig. 4.2 The Fourier transform of ~ 8.5 hrs of GONG Doppler imaging data. Power is shown as a gray scale as a function of the spatial wavevector $k_h = \sqrt{k_x^2 + k_y^2}$. Upper panel, before applying a phase-speed filter ; lower panel, after filtering.

In Fig. 4.2 the acoustic power is shown as a function of k_h and of frequency. The unfiltered data is equivalent to Fig. 1.7 which is from a much longer time series obtained with MDI and which therefore has lower level of noise. The filtered data shows that the horizontal phase speed filter cuts across the (k_h, ω)-diagram in such a way as to remove large sections of it, which are dominated by noise. Several clearly defined ridges of acoustic power remain which give a good response in the cross-correlation.

In Fig. 4.3 the cross-correlation function is shown using observational data obtained with the MDI instrument. The length of the time series is again ~ 8.5 hrs. During this time a selected rectangular patch on the solar

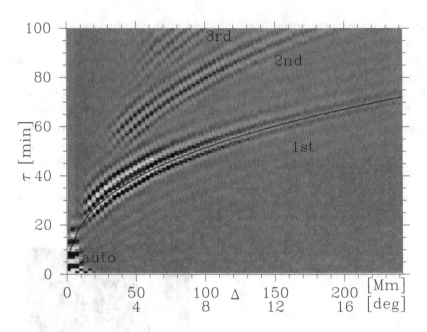

Fig. 4.3 The cross-correlation function as a gray scale for a range of horizontal separations Δ over the surface, and time delays τ. This function is obtained by averaging over many pixels, and reflects the mean temperature stratification below the photosphere in the Sun. The solid line is a prediction from a model, which corresponds to the solid line in Fig. 4.1 for the toy model. The first second and third ridges are well-defined in power. For each ridge there is a fine structure present, due to the fact that the filtering is done with a finite pass band, rather than cross-correlating a monochromatic signal (figure based on [Duvall et al. (1997)]).

surface is observed, taking care that the image is 'tracked' at the surface rotation rate of the Sun, which implies shifting the selected range of pixels from the instrument and carrying out interpolation of the Doppler images onto a fixed grid of 'virtual pixels'. This tracking is necessary, in particular for longer time series because otherwise in a cross-correlation function there is a loss of coherence. The analogy in a geo-seismology setting would be that the seismic detectors are dragged along at high speed during the collection of seismic data. Since a patch needs to be tracked at the surface rotation rate, the shift between successive Doppler images in the time series is not an integer number of pixels. Interpolation is necessary to prevent spurious periodic responses at frequencies that are integer multiples of $\Omega_{\text{surf}} R_\odot / s_{\text{pix}}$ where s_{pix} is the projected size of the pixels on the solar surface.

The tracked MDI time-series of data is Fourier transformed and filtered as outlined above. The cross-correlation function is calculated as a function of surface separation Δ and time-delay τ and an average of this is taken over all pixel pairs available at that separation. The averaging achieves that even the third-bounce can be seen in Fig. 4.3 but with this averaging the resulting diagram cannot be used as a diagnostic for horizontal variations: the diagram reflects the horizontally averaged conditions in the solar subphotosphere.

4.1.2 Averaging and masking

With current instrumentation of ground-based networks, or even MDI, the noise per pixel is still high enough that pixel by pixel cross-correlations, as outlined in the previous section, are dominated by noise. Instrumentation on planned satellite missions such as the Solar Orbiter may well reach sufficiently low noise levels per pixel that no further processing is necessary. With existing data however, there is a need for spatial averaging in order to improve the signal-to-noise ratio in the cross-correlations. A simple average over a contiguous set of pixels evidently provides an improvement in the signal-to-noise ratio. However, if such averaging is done over many pixels then the ability to resolve structures below the surface with tomographic techniques suffers, at which point the method loses much of its value. By designing 'masks' of weights for calculating weighted averages over contiguous pixels one can better balance the need for suppression of noise and the need for retaining spatial resolution in tomographic inversions.

Most averaging schemes that have been used so far select regions of pixels, within which the weight is set uniformly (see Fig. 4.4). One of these schemes, also used in [Duvall, et al. (1997)], is to correlate signal between a central circular patch of pixels and annuli concentric with this central patch. Directional information is largely lost in this way but variations around the horizontal mean of for instance the temperature can be investigated. Directional information can be obtained if only sections of the annulus (e.g. a quadrant) are used in the correlation. Another choice that can be found is to correlate arc to arc, in which the arc pairs are sections of an annulus centered on a point. By increasing the separation of the arcs, through increasing the radius of the annulus of which they are a part, deeper layers are being probed. In this case there is directional information as well as depth information through the centering of the arcs: for instance one arc pair for each annulus could be aligned East-West and a complementary pair would be aligned North-South.

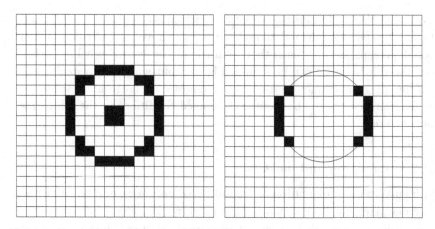

Fig. 4.4 Two widely used pixel weighting schemes for averaging. Left panel is centre-annulus averaging: signal is correlated between the average over the central 4 (or more) pixels and the average over an annulus of pixels symmetrically arranged around the central patch. Right panel is arc-to-arc averaging where the average is done over two arcs of pixels and the correlation is carried out between the two resulting signals.

Neither of these choices is optimised in terms of error propagation. Also not taken into account is the potential resolving power as a function of depth when using these averages as data in a tomographic inverse problem to determine horizontal temperature fluctuations or flows. For this reason, in [Hughes et al. (2006)] it is explored how one can devise strategies that will provide such masks in terms of weights for each pixel. With these weights, weighted averages can then be calculated for the data, that will provide a better starting point for subsequent inverse analysis. One can define a kernel $\mathbf{K}^{1,2}$ that represents the spatial sensitivity for the correlation between pixels in two spatially disjunct patches of pixels labelled 1 and 2. Without loss of generality the two patches can be assumed to have an equal number of pixels. There are two sets of weighting coefficients $\mathbf{c}_1$ and $\mathbf{c}_2$ corresponding to the two regions that are being cross-correlated, so that the cross-correlation of the averages corresponds to a spatial sensitivity kernel $\mathcal{K}(\mathbf{r}_0, \mathbf{r})$:

$$\mathcal{K}(\mathbf{r}_0, \mathbf{r}) \equiv \mathbf{c}_1(\mathbf{r}_0) \cdot \mathbf{K}^{1,2}(\mathbf{r}) \cdot \mathbf{c}_2(\mathbf{r}_0) \ . \qquad (4.17)$$

Even if the two patches are identical in shape and the indexing of each pixel within the two pixels is done symmetrically, the forward model that constitutes $\mathbf{K}$ is not symmetric i.e. the spatial sensitivity corresponding to correlating pixel i from patch 1 with pixel $j (\neq i)$ from patch 2 is not the same is that for the correlation of pixel j from patch 1 with pixel i from patch 2:

$$\mathbf{K}_{ij} \neq \mathbf{K}_{ji} . \tag{4.18}$$

By splitting up the coefficients into symmetric (c_s) and anti-symmetric (c_a) parts one can write:

$$\begin{aligned}\mathbf{c}_s &= \frac{1}{2}(\mathbf{c}_1 + \mathbf{c}_2) \\ \mathbf{c}_a &= \frac{1}{2}(\mathbf{c}_1 - \mathbf{c}_2) .\end{aligned} \tag{4.19}$$

When substituted into Eq. (4.17), the resulting equation is:

$$\begin{aligned}\mathbf{c}_1^T \cdot \mathbf{K} \cdot \mathbf{c}_2 &= (\mathbf{c}_s + \mathbf{c}_a)^T \cdot \mathbf{K} \cdot (\mathbf{c}_s - \mathbf{c}_a) \\ &= \mathbf{c}_s^T \cdot \mathbf{K} \cdot \mathbf{c}_s + \mathbf{c}_a^T \cdot \mathbf{K} \cdot \mathbf{c}_s - \mathbf{c}_s^T \cdot \mathbf{K} \cdot \mathbf{c}_a - \mathbf{c}_a^T \cdot \mathbf{K} \cdot \mathbf{c}_a .\end{aligned} \tag{4.20}$$

For any matrix A it is true that:

$$\mathbf{c}_s^T \cdot A \cdot \mathbf{c}_s = \left(\mathbf{c}_s^T \cdot A \cdot \mathbf{c}_s\right)^T = \mathbf{c}_s^T \cdot A^T \cdot \mathbf{c}_s \tag{4.21}$$

and similarly that:

$$\begin{aligned}\mathbf{c}_a^T \cdot A \cdot \mathbf{c}_a &= \mathbf{c}_a^T \cdot A^T \cdot \mathbf{c}_a \\ \mathbf{c}_s^T \cdot A \cdot \mathbf{c}_a &= \mathbf{c}_a^T \cdot A^T \cdot \mathbf{c}_s .\end{aligned} \tag{4.22}$$

In a similar fashion to the coefficients, one can also split up the matrix $\mathbf{K}$ into explicitly symmetric and anti-symmetric components:

$$\begin{aligned}\mathbf{K}_{s,ij} &= \mathbf{K}_{s,ji} \equiv \frac{1}{2}(\mathbf{K}_{ij} + \mathbf{K}_{ji}) \\ \mathbf{K}_{a,ij} &= -\mathbf{K}_{a,ji} \equiv \frac{1}{2}(\mathbf{K}_{ij} - \mathbf{K}_{ji}) .\end{aligned} \tag{4.23}$$

If for the matrix A in Eq. (4.21) and Eq. (4.22) the anti-symmetric matrix $\mathbf{K}_a$ is substituted, it is clear that it must render the combinations with equal coefficients equal to 0. Equally, substituting for A the symmetric matrix $\mathbf{K}_s$ must make the final equation of Eq. (4.22) disappear. Hence the terms in Eq. (4.20) can be re-written:

$$\begin{aligned}\mathbf{c}_s^T \cdot \mathbf{K} \cdot \mathbf{c}_s &= \mathbf{c}_s^T \cdot \mathbf{K}_s \cdot \mathbf{c}_s \\ \mathbf{c}_a^T \cdot \mathbf{K} \cdot \mathbf{c}_a &= \mathbf{c}_a^T \cdot \mathbf{K}_s \cdot \mathbf{c}_a \\ \mathbf{c}_a^T \cdot \mathbf{K} \cdot \mathbf{c}_s - \mathbf{c}_s^T \cdot \mathbf{K} \cdot \mathbf{c}_a &= 2\mathbf{c}_a^T \cdot \mathbf{K}_a \cdot \mathbf{c}_s\end{aligned} \tag{4.24}$$

and therefore:

$$\mathbf{c}_1^T \cdot \mathbf{K} \cdot \mathbf{c}_2 = \mathbf{c}_s^T \cdot \mathbf{K}_s \cdot \mathbf{c}_s - \mathbf{c}_a^T \cdot \mathbf{K}_s \cdot \mathbf{c}_a + 2\mathbf{c}_a^T \cdot \mathbf{K}_a \cdot \mathbf{c}_s \ . \quad (4.25)$$

In a SOLA localisation algorithm the functional to be minimised for the weights is:

$$\int \mathrm{d}x \left(\mathbf{c}_1^T \cdot \mathbf{K} \cdot \mathbf{c}_2 - T \right)^2 =$$

$$= \int \mathrm{d}x \left[\left(\mathbf{c}_s^T \cdot \mathbf{K}_s \cdot \mathbf{c}_s - \mathbf{c}_a^T \cdot \mathbf{K}_s \cdot \mathbf{c}_a - T_s \right) + \left(2\mathbf{c}_a^T \cdot \mathbf{K}_a \cdot \mathbf{c}_s - T_a \right) \right]^2$$

$$= \int \mathrm{d}x \left(\mathbf{c}_s^T \cdot \mathbf{K}_s \cdot \mathbf{c}_s - \mathbf{c}_a^T \cdot \mathbf{K}_s \cdot \mathbf{c}_a - T_s \right)^2 + \int \mathrm{d}x \left(2\mathbf{c}_a^T \cdot \mathbf{K}_a \cdot \mathbf{c}_s - T_a \right)^2$$

$$+ 2 \int \mathrm{d}x \left(\mathbf{c}_s^T \cdot \mathbf{K}_s \cdot \mathbf{c}_s - \mathbf{c}_a^T \cdot \mathbf{K}_s \cdot \mathbf{c}_a - T_s \right) \left(2\mathbf{c}_a^T \cdot \mathbf{K}_a \cdot \mathbf{c}_s - T_a \right) \quad (4.26)$$

where T is the target form for the spatial sensitivity. For all the masks of pixels that have been used thus far it is possible to devise a pixel indexing scheme such that there is a plane of symmetry, (whether it be flat or cylindrical) for the spatial structure of the kernels $\mathbf{K}_{s,a}(\mathbf{r})$. Making use of this, the third of the three integrals in Eq. (4.26) is the integral of the product of a function that is symmetric with respect to this plane of symmetry, and one that is antisymmetric:

$$\mathbf{K}_{s,ij}(\mathbf{r}) = \mathbf{K}_{s,ij}(\mathbf{r} - 2(\mathbf{r} \cdot \mathbf{n})\mathbf{n})$$
$$\mathbf{K}_{a,ij}(\mathbf{r}) = -\mathbf{K}_{a,ij}(\mathbf{r} - 2(\mathbf{r} \cdot \mathbf{n})\mathbf{n})$$
$$T_s = \frac{1}{2}\left(T(\mathbf{r}) + T(\mathbf{r} - 2(\mathbf{r} \cdot \mathbf{n})\mathbf{n})\right)$$
$$T_a = \frac{1}{2}\left(T(\mathbf{r}) - T(\mathbf{r} - 2(\mathbf{r} \cdot \mathbf{n})\mathbf{n})\right) \ . \quad (4.27)$$

Integrating that product over the entire volume yields by definition 0. If furthermore the target function is itself symmetric then the antisymmetric part of it T_a is 0. Eq. (4.26) then reduces to minimising:

$$\int \mathrm{d}x \left(\mathbf{c}_s^T \cdot \mathbf{K}_s \cdot \mathbf{c}_s - \mathbf{c}_a^T \cdot \mathbf{K}_s \cdot \mathbf{c}_a - T_s \right)^2 + \int \mathrm{d}x \left(2\mathbf{c}_a^T \cdot \mathbf{K}_a \cdot \mathbf{c}_s \right)^2 \ . \quad (4.28)$$

The second term is clearly minimal if all antisymmetric coefficients $\mathbf{c}_{a,i} = 0$, which also satisfies the constraint that $\sum \mathbf{c}_{a,i} = 0$. Since $\mathbf{K}_s$ is symmetric it can be written as $\mathbf{K}_s = U^T D U$ where U is a unitary matrix and D is diagonal. Define α as follows:

$$\alpha = \min_i |D_i| - \min_i D_i = \begin{cases} 0 & \text{if } \min_i D_i > 0 \\ 2\min_i |D_i| & \text{if } \min_i D_i < 0 \end{cases}.$$

With this definition the matrix defined by:

$$D + \alpha I \equiv PP \qquad (4.29)$$

is diagonal and positive definite, and therefore so is its 'square root' matrix P. Define a vector w by:

$$w \equiv \begin{pmatrix} w_s \\ w_a \end{pmatrix} = \begin{pmatrix} P & 0 \\ 0 & P \end{pmatrix} \begin{pmatrix} U & 0 \\ 0 & U \end{pmatrix} \begin{pmatrix} \mathbf{c}_s \\ \mathbf{c}_a \end{pmatrix}. \qquad (4.30)$$

It is clear that:

$$\mathbf{c}_s^T \cdot \mathbf{K}_s \cdot \mathbf{c}_s - \mathbf{c}_a^T \cdot \mathbf{K}_s \cdot \mathbf{c}_a - T_s = (\mathbf{c}_s^T \mathbf{c}_a^T) \begin{pmatrix} \mathbf{K} & 0 \\ 0 & -\mathbf{K} \end{pmatrix} \begin{pmatrix} \mathbf{c}_s \\ \mathbf{c}_a \end{pmatrix} - T_s$$

$$= (w_s^T w_a^T) \begin{pmatrix} DP^{-2} & 0 \\ 0 & -DP^{-2} \end{pmatrix} \begin{pmatrix} w_s \\ w_a \end{pmatrix} - T_s \qquad (4.31)$$

which can be written out in individual elements as:

$$\sum_i \beta_i \left(w_{s,i}^2 - w_{a,i}^2 \right) - T_s \qquad (4.32)$$

in which:

$$\beta_i \equiv d_i/p_i^2 . \qquad (4.33)$$

Now assume that an acceptable solution w has been found, minimising the first term of Eq. (4.28). Clearly a solution with $w_{a,i} = 0$ would provide the absolute minimal value for the second term, but it is assumed here merely that the minimisation of the first term has yielded a solution which satisfies the weaker constraint that:

$$|w_{a,i}| < |w_{s,i}| \quad \forall \, i . \qquad (4.34)$$

Now construct a $w_{s,i}^*$ such that:

$$\begin{aligned} w_{s,i}^* &= w_{s,i} \cosh \alpha_i - w_{a,i} \sinh \alpha_i \\ w_{a,i}^* &= -w_{s,i} \sinh \alpha_i + w_{a,i} \cosh \alpha_i = 0 \end{aligned} \qquad (4.35)$$

which implies that α_i satisfies:

$$\tanh \alpha_i = \frac{w_{a,i}}{w_{s,i}} . \tag{4.36}$$

With these definitions it is true that:

$$\begin{aligned}
\left(w_{s,i}^{*2} - w_{a,i}^{*2}\right) &= w_{s,i}^{*2} \\
&= w_{s,i}^2 \cosh 2\alpha_i + 2 w_{s,i} w_{a,i} \cosh \alpha_i \sinh \alpha_i \\
&\quad + w_{a,i}^2 \sinh 2\alpha_i \\
&= w_{s,i}^2 - w_{a,i}^2 .
\end{aligned} \tag{4.37}$$

This process can be carried out for each index i independently. Therefore, given a non-symmetric solution (w_s, w_a) of the minimisation of the first part of Eq. (4.28), that satisfies Eq. (4.34), by the above procedure one can always construct an equally satisfactory symmetric solution $(w_s^*, 0)$ of it, i.e. with all $w_{a,i}^* = 0$ and therefore all $\mathbf{c}_{a,i} = 0$. Since this also minimises the second part the symmetric solution must be superior. There is therefore no need to consider asymmetric weighting of the patches if the target function T is symmetric.

From this discussion it is clear that the centre-annulus cross-correlation scheme is not optimal. The arc to arc cross-correlation scheme does satisfy the required symmetry properties, so it can be made optimal by departing from uniform weights for each pixel within the arc. The appropriate weights are then obtained by solving an inverse problem as demonstrated in [Hughes et al. (2006)].

Note that in addition to using appropriate masks there are also ways to improve the extraction of parameters, such as the group travel time, from noisy cross-correlation functions for wave packets between two surface locations. In [Gizon & Birch (2004)] a method is implemented which determines time delays by minimising the functional:

$$\chi \equiv \int dt' f(\pm t') \left[\Gamma^\epsilon(\mathbf{r}_1, \mathbf{r}_2, t') - \Gamma^{\mathrm{ref}}(|\mathbf{r}_1 - \mathbf{r}_2|, t' \mp \epsilon t)\right]^2$$

$$\Gamma^\epsilon(\mathbf{r}_1, \mathbf{r}_2, t') \equiv \epsilon \Gamma(\mathbf{r}_1, \mathbf{r}_2, t') + (1-\epsilon)\Gamma^{\mathrm{ref}}(|\mathbf{r}_1 - \mathbf{r}_2|, t') \tag{4.38}$$

in which Γ is the cross-correlation function, and f is an isolation filter which helps to separate out the appropriate first-bounce wave-packet propagating from location $\mathbf{r}_1$ to $\mathbf{r}_2$. The reference cross correlation function Γ^{ref} can be determined either from a smooth reference model, or from an average over an area of quiet Sun. The parameter ϵ must lie between 0 and 1 and is dependent on the signal-to-noise ratio, such that $\epsilon = 1$ in the absence of noise and $\epsilon \downarrow 0$ as the noise level increases. In this limit it is shown in appendix B of [Gizon & Birch (2004)] that this fitting reduces to taking for the delay times $\tau_\pm$:

$$\tau_\pm = \mp \int dt'\, W(t) \left[\Gamma(\mathbf{r}_1, \mathbf{r}_2, t') - \Gamma^{\text{ref}}(|\mathbf{r}_1 - \mathbf{r}_2|, t')\right]$$
$$W(t) = \frac{f(\pm t)\partial_t \Gamma^{\text{ref}}(t)}{\int dt'\, f(\pm t')\left[\partial_t \Gamma^{\text{ref}}(t')\right]^2} \quad (4.39)$$

which is the same as is done to determine Doppler shifts of spectral lines (see Eqs. (1.9) and (1.11)), assuming uniform and uncorrelated errors among the samples of the cross-correlation function. In [Gizon & Birch (2004)] the propagation of measurement errors, for f-modes, into errors estimates for the time delays is also discussed in detail. The process of estimating time delays from Eq. (4.39) is evidently linear, which means that error propagation is a straightforward process, and can be generalised to apply to cross-correlation of p-mode signals as well.

4.2 The Forward Approach: Modelling

The conditions inside the Sun cover a very broad range of temperatures and densities. The state of the fluid is that of a highly ionised gas consisting of electrons, nuclei and photons. The equations governing the motion of this plasma are those of (magneto-)hydrodynamics (MHD). It is generally thought that the magnetic field strength is such, that it will dominate over the dynamics only in the outermost regions of the Sun, and that for much of the interior the contribution of the magnetic field can be ignored so that one need consider only the equations of fluid dynamics. Although this may be true for the mean (hydrostatic) structure of the Sun, the character of the solar rotation profile and the structure of the solar convection zone do require taking magnetic fields into account. For the purposes of modelling this requires numerical solution of the coupled system of differential equations governing conservation of mass, momentum and energy.

4.2.1 *Hydrodynamical models with radiation*

The dynamics of gas motion are described by differential equations for mass momentum and energy, in which advection is distinguished from source terms. *Advection* refers to material being dragged around by the flow and any given field at a particular location in space can change because of this, even if over the entire (closed) system there is no change. The *source terms* are the terms that express e.g. forces: generating momentum from the flow or removing it. The equation describing conservation of mass, in an Eulerian formulation, is:

$$\frac{\partial}{\partial t}\rho + \nabla \cdot (\rho \mathbf{v}) = 0 \qquad (4.40)$$

in which ρ is the material density and $\mathbf{v}$ is the velocity. There are no source terms here so the right-hand side of the equation is equal to 0. This same equation is also encountered in section 2.6.1 (Eq. (2.70)) for the derivation of the wave equation. Conservation of momentum is described by:

$$\frac{\partial}{\partial t}(\rho \mathbf{v}) + \nabla \cdot (\rho \mathbf{v}\mathbf{v}) = -\nabla p - \rho \nabla \Psi + \mathbf{F} \,. \qquad (4.41)$$

Here p is the gas pressure, related to density ρ and temperature T through an equation of state. Ψ is the gravitational potential and $\mathbf{F}$ is the sum of all other forces. the term $\mathbf{vv}$ is referred to as a tensor or dyadic ; a three by three matrix, which in this case is symmetric. The most convenient form of the equation of conservation of energy is as Eq. (2.77), with the terms slightly re-arranged:

$$\frac{\mathrm{d}p}{\mathrm{d}t} - c_s^2 \frac{\mathrm{d}\rho}{\mathrm{d}t} = \rho(\Gamma_3 - 1)\frac{\mathrm{d}q}{\mathrm{d}t} \,. \qquad (4.42)$$

In section 2.6.1 a perturbation analysis of these equations under adiabatic conditions produced the LAWE. In the outer layers of the Sun the adiabatic approximation must be reconsidered. Also, the influence of magnetic fields becomes a central issue if waves are to be used as probes of conditions in and around sunspots. The interaction between the radiation field and the hydrodynamics by itself introduces terms in the equations of conservation of momentum and energy. Detailed discussions can be found in [Mihalas (1978)] or [Mihalas & Mihalas (1984)]. Lmiting the discussion to non-relativistic flow so that terms $O(v^2/c^2)$ can be omitted the force term that needs to be included in the momentum equation is:

$$\mathbf{F}_r = -\left[\nabla \cdot \mathbf{P}_r^0 + \frac{1}{c^2}\left(\frac{\partial \mathcal{F}^0}{\partial t} - \mathbf{v}\nabla \cdot \mathcal{F}^0\right)\right] \qquad (4.43)$$

in which $\mathcal{F}^0$ is the radiation flux, which to this order can be evaluated in a coordinate frame locally co-moving with the fluid, and $\mathbf{P}_r^0$ is the radiation stress tensor which can be determined in the same frame. The contribution from radiation to the energy balance of non-relativistic fluids is:

$$\rho \frac{dq}{dt} = 4\pi \int_0^\infty d\nu \, \chi_\nu^0 \left(J_\nu^0 - S_\nu^0 \right) \tag{4.44}$$

in which χ_ν^0 is the monochromatic absorption coefficient of the material, S_ν^0 is the source function, and J_ν^0 is the mean intensity, all to be evaluated in a coordinate frame locally co-moving with the fluid. To complete the system of differential equations, the radiative transport equations for J_ν^0 need to be solved together with the equations of fluid dynamics. This is awkward because the radiative transport equations are integro-differential equations. There are fast methods to deal with radiative transfer (e.g. the Feautrier method and accelerated lambda iteration) which can be combined with hydrodynamical simulations. Details of such techniques fall beyond the scope of this book but can be found in [Mihalas & Mihalas (1984)], [Kalkofen (Ed.) (1988)] or [Castor (2004)].

4.2.2 The influence of magnetic fields

To include the effect of the magnetic field the force term **F** that appears in the equation of motion (4.41) is the *Lorentz* force:

$$F_L = \frac{1}{c} \mathbf{J}_c \times \mathbf{B} \tag{4.45}$$

in which $\mathbf{J}_c$ is the current density, c is the speed of light, and **B** is the magnetic field strength. For a self-consistent treatment of the magnetic field one needs to include Maxwell's equations (cf. [Jackson (1999)]):

$$\begin{aligned}
\nabla \cdot \mathbf{D} &= 4\pi \rho_c \\
\nabla \cdot \mathbf{B} &= 0 \\
\nabla \times \mathbf{E} + \frac{1}{c}\frac{\partial \mathbf{B}}{\partial t} &= 0 \\
\nabla \times \mathbf{H} - \frac{1}{c}\frac{\partial \mathbf{D}}{\partial t} &= \frac{4\pi}{c} \mathbf{J}_c \, .
\end{aligned} \tag{4.46}$$

Note that in most situations encountered in astrophysics $\mathbf{D} \equiv \mathbf{E}$ and $\mathbf{H} \equiv \mathbf{B}$ (in the Gaussian system the electric permittivity $\epsilon_c \equiv 1$ and the magnetic permeability $\mu_c \equiv 1$), that is to say that the fluid medium does not get electrically polarised or magnetised. Also, displacement currents and variations in the charge density can often be ignored here these effects come in as terms of order v^2/c^2. The current density $\mathbf{J}_c$ is related to the electric and magnetic fields through:

$$\mathbf{J}_c = \sigma\left(\mathbf{E} + \frac{\mathbf{v}}{c} \times \mathbf{B}\right) . \qquad (4.47)$$

The Lorentz force can be reformulated with the use of the fourth of equations (4.46) and with the vector identity (cf. appendix A):

$$\begin{aligned}\mathbf{B} \times (\nabla \times \mathbf{B}) &= \frac{1}{2}\nabla(\mathbf{B}\cdot\mathbf{B}) - (\mathbf{B}\cdot\nabla)\mathbf{B} \\ &= -\nabla\cdot\left(\mathbf{BB} - \frac{1}{2}|\mathbf{B}|^2\mathbf{I}\right) . \end{aligned} \qquad (4.48)$$

Here $\mathbf{I}$ is the unit dyadic. The expression for the Lorentz force becomes:

$$\begin{aligned} F_L &= \frac{1}{4\pi}(\nabla \times \mathbf{B}) \times \mathbf{B} \\ &= \nabla\cdot\left(\frac{\mathbf{BB}}{4\pi} - \frac{|\mathbf{B}|^2}{8\pi}\mathbf{I}\right) . \end{aligned} \qquad (4.49)$$

This shows that the effect of the magnetic field can be expressed in terms of a tension and a pressure term. With some manipulation of Maxwell's equations (4.46) the magnetic field is shown to obey a transport equation:

$$\frac{\partial \mathbf{B}}{\partial t} + \nabla \times (\mathbf{B} \times \mathbf{v}) = -\nabla \times (\eta\nabla \times \mathbf{B}) \qquad (4.50)$$

in which η is the resistivity Flows in rotating conducting media can give rise to large scale magnetic fields that are spontaneously generated out of infinitesimal fluctuations. The first formulation of this *dynamo action* was given in [Parker (1955)]. This formulation and subsequent work that uses the same techniques is now known as *mean field theory*. There is considerable and long-standing research effort. A comprehensive text book is [Childress & Gilbert (1995)], and recent reviews are [Schekochihin & Cowley (2006)] and [Brandenburg & Dobler (2002)]. An older but accessible discussion can also be found in [Hoyng (1990)]. A simple calculation of the diffusion time scale of a primordial magnetic field shows that the magnetic fields observed on the Sun, and also on magnetically active stars, must have been (re-)generated over the stellar main sequence life time. Therefore the modelling of the generation of magnetic fields is highly relevant to the Sun. A review more specifically of dynamo action in the Sun is [Ossendrijver (2003)].

4.2.2.1 Dynamo action

The starting point of dynamo theory is Eq. (4.50). Often the resistivity η will be assumed constant so that it can be written as:

$$\frac{\partial \mathbf{B}}{\partial t} = \nabla \times (\mathbf{v} \times \mathbf{B}) + \eta \nabla^2 \mathbf{B} \ . \tag{4.51}$$

If the velocity field $\mathbf{v}$ is assumed given then the equation is linear in the magnetic field and this is *kinematic dynamo theory*. As in all linear differential equations, an amplitude for $\mathbf{B}$ cannot be deduced from this system. The full problem where the equation of motion is also solved is *nonlinear dynamo theory*.

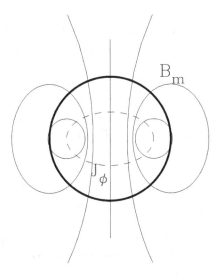

Fig. 4.5 An axially symmetric magnetic field of which selected meridional field lines B_m are shown. On the dashed line $B_m = 0$ and there is a non-zero current density J_ϕ.

One of the most important basic theorems in dynamo theory is Cowling's theorem:

- A stationary axisymmetric dynamo is impossible

The simplest proof of this is starts by imposing an initially axisymmetric magnetic field on a rotating star and proving that it cannot be stationary. The magnetic field is decomposed into a meridional component $\mathbf{B}_m$ and an azimuthal component $\mathbf{B}_\phi$. Because $\nabla \cdot \mathbf{B} = 0$ all the field lines are closed and then there must be at least one line on which the meridional component is exactly equal to 0. Because $\nabla \times \mathbf{B}_m \neq 0$ on this line N (dashed line in Fig. 4.5) there must be a component $\mathbf{J}_\phi$ of the current density there. With the help of Eq. (4.47) and also using that:

$$(\mathbf{v} \times \mathbf{B})_\phi = \mathbf{v}_m \times \mathbf{B}_m = 0 \qquad \text{on N,} \qquad (4.52)$$

it is clear that there then must be a finite $\mathbf{E}_\phi$. On the other hand:

$$\oint_N \mathbf{J}_\phi dl = \oint \mathbf{J}_\phi \cdot dl = \sigma \oint \mathbf{E}_\phi \cdot dl = \sigma \oint \mathbf{E} \cdot dl$$
$$= \sigma \int_S \nabla \times \mathbf{E} \cdot ds = -\frac{\sigma}{c} \int_S \frac{\partial \mathbf{B}}{\partial t} \cdot ds = 0 . \qquad (4.53)$$

The first line of identities in Eq. (4.53) is straightforward geometry. The second line follows from successively Stokes' theorem (cf. Appendix A), the third of Maxwell's equations (4.46), and the assumption of stationarity. This contradiction leads to the conclusion that a dynamo must produce a magnetic field that is either not axisymmetric or not stationary or is neither stationary nor axisymmetric.

The importance of this result for modelling of solar sub-surface layers is that the theorem implies that such modelling requires either a 3-dimensional approach or a time-dependent approach, even for the most simplified models. It is evident from the solar surface magnetic field that in fact the solar magnetic field is neither axisymmetric nor stationary, so that good models of the solar magnetic field behaviour are almost exclusively the domain of numerical simulations.

In mean field theory all the vector fields are split up into a mean field and a fluctuating component. Although this may seem the same as doing a perturbation analysis this is not the case. The fluctuating component is not assumed to be small. Next an *ensemble averaging* procedure is defined such that the mean quantities are unaffected. The average in this sense of the fluctuating components are by definition equal to 0, but the average of products of the fluctuating component of the fields does not have to be. For instance it can be shown (cf. [Knobloch (1978)] and [Stix (1983)]) that if the fluctuating velocity behaves as isotropic turbulence the following is true:

$$\langle \mathbf{v}_1 \times \mathbf{B}_1 \rangle \approx \alpha \mathbf{B}_0 - \beta \nabla \times \mathbf{B}_0 \equiv \alpha \mathbf{B}_0 - \frac{4\pi\beta}{c} \mathbf{J}_0 . \qquad (4.54)$$

The $\langle . \rangle$ denotes the taking of an ensemble average. The subscript 0 refers to the mean field and 1 to the fluctuating components. α and β are defined

as:
$$\alpha = -\frac{1}{3}\langle \mathbf{v_1} \cdot \nabla \times \mathbf{v_1}\rangle \tau_c$$
$$\beta = \frac{1}{3}\langle |\mathbf{v_1}|^2\rangle \tau_c \qquad (4.55)$$

where τ_c is the correlation time of the turbulence. The quantity $\nabla \times \mathbf{v_1}$ is sometimes referred to as the vorticity of the flow field, and $\mathbf{v_1} \cdot \nabla \times \mathbf{v_1}$ as the helicity. After some manipulation a closed equation for $\mathbf{B}_0$ can be derived:

$$\frac{\partial \mathbf{B}_0}{\partial t} = \nabla \times [\mathbf{v}_0 \times \mathbf{B}_0 + \alpha \mathbf{B}_0 - (\eta + \beta)\nabla \times \mathbf{B}_0] \ . \qquad (4.56)$$

Subsequently this equation is usually split up into two coupled equations: one for toroidal field and one for poloidal field. On the basis of this analysis usually three different types of dynamos are distinguished:

- $\alpha\Omega$ dynamo: the component of the shear flow term $\mathbf{v}_0 \times \mathbf{B}_0$ in the poloidal direction can be ignored. This means that toroidal field can only be generated from poloidal field through the 'alpha effect'. This yields dynamos that generally exhibit a periodic behaviour. The solar dynamo could be of this type.
- α^2 dynamo: the shear flow is ignored completely. This usually yields stationary solutions and might be representative of the magnetic field of the earth.
- $\alpha^2\Omega$ dynamo: the most complete of the three where all terms are assumed to be of comparable magnitude.

The mean field theory for dynamos is far from fully explored and there is still much active research in this field. Another technique for investigating magnetic field generation by convection and rotation is direct numerical MHD simulations (see e.g. [Rädler et al. (1990)], [Moss et al. (1991)] and [Brun et al. (2004)]). Such numerical work is subject to much the same restrictions such as in resolution that all simulations of turbulent convection are, but it is not limited to linear or quasi-linear approximations.

The importance of mean field theory for local helioseismology lies in that it provides a parameterised framework. Images of magnetic field and flow structures as produced by simulations and as inferred using tomographic techniques on the Sun may well appear similar, but such similarities must be quantified to properly test the models. The parameters of mean field theory are then appropriate quantities to attempt to measure. Examples are average values of flow and field helicity:

$$H_v \equiv \mathbf{v} \cdot \nabla \times \mathbf{v}$$
$$H_B \equiv \mathbf{A} \cdot \nabla \times \mathbf{A} \tag{4.57}$$

in which $\mathbf{A}$ is the electromagnetic vector potential related to the magnetic field by $\mathbf{B} \equiv \nabla \times \mathbf{A}$. Even for non-linear dynamo models, in which these quantities are not parameters, their measurement can provide constraints and guidance for improvement of the modelling.

4.2.3 Turbulence

In section 2.2 on global solar and stellar modelling it is discussed that convection zones are highly turbulent. For global modelling it is primarily of importance to determine the gradient of the pressure associated with turbulent flows and the energy transport mediated by turbulent convection. These quantities are low order moments of the turbulent spectrum for which one can derive transport equations. The problem, as mentioned in section 2.2., is that these transport equations involve terms that are higher order correlations, leading to what is an infinite hierarchy of equations. The starting point is very similar in spirit to that of dynamo theory. Temperature, density, velocity etc are expressed in terms of a mean field and a random fluctuating part with zero mean:

$$\begin{array}{ll} T = T_0 + \theta & \rho = \rho_0 + \delta\rho \\ p = p_0 + \delta p & \mathbf{v} = \mathbf{v}_0 + \delta\mathbf{v} \end{array} \tag{4.58}$$

where often the mean velocity is set $\mathbf{v}_0 = 0$. As in dynamo theory the fluctuations are not assumed to be small, and so non-linear terms in fluctuating variables are kept. Further, it is usual to solve these equations in the Fourier domain, i.e. for any random variable ψ is substituted:

$$\psi(\mathbf{x}, t) = \int d\mathbf{k}\, \widetilde{\psi}(\mathbf{k}, t) e^{i\mathbf{k}\cdot\mathbf{x}} \ . \tag{4.59}$$

This substitution implies that any spatial derivative $\nabla \cdot$ becomes an inner product with the vector $\mathbf{k}$. In the Fourier domain it is usual to envisage an eddy of a given scale, i.e. with a certain $\mathbf{k}$, to interact with eddies of larger and smaller scales and to separate out these two contributions, so that the time evolution of any random variable is then written formally as:

$$\frac{\partial}{\partial t}\widetilde{\psi}(\mathbf{k}, t) = f^<(\mathbf{k}, t) - \nu_d(\mathbf{k}, t) k^2 \widetilde{\psi}(\mathbf{k}, t) + f^{\text{ext}}(\mathbf{k}, t) \ . \tag{4.60}$$

Here f^{ext} is the contribution of external forcing of the turbulence, which would be 0 in free (decaying) turbulence, and $f^<$ expresses the forcing of turbulence at the scale corresponding to k by all larger (lower k) eddies. The forcing by the smaller scales (higher k) is expressed, in an analogy with viscous dissipation which occurs at very small scales, as a diffusive process. In normal space diffusion implies a behaviour $\propto \nu \nabla^2 \psi$ which in Fourier space becomes $\propto -\nu k^2 \tilde{\psi}$. The dynamical viscosity $\nu_d(\mathbf{k}, t)$ therefore depends on the true viscosity ν and on the cascading of energy in k-space:

$$\nu_d(\mathbf{k}, t) \equiv \nu + \nu_t(k) = \nu + \int_k^\infty \mathrm{d}k'\, V(k') \,. \tag{4.61}$$

One of the difficulties lies in the behaviour of the function $V(k)$. In [Canuto & Dubovikov (1996)] renormalisation group techniques are used to determine that:

$$\nu_d(\mathbf{k}, t) = \left[\nu^2 + \frac{2}{5} \int_k^\infty \mathrm{d}k'\, k'^{-2} E(k', t) \right]^{1/2} \tag{4.62}$$

where $E(k')$ is the energy density of the turbulence defined by:

$$E(k', t) \frac{1}{2} k^2 \int \mathrm{d}\mathbf{k}' \int \mathrm{d}\Omega_\mathbf{k} \left\langle \sum_i \delta \mathbf{v}_i(\mathbf{k}, t) \delta \mathbf{v}_i(\mathbf{k}', t) \right\rangle \tag{4.63}$$

which evidently depends on the fluctuating velocity one order higher than the velocity itself. The same is true for the turbulent forcing term $f^<$. A similar equation for E involves ensemble averages of third order products and so on. At some level in the hierarchy one therefore uses a closure model: expressing higher order correlations explicitly in terms of lower order correlations. In [Canuto & Dubovikov (1998)] such a closure scheme is presented, which closes the system at third order: correlations of products of three random field variables are expressed in terms of correlations of products of two random field variables.

Another approach is to abandon attempts to solve such a hierarchy, but instead to apply a numerical hydrodynamics code to this problem. Apart from issues of spatial resolution, there is also an issue of time resolution. In simulations of convection, propagating waves such as sound waves will be generated. These are not necessarily of interest for the modelling of the transport properties of the flow, but the time stepping in the hydrodynamical code will be limited by the period of these waves which is generally the shortest time scale in the problem. One way of dealing with this is to define fields of flow velocity and density fluctuations that are low-pass filtered, so that such propagating wave modes are eliminated. This concept is known as the anelastic approximation, introduced in [Gough (1969)]. In [Gilman & Glatzmaier (1981)] it is this approximation which is used as a basis for hydrodynamical calculations, and an extension to magnetohydrodynamical flows can be found in [Lantz & Fan (1999)].

One can also attempt to solve the hydrodynamical equations by explicitly splitting up the spectrum of turbulent motions into two domains: larger scales which are to be resolved explicitly and small scales which are not resolved, but which affect the larger scales through a dynamical viscosity, represented by a stress tensor. This is what is known as Large Eddy Simulations (LES). There are many recipes for constructing a viscosity stress tensor resulting from sub-grid scale turbulence. In principle one can even combine this approach with the closure models discussed above. In this case one would use a closure model just for the sub-grid scale turbulence. The reason to follow this path is that at the smaller scales, the turbulence could be close to homogeneous and isotropic even in stratified media, whereas the effects of stratification matter mostly for those scales of turbulence which are well-resolved.

In [McComb (1990)] a thorough discussion can be found of turbulence and techniques for modelling, although with an emphasis on incompressible media. Much more research effort has been focussed on incompressible flows than on compressible flows. A very comprehensive text that does cover aspects of compressible turbulence as well is [Monin & Yaglom (1975)]. Of particular interest for the discussion here is the fact that for free (i.e. no external forcing) homogeneous, isotropic turbulence certain correlations are conserved. Let α and β represent two homogeneous random scalar fields with zero means and A and B two homogeneous random vector fields such that:

$$\frac{\partial}{\partial t}\alpha = -\nabla \cdot A$$
$$\frac{\partial}{\partial t}\beta = -\nabla \cdot B \,. \qquad (4.64)$$

In free turbulence all the conservation equations (mass, momentum, and energy) can be expressed in this manner. One can form the following two point correlations:

$$\begin{aligned}Q_{\alpha\beta}(\mathbf{r},t) &= \langle \alpha(\mathbf{x},t)\beta(\mathbf{x}+\mathbf{r},t)\rangle \\ Q_{A_j\beta}(\mathbf{r},t) &= \langle A_j(\mathbf{x},t)\beta(\mathbf{x}+\mathbf{r},t)\rangle \\ Q_{\alpha B_j}(\mathbf{r},t) &= \langle \alpha(\mathbf{x},t)B_j(\mathbf{x}+\mathbf{r},t)\rangle \,.\end{aligned} \quad (4.65)$$

By combining the conservation equations one can show that these three correlations satisfy:

$$\frac{\partial}{\partial t} Q_{\alpha\beta} = \nabla \cdot \left(Q_{A_j\beta} - Q_{\alpha B_j} \right) \,. \quad (4.66)$$

Integrating over a spherical volume with radius R and using the divergence theorem (see appendix A) then yields:

$$\frac{\partial}{\partial t} \int_{r<R} d\mathbf{r}\, Q_{\alpha\beta} = \int_{r=R} dS \sum_j \left(Q_{A_j\beta} - Q_{\alpha B_j} \right) \frac{r_j}{r} \,. \quad (4.67)$$

If the correlations decrease more rapidly than r^{-3} the right hand side of (4.67) goes to 0 which means that:

$$\Lambda_{\alpha\beta} \equiv \int_0^\infty d\mathbf{r}\, Q_{\alpha\beta}(\mathbf{r},t) = \text{const.} \quad (4.68)$$

In homogeneous incompressible turbulence the correlation functions are usually assumed to decrease with distance according to a power law and more rapidly than r^{-3}. It is this *slow* because in incompressible fluids the propagation speed of perturbations is infinite. In compressible fluids the propagation speed is finite (the sound speed) and the decrease of the correlation function with distance can be assumed to be exponential which implies that one can expect Eq. (4.68) to hold. Because of the symmetry $Q_{\alpha\beta}(\mathbf{r}) = Q_{\alpha\beta}(-\mathbf{r})$ one can form 15 independent conserved quantities out of combinations of random fields (using the EOS). In isotropic turbulence there are in fact less than that, because there are further translational and rotational symmetries, which leaves four. Two involve just the fluctuating density field and the fluctuating velocity field:

$$\Lambda_{\rho\rho} = \int_0^\infty dr \, Q_{\rho\rho}(r,t) r^2$$

$$\Lambda_{vv} = \int_0^\infty dr \, \left(Q_{\|\|}(r,t) + 2Q_{\perp\perp}(r,t)\right) r^2 \, . \qquad (4.69)$$

The subscripts $\|$ and $\perp$ refer to the directions of the momentum vector along $\mathbf{r}$ and perpendicular to it. The other two involve the fluctuating part of the energy density of the fluid $\mathcal{E}$.

$$\Lambda_{\mathcal{E}\mathcal{E}} = \int_0^\infty dr \, Q_{\mathcal{E}\mathcal{E}}(r,t) r^2$$

$$\Lambda_{\rho\mathcal{E}} = \int_0^\infty dr \, Q_{\rho\mathcal{E}}(r,t) r^2 \qquad (4.70)$$

In the Sun the turbulence is not free, nor is it homogeneous and isotropic, so these conserved quantities may appear irrelevant to the modelling of turbulent convection in the Sun. However, one can argue that for small enough scales, the departures from homogeneity and isotropy are very small, and also that the forcing is spatially concentrated in layers very near the surface, where the optical depth decreases quickly. As discussed, some simulations rely on such assumptions to account for sub-grid scale structure. It is therefore of interest to measure these correlations as a function of depth and time in order to test the validity of such assumptions, which brings the discussion into the domain of tomography of random fields.

4.2.4 Wave propagation in inhomogeneous media

From the discussion in the previous sections it is clear that even if one can reproduce flows in the solar convection zone, using numerical magneto-hydrodynamical codes, the results are realisations of a stochastic process. The structures seen in these flows, and the topology of the magnetic fields, have to be compared with their counterparts on the Sun on a statistical basis ; as time-averages of moments of fluctuations in temperatures, velocities and magnetic fields, and as correlations between these quantities.

One expects that sound waves can be scattered by inhomogeneities, or that wave energy can be converted from one propagating mode to another, or dissipated as heat. Observations of the power in p-mode waves in the vicinity of perturbations with a clear signature on the surface such as sunspots (cf. [Braun et al. (1992)] [Braun (1995)]) certainly show that this occurs. It is the intention to use the wave field of sound waves or magneto-acoustic waves as probes of the properties of the stochastic fields of temperature, velocity and magnetic field in the solar convection zone. For this to be possible, one requirement is that one can make the distinction between the fluctuations due to waves, and other fluctuations. In practice this implies that only those fluctuations can be probed that are slow compared to the inverse of the frequency of the waves used as probes. Accepting this limitation, the next step is that one needs to be able to model the propagation of waves through an inhomogeneous medium. By assumption the medium is stationary, but the fluctuations in the medium are not necessarily small.

In the case at hand this means finding oscillatory solutions of the equations of motion in radiating, magnetised fluids. The basic equations and the properties of various types of solutions are presented in e.g. [Bogdan & Knölker (1989)]. A comprehensive treatment of the hydrodynamics of radiative fluids is [Mihalas & Mihalas (1984)] which also discusses wave propagation. Other relevant discussions can be found in [Dzhalilov et al. (1992)], [Dzhalilov et al. (1994)], [Zhugzhda et al. (1993)], and [Bogdan et al. (1991)]. Particular attention is paid to mode conversion in [Cally & Bogdan (1997)].

4.2.4.1 Example I: Scattering on a sphere in a homogeneous medium

As an example of the scattering of a wave field, one can consider a plane sound wave in a uniform medium scattering on a spherical bubble, that is stationary in this medium. Clearly this is a very simplified view of wave propagation in the solar convection zone, since velocity and magnetic fields are ignored. Also, the topology of the temperature fluctuations is not likely to be built up out of spherical bubbles. However, it does illustrate the difficulties that one can expect to encounter for more realistic situations. The problem of scattering of plane waves on spheres is a text-book problem in electromagnetism (cf. [Jackson (1999)]), where it is complicated by the fact that it involves the scattering of vector fields rather than scalar fields. In radiative transfer this same problem is often referred to as Mie scattering (cf. chapter 9 of [van de Hulst (1957)] or chapter 4 of [Bohren & Huffman (1983)]). However, much of the treatment carries over in a straightforward manner.

The pressure perturbation in an incident plane wave with wave vector **k**, in a medium with a uniform sound speed, can be written out as an expansion on spherical harmonics as follows:

$$p'_{\text{inc}} = A(\mathbf{k},\omega)e^{i(\mathbf{k}\cdot\mathbf{r}-\omega t)}$$
$$= A(\mathbf{k},\omega)e^{-i\omega t}\sum_l i^l(2l+1)\sqrt{\frac{\pi}{2}}\frac{J_{l+1/2}(kr)}{\sqrt{kr}}P_l(\cos\gamma) \quad (4.71)$$

where $k \equiv |\mathbf{k}|$, and γ is the angle between **k** and **r**:

$$\cos\gamma = \frac{\mathbf{k}\cdot\mathbf{r}}{kr}. \quad (4.72)$$

This expansion is convenient, because in order to calculate the total wave field, the incident wave field needs to be matched to the wave field inside the spherical gas pocket. By using this expansion the matching of the solutions reduces to simple conditions for the coefficients of the expansion. The wave field radiated by a spherical object at $\mathbf{r}'$ into a uniform medium must at large distances be approximately identical to the outgoing wave Green's function G of:

$$\left(\nabla^2 + k^2\right)G(\mathbf{r},\mathbf{r}') = -\delta(\mathbf{r}-\mathbf{r}') \quad (4.73)$$

which in spherical coordinates is:

$$G(\mathbf{r},\mathbf{r}') = \frac{e^{ik|\mathbf{r}-\mathbf{r}'|}}{4\pi|\mathbf{r}-\mathbf{r}'|}. \quad (4.74)$$

This is also sometimes referred to as the Green's function satisfying the Sommerfeld radiation condition. This Green's function can be expanded in the same way that the plane wave is:

$$\frac{e^{ik|\mathbf{r}-\mathbf{r}'|}}{4\pi|\mathbf{r}-\mathbf{r}'|} = \frac{i}{8\sqrt{rr'}}\sum_l J_{l+1/2}(kr_<)H^{(1)}_{l+1/2}(kr_>)(2l+1)P_l(\cos\theta) \quad (4.75)$$

in which:

$$\cos\theta \equiv \frac{\mathbf{x}\cdot\mathbf{x}'}{rr'}$$
$$r_< \equiv \min(r,r')$$
$$r_> \equiv \max(r,r'). \quad (4.76)$$

The function $H^{(1)}$ is referred to by several names: a Bessel function of the third kind or a Hankel function. The spherical bubble has radius a, and is assumed to have a uniform sound speed inside, which is different from that of the surrounding medium. A complete solution of the wave equation, valid inside this sphere where it is regular at the origin, and valid outside the sphere where it becomes an outgoing spherical wave, is:

$$p'_{\text{int}} = e^{-i\omega t} \sum_l a_l i^{l+1} (2l+1) \sqrt{\frac{\pi}{2}} \frac{J_{l+1/2}(k_i r)}{k_i r} P_l(\cos\theta) \quad r < a$$
$$p'_{\text{sca}} = e^{-i\omega t} \sum_l b_l i^{l+1} (2l+1) \sqrt{\frac{\pi}{2}} \frac{H^{(1)}_{l+1/2}(k_e r)}{k_e r} P_l(\cos\theta) \quad r > a \quad (4.77)$$

in which k_i and k_e are respectively the magnitude of the wave vector internal to the sphere and external to it. The two are related through the dispersion relation:

$$\omega^2 = k_{re}^2 c_{se}^2 = k_{ri}^2 c_{si}^2 \ . \quad (4.78)$$

At the interface between the two domains the total pressure perturbation outside must match the pressure perturbation inside. Also, the normal component of the displacement inside and outside must be equal, so that:

$$p'_{\text{inc}} + p'_{\text{sca}} = p'_{\text{int}}$$
$$\delta r_{n,\text{inc}} + \delta r_{n,\text{sca}} = \delta r_{n,\text{int}} \ . \quad (4.79)$$

The second of these expressions can be converted to a condition for the derivative of p' using Eq. (2.83), which in the absence of gravity ($\Psi_0 = \Psi' = 0$) is:

$$\delta r_n = -\frac{1}{i\omega\rho} \frac{\partial p'}{\partial r} \ . \quad (4.80)$$

Using the expansions Eq. (4.71) and Eq. (4.77) for p' these conditions reduce to linear equations between the unknown coefficients a_l, b_l in Eq. (4.77), and the amplitude of the incoming wave:

$$A(\mathbf{k}_e, \omega) j_l(k_e a) + b_l h_l^{(1)}(k_e a) = a_l j_l(k_i a)$$
$$A(\mathbf{k}_e, \omega) \frac{k_e j'_l(k_e a)}{\rho_e} + b_l \frac{k_e h_l^{(1)'}(k_e a)}{\rho_e} = a_l \frac{k_i j'_l(k_i a)}{\rho_i} \ . \quad (4.81)$$

In Eq. (4.81) spherical Bessel and Hankel functions $j_l(x)$ and $h_l^{(1)}(x)$ and their derivatives are introduced for convenience, and are defined by:

$$j_l(x) \equiv \frac{J_{l+1/2}(x)}{\sqrt{x}}$$

$$j_l'(x) \equiv \frac{lj_l(x) - xj_{l+1}}{x}$$

$$h_l^{(1)}(x) \equiv \frac{H_{l+1/2}^{(1)}x}{\sqrt{x}}$$

$$h_l^{(1)\prime}(x) \equiv \frac{lh_l^{(1)}(x) - xh_{l+1}^{(1)}}{x}. \tag{4.82}$$

The conditions of Eq. (4.81) can be solved for a_l and b_l to produce:

$$\begin{pmatrix} a_l \\ b_l \end{pmatrix} = A(\mathbf{k}_e, \omega) \mathbf{T}_l(k_e, \omega) \cdot \begin{pmatrix} j_l(k_e a) \\ \rho_i k_e j_l'(k_e a) \end{pmatrix} \tag{4.83}$$

$$\mathbf{T}_l(k_e, \omega) \equiv \frac{\begin{pmatrix} -\rho_i k_e h_l^{(1)\prime}(k_e a) & h_l^{(1)}(k_e a) \\ -\rho_e k_i j_l'(k_i a) & j_l(k_i a) \end{pmatrix}}{\rho_e k_i j_l'(k_i a) h_l^{(1)}(k_e a) - \rho_i k_e j_l(k_i a) h_l^{(1)\prime}(k_e a)}.$$

The matrix $\mathbf{T}_l(k_e, \omega)$ is the matrix which relates the amplitudes of the internal and external scattered field to the incoming field amplitude. Of particular interest for the purposes of measurements of the scattered wave field is the ratio $b_l/A(\mathbf{k}_e, \omega)$:

$$\frac{b_l(k_e, \omega)}{A(\mathbf{k}_e, \omega)} = \frac{\rho_i k_e j_l'(k_e a) j_l(k_i a) - \rho_e k_i j_l'(k_i a) j_l(k_e a)}{\rho_e k_i j_l'(k_i a) h_l^{(1)}(k_e a) - \rho_i k_e j_l(k_i a) h_l^{(1)\prime}(k_e a)}. \tag{4.84}$$

The incoming planar wave decomposition of Eq. (4.71) uses $j_l(k_e r)$ as base functions. An equivalent description can be obtained by making use of the incoming and outgoing spherical Hankel functions $h_l^{(1)}$ and $h_l^{(2)}$ from the identities (cf. [Gradshteyn & Ryzhik (1994)]):

$$j_l(x) = \frac{1}{2}\left(h_l^{(1)} + h_l^{(2)}\right)$$

$$y_l(x) = \frac{1}{2i}\left(h_l^{(1)} - h_l^{(2)}\right) \tag{4.85}$$

in which $y_l(x)$ is the spherical Bessel function of the second kind (more details can be found in eg. [Abramowitz & Stegun (1964)] or [Gradshteyn & Ryzhik (1994)]). The reason for making this substitution is that the $h_l^{(1)}$ and $h_l^{(2)}$ are mutually orthogonal (much as the sin and cos functions are), as are the functions j_l and y_l, whereas j_l and either of the h_l are not. In the decomposition of Eq. (4.85) the ratio of outgoing wave amplitude to incoming wave amplitude is given by $1 + 2b_l(k_e, \omega)$, which is a complex function, and can therefore be written as:

$$1 + 2b_l(k_e, \omega) \equiv R_l(k_e, \omega) e^{i\delta_l(k_e, \omega)} \qquad (4.86)$$

in which $R_l(k_e, \omega)$ is a purely real factor describing amplitude changes and $\delta_l(k_e, \omega)$ is purely real and describes phase changes. Under the conditions described here, the sphere does not absorb any energy and therefore $R_l(k_e, \omega) \equiv 1$. The angle $\delta_l(k_e, \omega)$ can be determined by evaluating:

$$\tan \delta_l(k_{re}, \omega) = \frac{\text{Im}[1 + 2b_l(k_{re}, \omega)]}{\text{Re}[1 + 2b_l(k_{re}, \omega)]} . \qquad (4.87)$$

For a few selected values of the parameters, the result is shown in Fig. (4.6). In the top panels the case is shown for the $l = 0$ component, where the contrast in sound speed is 10%, i.e. the temperature is $\sim 20\%$ higher (left) or lower (right) than the external medium. In each of these panels the results shown as dashed lines are calculated without density contrast between the bubble and the medium, which implies that the bubble is not in pressure equilibrium with its surroundings. The solid lines are cases for which there is pressure equilibrium $\rho_i c_{si}^2 = \rho_e c_{se}^2$. It is clear that for a bubble of a given size and temperature contrast the scattering angle increases in absolute value with increasing wavenumber, implying increasingly strong scattering. In the middle panels the temperature contrast is doubled. The dashed and solid lines have the same meaning as before. Evidently the scattering becomes stronger as well, and the finer-scale structure of the scattering angle also becomes more pronounced. At large distances from the bubble, the $l = 0$ component is most important in the scattering, because its amplitude decreases most slowly with increasing distance. However, in the case of the Sun, in particular for larger disturbances, the observational conditions are not necessarily 'far field' conditions. It is of some interest therefore to also show the behaviour for higher values of l. In the bottom left panel of Fig. (4.6) the same case is considered as in the top left panel: $c_{si}/c_{se} = 1.1$ and $\rho_i c_{si}^2 = \rho_e c_{se}^2$, but for increasing values of l. At higher l the scattering angle remains small up to larger values of the wavenumber, but then increases more steeply. Finally in the bottom right panel a case is considered in which the bubble does not have any temperature contrast, but is denser than the surrounding medium. Evidently such a bubble is not in pressure equilibrium with its surroundings. In this case one can see that there is not the same steep increase of scattering angle with wavenumber, but there is nevertheless a clear signature of the scatterer on the emerging wave field.

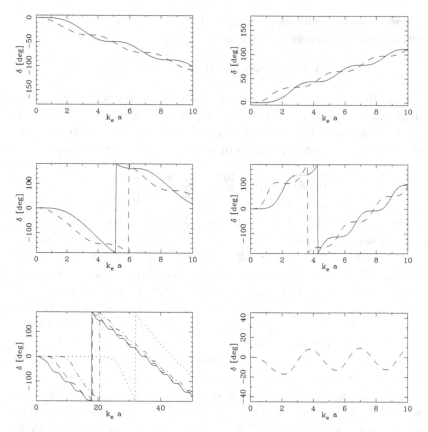

Fig. 4.6 The scattering angle δ shown as a function of $k_{re}a$ for various l, and various parameters. Top left panel: $l = 0$, $c_{si}/c_{se} = 1.1$, solid line: a bubble in pressure equilibrium with medium, dashed line: $\rho_i/\rho_e = 1$. Top right panel: $l = 0$, $c_{si}/c_{se} = 0.91$, solid and dashed lines as for top left panel. Middle left and right panels: $l = 0$, $c_{si}/c_{se} = 1.4$ and 0.714 resp., solid and dashed lines as for top left panel. Bottom left panel: $c_{si}/c_{se} = 1.1$, $l = 1$ (solid), $= 5$ (dashed), $= 10$ (dot-dash), $= 25$ (dotted). Bottom right panel: $l = 0$, $c_{si}/c_{se} = 1$, $\rho_i/\rho_e = 1.21$.

This illustrates that the change in propagation speed of the waves through the disturbance is not the only factor in the scattering process.

These diagrams can also be interpreted as the response obtained for waves of a given frequency, to bubbles with a range of radii. For typical parameters of solar oscillations in the sub-surface regions ($P \sim 5$ min, $c_{se} \approx 7$ km/s), one unit corresponds to 0.3 Mm in scale.

4.2.4.2 Example II: Scattering on an infinite cylinder in a homogeneous medium

The previous example is very highly idealised, although some of the qualitative features displayed are also seen in more realistic studies. One step towards this is to consider perturbations that are cylindrical in shape, which is closer to what one expects for instance from magnetic flux concentrations: flux tubes. The case of an infinite cylinder in a uniform atmosphere, is con-

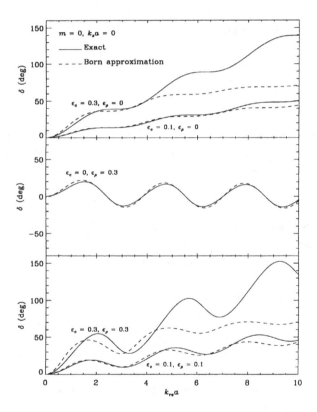

Fig. 4.7 The scattering angle δ shown as a function of $k_{re}a$ for normal incidence waves $k_z = 0$, and various cylinder parameters. Solid lines are the exact solution as shown here, dashed lines are from an approximate calculation using the Born approximation. (Figure from [Fan et al. (1995)] reproduced by permission of the AAS).

sidered in the appendix of [Fan et al. (1995)]. In this case the cylinder is aligned with the z-axis, and has radius a. External to the cylinder the sound speed and density are c_{se} and ρ_e. The internal sound speed and density are c_{si} and ρ_i. As above the boundary conditions require continuity of the pressure and of the component of the velocity perturbations normal to the interface. In this configuration the coordinate system of choice is cylindrical (r, ϕ, z). The pressure perturbation p' of the resulting total wave field (incoming + scattering) outside the cylinder in this case satisfies:

$$p' = e^{i(k_z z + l\phi + \omega t)} \sum_l \int_{-\infty}^{\infty} dk_z \int_0^{\infty} dk_r \; A_l(k_{re}, k_{ze}) \left[J_l(k_{re} r) \right.$$
$$\left. + b_l(k_{re}, k_z) H_l^{(2)}(k_{re} r) \right] \quad (4.88)$$

where b_l is determined by the interface conditions:

$$b_l(k_{re}, k_z) = \frac{\rho_i k_{re} J_l'(k_{re} a) J_l(k_{ri} a) - \rho_e k_{ri} J_l(k_{re} a) J_l'(k_{ri} a)}{\rho_e k_{ri} H_l^{(2)}(k_{re} a) J_l'(k_{ri} a) - \rho_i k_{re} H_l^{(2)'}(k_{re} a) J_l(k_{ri} a)} \quad . \quad (4.89)$$

Here the vertical wavenumber k_z is identical inside and outside the cylinder. k_{ri} and k_{re} are the radial wavevectors internal and external to the cylinder. and the dispersion relation is:

$$\omega^2 = c_{si}^2 \left(k_{ri}^2 + k_z^2 \right) = c_{se}^2 \left(k_{re}^2 + k_z^2 \right) \; . \quad (4.90)$$

Plotted in Fig. 4.7 is the phase shift δ between incoming and outgoing waves of mode (l, k_{re}, k_z) (note that in the notation of [Fan et al. (1995)] m is used instead of l), which is defined in the same was as in Eq. (4.87).

4.2.4.3 More general scattering problems

In either of these examples an incoming wave with wave-number k_e and azimuthal degree l gives rise to a scattered wave with the same external wavenumber and azimuthal degree: there is no mechanism that converts energy from one mode to another. A different way of expressing this is that if one expresses the scattering process for all l combined in matrix form $T^{l \to l'}(\omega)$, this matrix is diagonal:

$$T^{l \to l'}(\omega) = \delta_{ll'} b_l(\omega) \; . \quad (4.91)$$

For more general cases than these two examples, there is no reason to assume that the scattering matrix remains diagonal. In the main paper of [Fan et al. (1995)], the problem of scattering on a cylinder-symmetric perturbation in a stratified atmosphere is considered. The axisymmetry assumed in this problem, produces a result in which the scattering matrix is still diagonal in terms of the azimuthal degree l. The stratification together with appropriate boundary conditions produces a discrete spectrum of vertical wave functions which can be classified in terms of the order n, or equivalently the vertical wave vector k_{zn}. The scattering matrix for identical values $l = l'$ but $n \neq n'$ is generally not equal to zero. An incoming wave with a certain vertical wave vector can produce scattered waves with different vertical wave vectors.

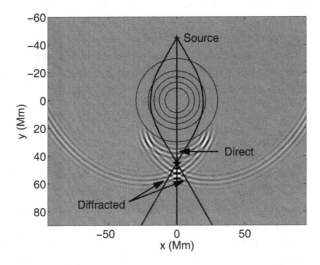

Fig. 4.8 A finite difference simulation of a wave field emanating from a point source and scattering on a spherical inhomogeneity. The solid circles are contours of equal sound speed. The thick lines correspond to ray paths. (Figure from [Birch et al. (2001)] reproduced by permission of the AAS)

The perturbations studied by [Fan et al. (1995)] are still not very realistic approximations of magnetic flux tubes since the effect of the magnetic field on the waves is ignored. More realistic studies of the scattering of waves by magnetic flux concentrations can be found in [Rosenthal (1995)] and [Bogdan & Knölker (1991)]. One also expects some degree of increased vorticity to be associated with convection: the overturning motions of eddies. A description of the scattering of sound waves on zones of increased vorticity $\nabla \times v$ can be found in [Bogdan (1989)]. An example of an explicit finite difference calculation of the wave field is shown in [Birch et al. (2001)] (see Fig. 4.8) as part of a comparison between various approximation methods for calculating wave fields in the presence of inhomogeneities. Certainly for more complicated perturbation topologies this is more appropriate. However one should keep in mind that for testing models of convective turbulence one needs to obtain statistical measures of its properties. Finite difference calculations would have to be carried out on an ensemble of realisations of simulations to provide such measures: i.e. one would have to carry out a Monte Carlo simulation.

In parallel with such efforts one can make progress in terms of tomographic imaging by making use of approximate methods for solving the wave field which can be the basis of inverse problems. The most commonly used approximations are the ray approximation, the Born approximation and the Rytov approximation. Each of these approximations has been used as a basis for inversions to image inhomogeneities in the solar subsurface layers. The discussion of each of these methods is therefore deferred to the appropriate sections (4.3.3-4.3.5).

4.3 Inverse Methods

The phase or group velocity of signals, measured at the surface of the Sun, is determined by conditions below the surface. Regardless of whether one correlates components of line-of-sight velocities between two patches or reconstructs the horizontal phase velocity averaged over a surface patch, the sensitivity to conditions below the surface is neither point-like ($\delta(\mathbf{r}-\mathbf{r}')$) nor uniform. In order to localise the cause of particular features in the observed quantities one therefore needs to solve an inverse problem. The precise character of the inverse problem depends on the type of data processing. The most commonly used methods are discussed here.

4.3.1 *Ring diagrams*

The ring diagram technique makes use of the location of the peaks of acoustic power as a function of the horizontal components of the wave vector. As outlined in section 4.1 a rectangular patch of solar surface is selected and tracked along with the surface rotation rate for a length of time which can range from 8 h to a few days. The resulting stack of Doppler images is then Fourier transformed in the two spatial directions and in time. The acoustic power within this data-cube in the Fourier domain is concentrated along curved surfaces that, when cut at constant frequency, appear as a set of nested rings of high acoustic power, corresponding to different radial orders of the oscillations. In Fig. 4.9 a diagram is shown for a simulation of a wave

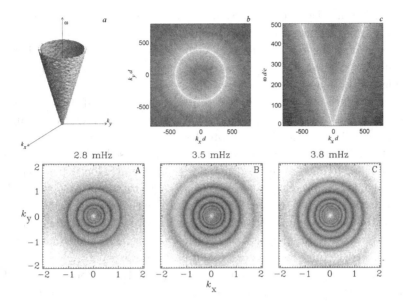

Fig. 4.9 Top row: artificial ring diagrams for a model with uniform density and sound speed. The power is shown as a surface in (k_x, k_y, ω)-space (left panel) and as cuts at constant frequency ω (middle panel) and also at constant $k_y = 0$ (right panel) (from [Hindman et al. (2005)]). Bottom row: ring diagrams for three selected frequencies, computed from observations of patches ~ 15 deg across for a duration of ~ 27 h (from [Haber et al. (2002)]). Both figures reproduced by permission of the AAS.

field in an atmosphere with uniform sound speed and density. In [Hindman et al. (2005)] this is used as a reference model to calculate the sensitivity of these ring diagrams to (horizontal) flow fields. For a uniform flow field **v** one can calculate the frequencies with the usual dispersion relation $\omega = kc_s$ in a frame co-moving with the fluid, and a simple Galilean transform between this coordinate frame and a 'laboratory' coordinate frame then produces a Doppler frequency shift $\Delta\omega = \mathbf{k} \cdot \mathbf{v}$. In ring diagrams (top middle panel of Fig. 4.9) the rings are therefore shifted so that they are no longer centered on $(0,0)$. From the bottom row panels of Fig. 4.9 (cf. [Haber et al. (2002)]) one can see that the observed data produces ring diagrams which appear very similar to the top row, indicating flow velocities that are small compared to the sound speed c_s. A detailed analysis of observed ring diagrams and inversion results can be found in [Haber et al. (2002)] and [González Hernández et al. (2006)]. In these two papers ring diagram analysis is used in particular to trace the meridional flow, its asymmetry between northern and southern hemisphere of the Sun, and its time variability. The global mode analysis can be used for changes in rotation rate with epoch, but by only determines the symmetric component of flows. From this it is evident that the ring diagram technique, based on the same perturbation analysis, provides valuable additional information.

The influence of velocity fields with a more complicated depth dependence, in a plane-parallel polytropic atmosphere, can be found in [Gough & Toomre (1983)]. However, the problem can also be viewed as a variant of the influence of rotation on frequencies discussed in section 3.3. In Eq. (3.63) the frequency perturbation due to a general velocity field is presented, which is subsequently applied to a purely rotational velocity field. For the application at hand Eq. (3.63) can also be the starting point. Some additional manipulation is necessary however, in order to make the connection between the data analysis and the theory. Several methods have been proposed to determine the frequency shift $\delta\omega$ from the (k_x, k_y, ω) data cube (cf. [Patron et al. (1997)], [Schou & Bogart (1998)]). Although details of parametrisation can differ, a function is fitted to the data which has the form:

$$P(k_x, k_y, \omega) = \frac{a}{1 + (\omega - \omega_0 - k_x U_x - k_y U_y)^2/\Gamma^2} + bN^2(k) \qquad (4.92)$$

in which $N^2(k)$ is the noise background which can be e.g. a power-law: $N^2(k) \propto k^\alpha$. The parameters fitted for are $a, b, \omega_0, \Gamma, U_x$, and U_y. This profile is the usual Lorentzian shape encountered in Eq. (1.41). One could envisage fitting asymmetric profiles as in Eq. (1.42) but because of the higher noise level in these comparatively short time series, the influence of any asymmetry is negligible. The frequency perturbation $\delta\omega$ is thus expressed here in terms of the fitting parameters U_x and U_y:

$$\delta\omega = k_x U_x + k_y U_y \ . \qquad (4.93)$$

If one fixes $k = \sqrt{k_x^2 + k_y^2}$, corresponding to taking a cylindrical cut through the data cube, the fitting is equivalent to determining the frequency perturbation at fixed l. For each n there is a separate ridge of power and therefore the fitting produces $\delta\omega$ at fixed n and l, which is then a function of angle ϕ_k in a polar representation of the (k_x, k_y)-plane.

To make the connection with (3.63) for this type of flow the same spherical coordinate system can be chosen which has its pole located at the rotational pole. It is expected that in the Sun radial velocities are small compared to horizontal velocities, apart perhaps from downflows directly associated with sunspots. For this reason in the analysis here it is assumed that $v_r = 0$ identically. The velocities in $\widehat{e}_x$ and $\widehat{e}_y$ directions map directly from the $\widehat{e}_\phi$ and $\widehat{e}_\theta$ directions so that:

$$\mathbf{v}_0 \equiv -v_y(r,\theta,\phi)\widehat{e}_\theta + v_x(r,\theta,\phi)\widehat{e}_\phi \ . \tag{4.94}$$

The effect of the $\widehat{e}_\phi$ component is similar to that of rotation, which is already shown in Eq. (3.64). Instead of Ω one needs to use $v_x(r,\theta,\phi)/r\sin\theta$. Further one needs to adjust the integration limits to account for the fact that only a patch of solar surface is observed and therefore a limited volume is probed. For local area helioseismology the horizontal wavenumbers used are always high and correspond to l values well in excess of 100. One consequence of this is that one can restrict the volume integral for the inner products of eigenfunctions, to be evaluated in the variational principle, without this invalidating to any significant extent the orthogonality properties of the eigenfunctions. This results in an expression for the frequency perturbations:

$$E_{nl}\delta\omega = m \int_0^R dr \int d\cos\theta \int d\phi\, \rho_0 r^2 \frac{v_x(r,\theta,\phi)}{r\sin\theta} \left\{ \widetilde{\delta r}_r^2 P_l^{m\,2} - 2P_l^{m\,2} \widetilde{\delta r}_r \widetilde{\delta r}_h \right.$$
$$\left. + \widetilde{\delta r}_h^2 \left[\left(\frac{dP_l^m}{d\theta}\right)^2 + \frac{m^2}{\sin^2\theta} P_l^{m\,2} \right] - 2P_l^m \widetilde{\delta r}_h^2 \cot\theta \frac{dP_l^m}{d\theta} \right\} \tag{4.95}$$

where the limits of integration for the θ and ϕ directions are omitted on purpose, since they are set by the size of the patch which is being analysed. It is not unusual in this type of analysis to assume that the velocity is a function of depth only. Horizontal resolution is achieved by applying the analysis to a number of patches, possible partially overlapping, which together tile the part of the solar surface of interest. In this case, with appropriate normalisation applied to account for the integration over a patch rather than the full Sun, Eq. (4.95) reduces to:

$$U_x^{nl} \equiv \frac{\delta\omega}{k_x} = \int_0^R dr\, K_{nl}^{\text{ring}}(r) v_x(r) \tag{4.96}$$

in which the kernel $K^{\text{ring}\,nl}(r)$ is given by

$$K^{\text{ring}\,nl}(r) = \frac{\rho(r)r}{E_{nl}}\left[g_1^{lm}\left(\widetilde{\delta r}^2_{r\,nl}(r) - 2\widetilde{\delta r}_{r\,nl}(r)\widetilde{\delta r}_{h\,nl}(r) + L^2\widetilde{\delta r}^2_{h\,nl}(r)\right)\right.$$
$$\left. - g_2^{lm}\widetilde{\delta r}^2_{h\,nl}(r)\right] \quad (4.97)$$

and the $g_{1,2}$ are normalisation factors:

$$g_1^{lm} \equiv \int d\theta\, [P_l^m(\cos\theta)]^2$$
$$g_2^{lm} \equiv \int d\theta \frac{1}{2}\sin^2\theta \frac{d^2 P_l^{m\,2}}{d\cos\theta^2}. \quad (4.98)$$

The expression for the v_y component has to be identical since the perturbation is done with respect to a spherically symmetric background. Because of this symmetry, rotation of the coordinate system around an axis perpendicular to the solar surface at the centre of the patch, must leave the expressions invariant, which implies:

$$U_y^{nl} = \int_0^R dr\, K^{\text{ring}\,nl}(r) v_y(r) \quad (4.99)$$

with the same kernel $K^{\text{ring}\,nl}$. Good depth resolution can be obtained by combining data for a range of n and l, using the same inverse methods as are discussed in chapter 3.

Evidently the assumption that the rotation rate is horizontally uniform over a tile does not represent reality, since the results for different tiles on the Sun show a varying flow field. The influence of this on the inversion results is explored in [Hindman et al. (2005)] from which the upper panels of Fig. 4.9 are taken. As an example the case is considered in which there is a narrow stream of high velocity plasma running through a tile. Such jets might exist as solar counterparts to the jet-stream in the Earth's atmosphere. It is shown that in this case the rings are not only displaced, but also become smeared out with a double peak in power. One peak will be roughly circular in shape, whereas the other will appear ellipsoidal in shape. As the flow speed of the feature increases to a significant fraction of the local sound speed, this result becomes more pronounced. As is shown in the appendix of [Hindman et al. (2005)] the shape of the profile of power in cuts through rings can yield additional information on the level of horizontal inhomogeneity, and is therefore of additional diagnostic value.

4.3.2 Time-distance techniques

The first applications of time distance helioseismology can be found in [Duvall et al. (1993b)] in which cross-correlations of intensity fluctuations on the solar surface are calculated. Formal descriptions of appropriate analysis techniques, within the context of helioseismology, have become available following this pioneering work. The first relevant paper is [D'Silva (1996)] in which the ray approximation is used to interpret the cross-correlation function. An example of a normalised cross-correlation function for the

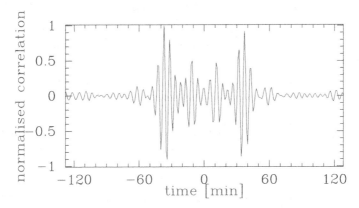

Fig. 4.10 An example of a cross-correlation function in the time domain computed from a time series of 512 Doppler images ($\sim$ 8.5 hrs of 128x128 pixels corresponding to $\sim$ 195x195 Mm on the surface, after appropriate filtering, and for cross-correlation between arcs (right hand panel of Fig. 4.4).

arc-to-arc configuration shown in the right hand panel of Fig. 4.4. This is what in geophysical settings would be referred to as a seismic trace. It is traces such is this which need to be interpreted either directly or in terms of the time of the peak of the envelope, or of the 'leading edge' of the wave package if such can be identified robustly. The trace shown in Fig. 4.10 is symmetrical to a high degree because the arc-to-arc correlations are averaged over an entire patch. Individual correlations show less clearly definable wave packets, and are generally much less symmetric. Three analysis methods are discussed in the following sections: the ray approximation, the Born approximation, and the Rytov approximation. An overview of in particular the first two methods and applications can also be found in the review papers [Kosovichev et al. (2000)] and [Gizon & Birch (2005)].

4.3.3 The ray approximation

In [D'Silva (1996)] the ray approximation as applied to local helioseismology is described in detail. Starting point is to consider the quantity observed at the surface v and its Fourier transform with respect to time:

$$v(t,\mathbf{r}) = \frac{1}{2\pi}\int d\omega\, \tilde{v}(\omega,\mathbf{r})e^{-i\omega t}\,. \tag{4.100}$$

Cross-correlation of this signal with a signal at some displacement $\Delta \mathbf{r}$ is:

$$\Gamma(\Delta t, \Delta \mathbf{r}, \mathbf{r}) = \int dt\, v(t+\Delta t, \mathbf{r}+\Delta \mathbf{r})v(t,\mathbf{r})^*$$

$$= \frac{1}{2\pi}\int d\omega\, \tilde{v}(\omega, \mathbf{r}+\Delta \mathbf{r})e^{-i\omega\Delta t}\tilde{v}(\omega,\mathbf{r})^* e^{-i\omega t} \tag{4.101}$$

where the second equality is due to the fact that in the Fourier domain the cross-correlation is the product of the Fourier transform of the quantities being correlated. The next step is to treat the wave packet as locally plane waves so that the phase difference between the observable at two different locations can be expressed as a line integral:

$$\tilde{v}(\omega, \mathbf{r}+\Delta \mathbf{r}) = \tilde{v}(\omega, \mathbf{r})\exp\left[i\int_{\mathbf{r}}^{\mathbf{r}+\Delta\mathbf{r}} \mathbf{k}(\omega,\mathbf{r})\cdot d\mathbf{r}'\right]\,. \tag{4.102}$$

With this the expression for Γ reduces to:

$$\Gamma(\Delta t, \Delta \mathbf{r}, \mathbf{r}) = \frac{1}{2\pi}\int d\omega\, |\tilde{v}(\omega,\mathbf{r})|^2$$

$$\times \exp\left[-i\omega(t+\Delta t) + i\int_{\mathbf{r}}^{\mathbf{r}+\Delta\mathbf{r}} \mathbf{k}(\omega,\mathbf{r}')\cdot d\mathbf{r}'\right]\,. \tag{4.103}$$

The absolute time t appearing on the right-hand side of this expression is a constant of integration. By adjusting the integration limits in the line integral over $\mathbf{r}$ the dependence on t disappears:

$$\Gamma(\Delta t, \Delta \mathbf{r}, \mathbf{r}) = \frac{1}{2\pi}\int d\omega\, |\tilde{v}(\omega,\mathbf{r})|^2$$

$$\times \exp\left[-i\Delta t\left(\omega - \frac{1}{\Delta t}\int_0^{\Delta \mathbf{r}} \mathbf{k}(\omega,\mathbf{r}')\cdot d\mathbf{r}'\right)\right]\,. \tag{4.104}$$

This integral is a generalised Fourier integral, i.e. of the type:

$$I(\tau) = \int d\omega\, f(\omega) e^{i\tau\psi(\omega)} \,. \tag{4.105}$$

One can apply the Riemann-Lebesgue lemma to integrals of this type to show that in the limit of large τ the value of all integrals of this type goes asymptotically to 0, as long as the function $|f(t)|$ is integrable, and $\psi(\omega)$ is continuously differentiable and not constant on any subinterval in the range of integration. The most slowly decreasing contribution to the integral is obtained by integration by parts and is (cf. [Bender & Orszag (1978)]):

$$I(\tau) \approx \frac{f(\omega)}{i\tau\psi'(\omega)} e^{ix\psi(\omega)} \bigg|_{\omega=\omega_-}^{\omega=\omega_+} . \tag{4.106}$$

If, however, there is a stationary point for ψ in the range of integration, the integration by parts fails. The leading order behaviour, i.e. the most slowly decreasing term, can then be obtained as follows. Without loss of generality (cf. [Bender & Orszag (1978)]) the integration can always be split up in such a way that there is no more than a single point ω_s at which $\psi'(\omega_s) = 0$. At this point higher order derivatives could in principle vanish as well, and so in the most general case the first non-zero derivative may be of order p:

$$\psi'(\omega_s) = \cdots = \psi^{(p-1)}(\omega_s) = 0$$
$$\psi^{(p)}(\omega_s) \neq 0 \,. \tag{4.107}$$

In this case the integral becomes:

$$I(\tau) \approx f(\omega_s) e^{i\tau\psi(\omega_s) \pm i\pi/2p} \left[\frac{p!}{\tau|\psi^{(p)}(\omega_s)|}\right]^{1/p} \frac{\Gamma(1/p)}{p} \tag{4.108}$$

in which Γ is the Gamma-function (not to be confused with the cross-correlation), and the sign of the factor $i\pi/2p$ in the exponential is the same as the sign of $\psi^{(p)}(\omega_s)$. A stationary point of ψ is given by:

$$0 = \psi'(\omega) = 1 - \frac{1}{\Delta t} \int_0^{\Delta \mathbf{r}} \frac{\partial \mathbf{k}(\omega, \mathbf{r}')}{\partial \omega} \cdot d\mathbf{r}' \tag{4.109}$$

which can also be written in a differential form:

$$\frac{d\mathbf{r}}{dt} = \frac{\partial \omega}{\partial \mathbf{k}} \equiv v_{\text{group}} \,. \tag{4.110}$$

This result means that the dominant behaviour for Γ is determined by waves satisfying Eq. (4.110). In this case a packet of sound waves, with frequencies limited to a band centered on some reference frequency ω_0, are considered to propagate along a path given by the magnitude and direction of their group velocity. In the limiting case of a non-dispersive medium $\omega = c_s \mathbf{k}$, and the phase velocity $\omega/\mathbf{k}$ is identical to the group velocity. The rays calculated in the toy model in section 4.1.1 are then identical to the rays considered here. In [D'Silva (1996)] the wave packet is assumed to have a Gaussian power spectrum centered on ω_0:

$$|\tilde{v}(\omega, \mathbf{r})|^2 = A_c \exp\left[-b(\omega - \omega_0)^2\right] . \tag{4.111}$$

In [D'Silva (1996)] it is demonstrated that the peak of the envelope of the cross-correlation function is located at a time delay Δt determined by Eq. (4.109). It is also shown that the individual maxima of Γ correspond to a phase travel time given by:

$$\omega_0 \Delta t_{\rm ph} = \left(2n \pm \frac{1}{4}\right)\pi + \int_0^{\Delta \mathbf{r}} \mathbf{k}(\omega, \mathbf{r}') \cdot d\mathbf{r}' \tag{4.112}$$

where the path of the integral is given by Eq. (4.110), and the sign of the factor $1/4$ is the same as the sign of $\psi''(\omega_0)$ which is assumed not to vanish. Note that in practice one obtains the observational group (and phase) travel times by fitting a function of the form given in Eq. (4.108), with $f(\omega_s)$ given by Eq. (4.111). This functional form is also referred to as a Gabor wavelet.

The relationship obtained between the group velocity along a ray and the location in time of the peak of the envelope of the cross-correlation function, also provides the basis of an inverse problem for that group velocity. If one considers the difference between the group travel time obtained from observations, and the travel time calculated for an appropriate reference model this satisfies:

$$\Delta t_{\rm grp} - \Delta t_{\rm grp\ ref} = \int_0^{\Delta \mathbf{r}} \left[\frac{\partial \mathbf{k}(\omega, \mathbf{r}')}{\partial \omega} - \left.\frac{\partial \mathbf{k}(\omega, \mathbf{r}')}{\partial \omega}\right|_{\rm ref}\right] \cdot d\mathbf{r}' . \tag{4.113}$$

Evidently the difference in travel time between the true Sun and a reference model is a measure of the difference in group velocity along the ray of propagation. If the perturbations are (assumed to be) small this is normally explicitly indicated by using the notation:

$$\delta \Delta t_{\text{grp}} = \int_0^{\Delta r} \delta \frac{\partial \mathbf{k}(\omega, \mathbf{r}')}{\partial \omega} \cdot d\mathbf{r}' \ . \tag{4.114}$$

The cross-correlation function has maxima at positive and negative time delays, corresponding to wave packets travelling in either direction along the ray. The dispersion relation for (magneto-)acoustic waves in the solar convection zone is very similar to that given by Eq. (2.129), but now local flow and magnetic fields are accounted for:

$$(\omega - \mathbf{k} \cdot \mathbf{v}_0)^2 = \omega_c^2 + k^2 c_f^2 \tag{4.115}$$

in which c_f is the fast magneto-acoustic speed:

$$c_f^2 = \frac{1}{2} \left(c_s^2 + v_A^2 + \sqrt{(c_s^2 + v_A^2)^2 - 4 c_s^2 (\mathbf{k} \cdot \mathbf{v}_A)^2 / k^2} \right)$$

$$\mathbf{v}_A = \frac{\mathbf{B}}{\sqrt{4\pi\rho}} \ . \tag{4.116}$$

Using this dispersion relation, the difference between the corresponding travel times is related to flows (including rotation):

$$\frac{|\delta \Delta t_{\text{grp}} +| - |\delta \Delta t_{\text{grp}} -|}{2} \approx \int_{\text{ray}} \frac{\mathbf{v}_0}{c_s^2} \cdot d\mathbf{r} \tag{4.117}$$

in which $\mathbf{v}_0$ is the flow velocity. The mean of the travel time differences is sensitive to structural differences between the Sun and the model (cf. [Kosovichev & Duvall (1997)]). Limiting structural differences to those expressible in terms of differences in sound speed or differences in magnetic field the expression is:

$$\frac{|\delta \Delta t_{\text{grp}} +| + |\delta \Delta t_{\text{grp}} -|}{2} \approx -\int_{\text{ray}} \left[\frac{\delta c_s}{c_s} + \frac{1}{8\pi} \left(\frac{\mathbf{B} \cdot \mathbf{B}}{\rho c_s^2} - \frac{(\mathbf{k} \cdot \mathbf{B})^2}{\rho k^2 c_s^2} \right) \right] \frac{\mathbf{k}}{\omega} \cdot d\mathbf{r} \ . \tag{4.118}$$

Eqs. (4.117) and (4.118) both specify inverse problems since a line integral is involved. By solving the inverse problem it is possible to build up a tomographic reconstruction of the flows and structural differences in three dimensions. As pointed out in [Kosovichev et al. (2000)] the magnetic field introduces a directional dependence in the structural differences, so that any anisotropy in structural differences that is detected will allow a separation of the effect of sound speed differences from those of magnetic field differences.

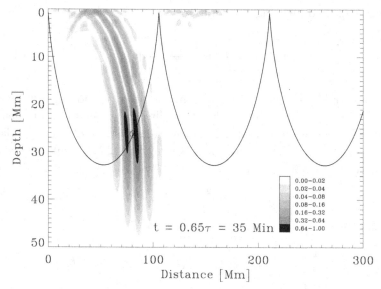

Fig. 4.11 Contour plot of the normalised energy density of a wave packet of waves with a phase speed around 65 km s^{-1}, propagating from a location on the surface after a time of $\sim$ 35 min. The solid line is the ray, with the asterisk the expected location of the wave packet on the basis of the method of stationary phase. (figure from [Bogdan (1997)] reproduced by permission of the AAS).

In practice this has not been achieved due to the limitations imposed by noise.

Physically less satisfactory is the implication of this result, that a measurement is sensitive to variations in the Sun on 1-dimensional lines: i.e. objects that have measure 0 in three dimensions. If one were to rewrite the integral in terms of a volume integral, the kernel would be a the product of a Dirac delta function centered on the ray and a more usual sensitivity function along the ray. To some extent this problem can be circumvented by working instead with 'ray bundles' or 'fat rays': collections of rays that together fill a 'tube'. Various methods for collecting rays and constructing appropriate non-singular kernels in a geophysical setting can be found in [Marquering et al. (1999)].

Another way of approaching this problem is to consider the character of the approximation in the method of stationary phase which leads to this unphysical sensitivity function. The method of stationary phase provides the leading asymptotic behaviour for $\tau \to \infty$. To ensure that one is operating in the asymptotic regime, it is required that:

$$\omega\tau \gg 1 \ . \tag{4.119}$$

The acoustic power in the Sun is concentrated around frequencies corresponding to periods of the order of 5 min. As can be seen from Fig. 4.10 the time delays are at most hours and usually less. It is therefore difficult to argue that this requirement is met. This problem is first pointed out in [Bogdan (1997)], in which it is shown that for realistic conditions, the energy in wave packets is only approximately centered on the ray path. Furthermore there is a wide range perpendicular to the wave path over which the energy in the wave packet is considerable. The result of this calculation is physically much more acceptable, in that the time delay is sensitive to conditions over a finite volume sampled by the wave field. The ray approximation can be used as an indication of the depth of penetration of the sampling, and a basis on which to decide parameters of the filters discussed in section 4.1.1. As a basis of an inverse problem it is now being replaced by the methods of the next sections.

4.3.4 The Born approximation

As is seen clearly from Eq. (2.111) the wave equation for solar oscillations in the Cowling approximation can be cast into the form:

$$\mathcal{L}\xi + \omega^2 \xi = F_\omega(\mathbf{r}_s) \tag{4.120}$$

in which $\mathcal{L}$ is a differential operator and F_ω is a source function: determined by the mechanism driving the oscillations (see also section 2.10). A general solution to this equation, making use of the Green's function solution G_ω of the wave equation, is:

$$\xi(\mathbf{r}, \mathbf{r}_s, t) = \int d\omega \ G_\omega(\mathbf{r}, \mathbf{r}_s) F_\omega(\mathbf{r}_s) e^{-i\omega t} \ . \tag{4.121}$$

The cross-correlation Γ of two points on the surface, responding to the same source, in this notation is:

$$\begin{aligned}
\Gamma(\mathbf{r}, \mathbf{r}', \Delta t) &\equiv \int d\mathbf{r}_s \int dt \ \xi(\mathbf{r}', \mathbf{r}_s, t + \Delta t) \xi^*(\mathbf{r}, \mathbf{r}_s, t) \\
&= \frac{1}{2\pi} \int d\mathbf{r}_s \int d\omega \ G_\omega(\mathbf{r}', \mathbf{r}_s) |F_\omega(\mathbf{r}_s)|^2 G_\omega(\mathbf{r}, \mathbf{r}_s) e^{-i\omega t}
\end{aligned} \tag{4.122}$$

where the second equality again makes use of the convolution theorem of Fourier transforms. It is useful to consider the Fourier transform of the cross-correlation function $\tilde{\Gamma}$ for which:

$$\tilde{\Gamma}(\mathbf{r},\mathbf{r}',\omega) = \int d\mathbf{r}_s G_\omega(\mathbf{r}',\mathbf{r}_s)|F_\omega(\mathbf{r}_s)|^2 G_\omega(\mathbf{r},\mathbf{r}_s) \ . \qquad (4.123)$$

In an analogy with designations used in quantum theory and also in the field of turbulence the Green's function can be referred to as 'propagators' since they play the role of propagating the power in the forcing function/source $|F_\omega(\mathbf{r}_s)|^2$ from its position $\mathbf{r}_s$ to the position of the 'detectors': the locations on the solar surface for which the signal is cross-correlated.

In the field of geophysics a theorem referred to as Claerbout's conjecture (cf. [Rickett & Claerbout (2000)]) states that: "By cross-correlating noise traces recorded at two locations on the surface, it is possible to construct the wavefield that would be recorded at one of the locations if there was a source at the other". A proof of this conjecture, valid in geophysical settings, can be found in [Wapenaar et al. (2004)]. From Eq. (4.123) it is clear that this conjecture must hold as long as the propagator, the Green's function, is symmetric and satisfies a convolution property:

$$\begin{aligned} G_\omega(\mathbf{r},\mathbf{r}') &= G_\omega(\mathbf{r}',\mathbf{r}) \\ G_\omega(\mathbf{r},\mathbf{r}') &= \int d\mathbf{r}'' \ G_\omega(\mathbf{r},\mathbf{r}'')G_\omega(\mathbf{r}'',\mathbf{r}') \end{aligned} \qquad (4.124)$$

which has to be valid for any $\mathbf{r}$, and $\mathbf{r}'$. If one uses these two properties of Eq. (4.124) in Eq. (4.123) then it is evident that one can pretend that the power $|F_\omega(\mathbf{r}_s)|^2$ is generated at either $\mathbf{r}$ or $\mathbf{r}'$ instead of at $\mathbf{r}_s$ and then is propagated to either $\mathbf{r}'$ or $\mathbf{r}$ respectively. In [Kosovichev & Duvall (1997)] it is shown that the Green's function in a spherically symmetric model of the Sun indeed does satisfy these properties. More generally the first condition is satisfied as long as the operator $\mathcal{L}$ is self-adjoint. This indicates under what circumstances Claerbout's conjecture may fail: e.g. in a medium in which the propagation speed of waves is anisotropic and in particular is not invariant under a sign change of the wave vector $\mathbf{k}$. If one examines the dispersion relation for magneto-acoustic plane waves Eq. (4.116) it appears that a non-zero flow $\mathbf{v}_0$ can cause problems for the first of the conditions Eq. (4.124). However, a uniform flow in a plane parallel atmosphere, or uniform rotation in a spherical atmosphere, does not. A Galilean transform in this case eliminates the flow term. In the case of the Sun, the rotation is slow and the 'receivers' track with the surface rotation by construction. The inertial forces introduced by working in a co-rotating frame can be neglected and so it is only the differential part of the rotation which affects the isotropy of propagation, as well as convective motions. Effects of anisotropy are therefore not expected to be substantial. Attenuation of the magneto-coustic waves by for instance radiative damping or mode conversion from e.g. magneto-acoustic to Alfvén waves, is a problem as well, because it affects the second of conditions Eq. (4.124).

In what follows it is assumed that, to lowest order, in the solar atmosphere the two conditions Eq. (4.124) do hold, i.e. there is an operator $\mathcal{L}_0$ with its Green's function G_0 so that:

$$\mathcal{L}_0 G_0 + \omega^2 G_0 = \delta(\mathbf{r} - \mathbf{r}_s) \tag{4.125}$$

and G_0 satisfies Eq. (4.124). The difference with the full operator $\mathcal{L}$ is a perturbed operator $\mathcal{L}_1$ so that:

$$(\mathcal{L}_0 + \mathcal{L}_1)(G_0 + G_1) + \omega^2 (G_0 + G_1) = \delta(\mathbf{r} - \mathbf{r}_s) . \tag{4.126}$$

The difference between these two is:

$$\mathcal{L}_0 G_1 + \omega^2 G_1 = \mathcal{L}_1 (G_0 + G_1) \tag{4.127}$$

which is known as the Lippman-Swinger equation. The Born approximation is obtained by neglecting on the right-hand side the term G_1. This approximation means that Eq. (4.127) in this case is straightforward to solve in terms of the zero order Green's function G_0 since the source term of the right-hand side no longer depends on G_1.

Based on the approach presented for a geophysical setting in [Zhao & Jordan (1998)], in [Birch & Kosovichev (2000)] the operator $\mathcal{L}_1$ is presented for the case when only sound speed perturbations are present in the medium. Contrary to the discussion above, in [Birch & Kosovichev (2000)] the Green's function is treated in the time domain rather than the Fourier domain. In [Birch & Kosovichev (2000)] the impulsive source function is taken as $S = A\nabla \delta(\mathbf{r} - \mathbf{r}_s)\delta(t - t_s)$ so that the equations corresponding to Eqs. (4.125) and (4.127) are:

$$\begin{aligned}\left(\rho_0 \frac{\partial^2}{\partial t^2} + \mathcal{L}_0\right)\xi_0 &= A\nabla \delta(\mathbf{r} - \mathbf{r}_s)\delta(t - t_s) \\ \left(\rho_0 \frac{\partial^2}{\partial t^2} + \mathcal{L}_0\right)\xi_1 &= -\mathcal{L}_1 \xi_0\end{aligned} \tag{4.128}$$

with the differential operators:

$$\mathcal{L}_0 \boldsymbol{\xi} = -\nabla \left[c_{s\,0}^2 \rho_0 \nabla \cdot \boldsymbol{\xi} - \xi_r \frac{\partial P_0}{\partial r} \right] + \nabla \cdot (\rho_0 \boldsymbol{\xi}) \, \mathbf{g}_0$$
$$\mathcal{L}_1 \boldsymbol{\xi} = -\nabla \left[\delta c_s^2 \rho_0 \nabla \cdot \boldsymbol{\xi} \right] \tag{4.129}$$

in which $\boldsymbol{\xi}$ is the displacement vector and $\mathbf{g}_0$ the gravitational acceleration. Since the operators are separable in spatial and time variables, the solutions must be separable as well. The spatial part of the Green's function can be obtained from global normal mode summation:

$$\mathbf{G}(\mathbf{r}, \mathbf{r}_s) = \sum_{nlm} \frac{\xi^{nlm}(\mathbf{r}) \xi^{nlm}(\mathbf{r}_s)}{\omega_{nlm}^2} . \tag{4.130}$$

Note that in general $\mathbf{G}$ is a tensor since the displacement eigenfunctions ξ^{nlm} are vectorial. The normal mode summation can be simplified somewhat by making use of the summation theorem of spherical harmonic functions (cf. [Gradshteyn & Ryzhik (1994)]):

$$\frac{2l+1}{4\pi} P_l(\mathbf{x}_1 \cdot \mathbf{x}_2) = \sum_{m=-l}^{l} P_l^m(\mathbf{x}_1 \cdot \mathbf{x}) P_l^m(\mathbf{x}_2 \cdot \mathbf{x}) e^{im(\phi_1 - \phi_2)} \tag{4.131}$$

where the $\phi_{1,2}$ are the azimuthal coordinate in a spherical system corresponding to the vectors $\mathbf{x}_1$ and $\mathbf{x}_2$ respectively.

The time-dependent part for each of the normal modes is then given by:

$$g(t, t_s) = H(t - t_s) e^{-i\omega_{nlm}(t - t_s)} . \tag{4.132}$$

With this Green's function the first order problem can then be solved, since the operators on the left-hand side of the two equations (4.128) are identical. If the term on the right-hand side of the second of Eqs. (4.128) is also expanded on normal modes with the time dependence for each mode of $e^{-i\omega_{nlm}t}$ then the time-dependency of the ξ_1 is obtained from:

$$\int_0^\infty dt' \, g(t,t') e^{-i\omega_{nlm}t'} = \int_0^\infty dt' \, H(t-t') e^{-i\omega_{n'l'm'}(t-t')} e^{-i\omega_{nlm}t'}$$
$$= \int_0^t dt' \, e^{-i\omega_{n'l'm'}(t-t')} e^{-i\omega_{nlm}t'}$$
$$= i \frac{e^{-i\omega_{nlm}t} - e^{-i\omega_{n'l'm'}t}}{\omega_{nlm} - \omega_{n'l'm'}} . \tag{4.133}$$

The spatial part of the solution is expressed by evaluating:

$$-\int d\mathbf{r}'\ \mathbf{G}(\mathbf{r}',\mathbf{r}) \cdot \mathcal{L}_1 \xi_0(\mathbf{r}') \qquad (4.134)$$

where ξ_0 is also obtained using this Green's function:

$$\xi_0 = A \int d\mathbf{r}'\ \mathbf{G}(\mathbf{r}',\mathbf{r}) \cdot \nabla \delta(\mathbf{r} - \mathbf{r}')\ . \qquad (4.135)$$

Solutions are given explicitly in terms of the radial displacement eigenfunctions and products of Legendre polynomials $P_l P_{l'}$ in [Birch & Kosovichev (2000)], for that choice of source function. The kernel relating travel time perturbations to differences in sound speed between a reference model and a model (or the Sun) with sound speed inhomogeneities is derived from this displacement. As is pointed out in [Birch & Kosovichev (2000)], the kernel thus obtained is not symmetric between source and receiver, which is in part due to the approximation that the source is impulsive and localised. This is undesirable since the cross-correlation process is invariant when interchanging the two points or patches in the correlation. In [Birch & Kosovichev (2000)] an ad-hoc replacement of the source and receiver parts of the response is introduced which restores the symmetry.

Assuming instead a distributed source function should solve this problem in a physically more satisfactory way. In [Gizon & Birch (2002)] the filtering procedures are included in the analysis to obtain kernels relating travel time perturbations to inhomogeneities in the Sun. General expressions for the case of distributed source functions are given: i.e. in Eq. (4.135) the term $\nabla\delta(\mathbf{r}-\mathbf{r}')$ is replaced by the general forcing function $\mathbf{F}(\mathbf{r}_s,t)$. Also, allowance is made for the source function itself to be perturbed which means that in Eq. (4.134) a term $\delta\mathbf{F}$ has to be added to the term $\mathcal{L}_1\xi_0(\mathbf{r}')$. Appendix C of [Gizon & Birch (2002)] gives expressions for the kernels, relating travel time perturbations to perturbations in source strength and damping rate, for surface gravity waves. Note that in order to determine the travel time from the observations it is better to use the zero-order cross-correlation as a reference wavelet, rather than the Gabor wavelet. The reason for this is that the Gabor wavelet, with its close connection to the ray picture of wave propagation, ignores the wavefield character of waves propagating in an unperturbed medium, and therefore the imperfections of that picture can propagate and produce systematic errors in the travel time perturbation measurements, and then be interpreted as real perturbations. The construction of the reference wavelet is straightforward operationally, and the fitting process to extract travel time perturbations is achieved through a standard least-squares procedure. Details of the technique can also be found in [Gizon & Birch (2002)] and in [Gizon & Birch (2004)].

One of the issues that these treatments highlight is that the kernels are evidently dependent on the properties of the source function which are not very well known. If the weighting kernels for travel time perturbations are uncertain, this introduces an additional source of errors in inversions of the travel time perturbations for structural inhomogeneities. Another issue is that of strong scattering. As is shown in section 4.2.4 the Born approximation fails if the perturbations become too large, since then the scattering is too strong and the term $\mathcal{G}_1$ on the right-hand side in Eq. (4.127) should not be dropped. In this case one can retain higher order correction terms in a Born series, constructed recursively but one should keep in mind that for strong/multiple scattering such series do not necessarily converge very quickly. It is also possible to apply renormalisation group techniques, as is done in quantum physical applications and also in turbulence modelling. In principle this achieves the same as higher order Born series approximations, but convergence is greatly improved by re-arranging the terms in the series appropriately, so that one obtains a problem that appears formally linear with a renormalised propagator $\widetilde{\mathbf{G}}$ and and a renormalised source function $\widetilde{\mathbf{S}}$ which include in principle the scattering terms to all orders. In [McComb (1990)] the application of these methods to turbulence modelling is discussed in great detail, and much of that discussion carries over into the problem at hand. Neither of these methods has been applied as yet.

As mentioned in section 4.2.4.3, one of the aims of tomographic techniques is to provide quantitative tests of stellar convection theories and simulations, through providing measures of correlations and structure functions of the turbulent velocity field in the solar convection zone. In this context it is of particular interest to mention the work reported in [Skartlien (2002b)]. There the perturbed operator $\mathcal{L}_1$ is expanded in terms of correlations of increasing order of the fluctuating density and velocity fields in the solar convection zone (cf. section 4.2.3). Much of the discussion in that paper is focussed on the collective effect of weak scattering on averaged profiles of power as a function of frequency of the resonant modes (see also section 2.10). The same strategy in principle can be extended to interpret perturbations of cross-correlation functions in terms of the correlations of fluctuating quantities and structure functions of turbulence.

4.3.5 The Rytov approximation

The Rytov approximation is of the same order as the first order Born approximation and therefore suffers from limitations that are very similar to it. It does have somewhat different properties however, with the result that in some cases it has been shown to provide some improvements of the first order Born approximation. In the Rytov approximation the perturbation is not considered in terms of the scattered wavefield, but in terms of a complex phase. Details can be found in the text-book [Ishimaru (1978)]. Much as the Born approximation, the Rytov approximation has been applied in geophysical settings (cf. [Snieder & Lomax (1996)]) and is now being adapted for application in time-distance helioseismology. The first step in this direction can be found in [Jensen & Pijpers (2003)]. Starting point is the wave equation in the Cowling approximation, as is the case for all local helioseismology techniques. In [Jensen & Pijpers (2003)] the dependent variable is chosen to be that defined in Eq. (2.130), with the associated additional approximation that terms $O(1/r)$, i.e. effects of sphericity, are considered small. As in the previous section, the analysis concentrates on those modes that do not propagate below the convective layer, and the term in the buoyancy frequency N^2 is also set to 0. In [Jensen & Pijpers (2003)] magnetic fields and velocity fields are omitted from the analysis as well. This is not part of the Rytov approximation scheme, but it simplifies the analysis so that it is clearer what the role of the Rytov approximation is. The waves are not expanded onto spherical harmonics so that the wave equation from Eq. (2.111) with (2.129), and the additional approximations outlined above becomes:

$$\left[\nabla^2 + \frac{\omega^2 - \omega_c^2}{c_s^2}\right] \Psi = 0 \tag{4.136}$$

in which the dependent variable $\Psi = c^2 \rho^{1/2} \nabla \cdot \boldsymbol{\delta r}$ replaces X to make it clear that some additional approximations are made compared with Eq. (2.129). Considering only sound-speed perturbations, to first order, produces:

$$\begin{aligned} \left[\nabla^2 + K_0^2\right] \Psi_0 &= 0 \\ \left[\nabla^2 + K_0^2\right] \Psi &= 2K_0^2 \frac{\delta c_s}{c_{s\,0}} \Psi \end{aligned} \tag{4.137}$$

with K_0^2 defined by:

$$K_0^2 = \frac{\omega^2 - \omega_c^2}{c_{s\,0}^2} \,. \tag{4.138}$$

The wave field Ψ is now expressed in terms of a complex phase so that incoming waves and scattered waves are expressed in terms of a sum of phases. With the subscript 0 referring to an incoming (unperturbed) wave field the following definitions are used:

$$\Psi \equiv e^{\Phi} = e^{\Phi_0 + \Delta\Phi} = \Psi_0 e^{\Delta\Phi} \ . \tag{4.139}$$

The real part of the complex phase difference $\Delta\Phi$ represents amplitude changes in the wave field, and the imaginary part represents phase changes in the field. Inserting these definitions into Eq. (4.137) yields:

$$(\nabla\Phi)^2 + \nabla^2\Phi + K_0^2 = 2K_0^2 \frac{\delta c_s}{c_{s\,0}} \tag{4.140}$$

and therefore:

$$(\nabla\Phi_0)^2 + 2\nabla\Phi_0 \cdot \nabla\Delta\Phi + (\nabla\Delta\Phi)^2 + \nabla^2\Phi_0 + \nabla^2\Delta\Phi + K_0^2 = 2K_0^2 \frac{\delta c_s}{c_{s\,0}} \ . \tag{4.141}$$

Since Ψ_0 is the solution of the unperturbed equation, terms can be eliminated from Eq. (4.141) so that it can be re-arranged as:

$$\begin{aligned}\Psi_0 \left[2K_0^2 \frac{\delta c_s}{c_{s\,0}} - (\nabla\Delta\Phi)^2 \right] &= 2\Psi_0 \nabla\Phi_0 \cdot \nabla\Delta\Phi + \Psi_0 \nabla^2\Delta\Phi \\ &= 2\nabla\Psi_0 \cdot \nabla\Delta\Phi + \Psi_0 \nabla^2\Delta\Phi \\ &\quad + \Delta\Phi \left(\nabla^2 \Psi_0 + K_0^2 \Psi_0 \right) \\ &= \left[\nabla^2 + K_0^2 \right] \Psi_0 \Delta\Phi \ . \end{aligned} \tag{4.142}$$

The Rytov approximation consists of assuming that the phase perturbation is small in the sense that:

$$\frac{\nabla\Delta\Phi}{K_0} = O\left(\frac{\delta c_s}{c_{s\,0}}\right) \tag{4.143}$$

so that on the left hand side of Eq. (4.142) the square of this term is removed since it is of second order. In this case Eq. (4.142) becomes an inhomogeneous wave equation for $\Psi_0 \nabla\Phi$ and can be solved using Green's functions. With the Green's function $G_0(\mathbf{r}', \mathbf{r}_1)$ defined by:

$$\left[\nabla^2 + K_0^2 \right] G_0 = \delta(\mathbf{r}' - \mathbf{r}_1) \ . \tag{4.144}$$

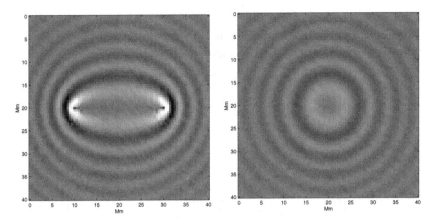

Fig. 4.12 Gray scale plot of the monochromatic Rytov kernel for the complex phase shift, in a homogeneous medium and for a point source for the wave field. The left-hand panel shows a horizontal cut, the right-hand panel shows a vertical cut (figure from [Jensen & Pijpers (2003)] reproduced by permission of A&A).

The expressions for Ψ_0 and $\Delta\Phi$ are:

$$\Psi_0(\mathbf{r}) = \int d\mathbf{r}' G(\mathbf{r}',\mathbf{r}) F_\omega(\mathbf{r}')$$
$$\Delta\Phi(\mathbf{r}_1) = \int d\mathbf{r}' 2K_0^2 \frac{G_0(\mathbf{r}',\mathbf{r}_1)\Psi_0(\mathbf{r}')}{\Psi_0(\mathbf{r}_1)} \frac{\delta c_s}{c_{s\,0}} . \quad (4.145)$$

In Fig. 4.12 a Rytov kernel is shown for the case when $F_\omega(\mathbf{r}') = \delta(\mathbf{r}' - \mathbf{r}_2)$ in a homogeneous medium. This is the kernel for the complex phase shift, i.e. in the second equality of Eq. (4.145) the function that appears with the sound speed perturbation under the integral sign. Note that Rytov kernels, as well as Born kernels (cf. [Marquering et al. (1999)]), show very low (or zero) sensitivity along the ray path itself, which has come to be referred to as the banana-doughnut paradox. This zero sensitivity is only true for phase perturbations. There is non-zero amplitude for amplitude perturbations. If one considers a perturbation placed on the ray path this can be considered to act as a point source, from Huygens' principle. The wavefield emitted arrives in phase with the scattered wavefield and therefore does not give rise to a phase perturbation, although the amplitude can be changed.

The next step to take, as with the Born approximation, is to construct the cross-correlation function in the Fourier domain $\widetilde{\Gamma}$ of the line-of-sight component of the velocity, including the effects introduced by the filtering and the averaging procedures discussed in sections 4.1.1 and 4.1.2. It is straightforward to show that, for adiabatic waves in the Cowling approximation, the surface velocity field is related to Ψ as:

$$\mathbf{v} = \frac{i\omega}{\rho_0}\nabla\left(\rho^{1/2}\Psi\right) . \qquad (4.146)$$

Therefore the correlation between points at locations $\mathbf{r}_1$ and $\mathbf{r}_2$ is given by:

$$\begin{aligned}\widetilde{\Gamma}(\omega,k_h) &= H(\omega,k_h)\nabla\left(\rho_0^{1/2}\Psi_0 e^{\Delta\Phi}\right)\bigg|_{r_1}\nabla\left(\rho_0^{1/2}\Psi_0^* e^{\Delta\Phi^*}\right)\bigg|_{r_2}\\ &= H(\omega,k_h)e^{\Delta\Phi(\mathbf{r}_1)+\Delta\Phi(\mathbf{r}_2)^*}\bigg[\nabla\left(\rho_0^{1/2}\Psi_0\right)\bigg|_{r_1}\nabla\left(\rho_0^{1/2}\Psi_0\right)\bigg)\bigg|_{r_2}\\ &+\rho_0^{1/2}\Psi_0(\mathbf{r}_1)\nabla\Delta\Phi|_{r_1}\nabla\left(\rho_0^{1/2}\Psi_0^*\right)\bigg|_{r_2}\\ &+\rho_0^{1/2}\Psi_0(\mathbf{r}_2)^*\nabla\Delta\Phi|_{r_2}\nabla\left(\rho_0^{1/2}\Psi_0\right)\bigg|_{r_1}\\ &+\rho_0\Psi_0(\mathbf{r}_1)\Psi_0(\mathbf{r}_2)\nabla\Delta\Phi|_{r_1}\nabla\Delta\Phi|_{r_2}\bigg] \qquad (4.147)\end{aligned}$$

where the amplitude $H(\omega, k_h)$ is introduced to account for the prefactors to the divergence term in Eq. (4.146) and also any additional filtering and/or averaging as outlined in sections 4.1.1 and 4.1.2. If one were to include non-adiabatic effects, which can be important in the surface layers, then H must also include an appropriate phase factor which can be frequency and wavelength dependent. As long as these effects are the same in the true Sun and in the model or quiet Sun from which a reference cross-correlation $\widetilde{\Gamma}_0$ can be constructed, the non-adiabaticity is not a major issue however. It is useful to be able to describe the cross-correlation in a way similar to the wave field, i.e. in terms of the cross-correlation in the absence of any perturbers and a complex phase factor. This can be done from the above equation as follows:

$$\begin{aligned}\widetilde{\Gamma} &= \widetilde{\Gamma}_0 e^{\Delta\Phi(\mathbf{r}_1)+\Delta\Phi(\mathbf{r}_2)^*}\bigg[1\\ &+ \frac{\rho_0^{1/2}\Psi_0}{\nabla\left(\rho_0^{1/2}\Psi_0\right)}\bigg|_{r_1}\nabla\Delta\Phi|_{r_1} + \frac{\rho_0^{1/2}\Psi_0^*}{\nabla\left(\rho_0^{1/2}\Psi_0^*\right)}\bigg|_{r_2}\nabla\Delta\Phi^*|_{r_2}\\ &+ \frac{\rho_0^{1/2}\Psi_0}{\nabla\left(\rho_0^{1/2}\Psi_0\right)}\bigg|_{r_1}\frac{\rho_0^{1/2}\Psi_0^*}{\nabla\left(\rho_0^{1/2}\Psi_0^*\right)}\bigg|_{r_2}\nabla\Delta\Phi|_{r_1}\nabla\Delta\Phi^*|_{r_2}\bigg]\\ &\approx \widetilde{\Gamma}_0 e^{\Delta\Phi(\mathbf{r}_1)+\Delta\Phi(\mathbf{r}_2)^*} \qquad (4.148)\end{aligned}$$

where the approximate equality results from neglecting the terms of first and second order in $\Delta\Phi$ within the brackets, which is a consistent use of the Rytov approximation. By defining a complex phase factor for Γ it becomes clear that the cross-correlation function behaves much the same as the wave field itself:

$$\widetilde{\Gamma} = \widetilde{\Gamma}_0 e^{\Delta\Phi_\Gamma}$$
$$\Delta\Phi_\Gamma \equiv \Delta\Phi(\mathbf{r}_1) + \Delta\Phi(\mathbf{r}_2)^* \tag{4.149}$$

and therefore construction of a kernel relating $\widetilde{\Gamma}$ to sound speed perturbations is readily done from the corresponding kernels for the wavefield Ψ.

The method based on the Rytov approximation can be generalised to also invert for velocity fields and magnetic fields, using an appropriate wave equation as a starting point. The limitations of this method are much the same as those for the Born approximation: if the character of the source function is poorly known this will reflect itself in the Green's function. Since the source function is needed to calculate the Green's function, using an incorrect source function produces a Green's function that is inappropriate for the actual Sun, and hence the kernel is then inappropriate as well. If the scattering is strong then higher order terms can no longer be neglected.

4.3.6 Acoustic holography

The method referred to as acoustic holography is presented in the context of local helioseismology in [Lindsey & Braun (1990)], as a method for imaging surface features such as sunspots while they reside on the far side of the Sun. The method is further clarified in [Lindsey & Braun (1997)], where it is discussed more generally as a tomographic method. Although referred to as holography there are some notable differences between optical holography and acoustic holography, as outlined in [Lindsey & Braun (1997)]. One important difference is that in helioseismic acoustics both amplitude and phase are recorded by the time-resolved Doppler imaging, and therefore a reference beam is not required to provide a phase reference, which is fortunate as none is available. This also means that there is no need for working within a very narrow frequency band, which is done in optical holography to retain coherence.

The method as outlined in [Lindsey & Braun (1997)] can in fact also be recognised in terms of methods that are familiar within the fields of geoseismology or underwater water acoustics, which is back-propagation or downward migration: the signal recorded at the surface, the solar surface in this case, is propagated in time reverse using a wave equation in a linear approximation such as the Born approximation. An example of this in underwater acoustics can be found in [Jackson & Dowling (1991)]. Evidently such a method will break down if there is strong scattering in the medium or if there are caustics in the inverse propagation, beyond which the lack of unicity precludes further migration of the wave field.

An explicit example of how this can work is given in [Lindsey & Braun (1997)] where some measure of an acoustic wave field, such as the radial component of the velocity, is decomposed in terms of the normal modes:

$$\psi(\mathbf{r},t) = \sum_{nlm} a_{nlm} \widetilde{\delta r}_{r\ nl}(r) Y_l^m(\theta,\phi) e^{i\omega_{nlm}t} . \qquad (4.150)$$

If only a section of the solar surface is observed there is some indeterminacy for the coefficients a_{nlm} but over the range of l values normally used in local helioseismology this is not a severe restriction. The acoustic power P can be found by summing the power per mode over all modes:

$$P(\mathbf{r}) = \sum_{nlm} \left| a_{nlm} \widetilde{\delta r}_{r\ nl}(r) Y_l^m(\theta,\phi) \right| . \qquad (4.151)$$

In the limit of medium to high $l > 100$ the Y_l^m are asymptotically identical to the functions $e^{i\mathbf{k}_h \cdot \mathbf{r}}$, apart from a phase fractor (cf. [Abramowitz & Stegun (1964)]) so that the expression for ψ reduces to:

$$\psi(\mathbf{r},t) = \int d\omega \int d\mathbf{k}_h\ a(\omega,\mathbf{k}_h) \widetilde{\delta r}_r(\omega,\mathbf{k_h},r) e^{i(\mathbf{k}_h \cdot \mathbf{r} - \omega t)} . \qquad (4.152)$$

In either case the $a_{nlm}\widetilde{\delta r}_r(R_\odot)$ or the $a(\omega,\mathbf{k}_h)\widetilde{\delta r}_r(\omega,\mathbf{k}_h,R_\odot)$ respectively are obtained from the surface signal, in the latter case using a straightforward Fourier transform of the wave signal at the surface. Using the known behaviour of the $\widetilde{\delta r}_r(r)$ as a functions of r evidently one can then obtain the acoustic power at any depth. This means that locations which act as sources or sinks of acoustic power can be located by 'scanning through' depth: the location of a source or sink is there where it shows up most clearly as a source or sink of acoustic power. The interpretation of such behaviour in terms of the physical character of the perturbations is then a separate modelling step. In [Lindsey & Braun (1997)] some examples of this method are shown. From these results it is expected that if at some location the wave field is scattered rather than emitted or absorbed there will be a rather less clear signal to isolate in this way, or even none at all.

However the presence of scatterers can be determined from correlation of what is named egressing H_+ and ingressing H_- wave fields defined as:

$$H_\pm(\mathbf{r},t) = \int dt \int d\Omega \; \psi(\mathbf{r},t) G_\pm(\mathbf{r},t,\mathbf{r}',t') \qquad (4.153)$$

where the Green's functions G_+ and G_- represent the outward and inward propagation of a pulse. In [Lindsey & Braun (1997)] these Green's functions are expressed as:

$$G_\pm(\mathbf{r},t,\mathbf{r}',t') = \delta\left((t-t') \pm T(|\mathbf{r}_h - \mathbf{r}'_h|, z)\right) A(|\mathbf{r}_h - \mathbf{r}'_h|, z) \qquad (4.154)$$

in which T is the ray travel time obtained from a ray path integration of Eq. (4.13), and A is the amplitude of the pulse. For convenience the spatial coordinates are written as a horizontal vector $\mathbf{r}_h$ and a depth z below the surface. The correlation between H_+ and H_- is defined in the usual way, and signatures of perturbations would show up in such a correlation function in the same way that it does in standard time-distance techniques (see section 4.3.2 ff.).

Clearly the Green's functions defined in this way correspond to using the method of stationary phase. The method can be generalised easily by replacing these Green's functions with those from improved approximations such as the Born approximation, and also taking into account distributed sources, which is done in [Skartlien (2001)]. In [Skartlien (2001)] some explicit examples are shown for a uniform plane parallel atmosphere, in which the exact Green's function has a simple analytical form, in order to demonstrate the procedure. This paper also pays some attention to the smearing effect introduced by a variety of observational limitations (limited field-of-view, limited wavelength range of the acoustic waves, damping, etc.). A further development of the technique is described in [Skartlien (2002a)] where the back-propagation of the wave field is done using more realistic Green's functions. Two different methods are shown for the holographic reconstruction of the full wave-field inside the Sun, which differ primarily in their use of the Green's function. It is then this full 3-dimensional wave field, rather than the 2-dimensional section of it at the surface, which is related to the perturbations of the smooth reference model of the Sun through a volume integral. Rendering of these perturbations in 3 dimensions is then a standard inverse problem to which one can apply any of the techniques discussed in sections 3.5 and 3.7.

One issue discussed in [Skartlien (2002a)] which is of particular importance, is that in order to carry out the back-propagation both the value of ψ and its radial gradient at the surface are required. This can be understood by regarding the back-propagation problem as solving an initial value problem for a second order differential equation, where the initial conditions are the wave field, and its derivative. If the waves were adiabatic then one could apply the steps taken in the derivation of the LAWE in the Cowling approximation (see section 2.7) to show that both are determined from observations of only one field such as the radial velocity. However, at the surface where the measured Doppler shifts, which sample the wave field, originate, the waves are no longer to a good approximation adiabatic. If instead the waves are treated as isothermal, it is also possible to obtain both ψ and $\partial\psi/\partial r$ from a single observable. In intermediate cases one needs either a description of the non-adiabatic behaviour of the waves to provide the required boundary conditions, or one needs to combine two independent observables: i.e. the Doppler shift from two or more spectral lines with slightly separated layers of line formation, or the combination of velocity and intensity. A discussion of this is found in appendix E of [Skartlien (2002a)].

4.3.7 Seismology of magnetic loops

This section appears somewhat out of place with the rest of this book, since the structures under discussion are not interior to the Sun. Especially at UV and X-ray wavelengths the Sun shows well defined loop structures which extend up from the photosphere into the hot outer regions of the Sun: chromosphere and corona. The footpoints of these loops appear anchored in the photosphere of the Sun. Recent time resolved observations are now available from eg. the TRACE instrument on board SOHO. The footpoints of these loops can be identified with strong magnetic features such as Sun spots which are also evident in the visible range of the spectrum, which suggests that these loops are concentrations of strong magnetic field. The physics of magnetic structures in low density plasmas is a vast subject, beyond the scope of this book. For an overview of the field one can see for instance [Priest & Hood (Eds.) (1991)].

The reason for including this topic is that recently both from theoretical considerations, and also from the observations of these loops (cf. [Aschwanden et al. (1999)]), it has become clear that such loops can oscillate and that these oscillations could be used as tools to investigate the physics that governs the structure and evolution of such loops, thereby shedding light on the interaction of magnetic fields and matter and the processes responsible for heating the corona.

The theoretical description of oscillations in cylindrical flux tubes predates the high resolution TRACE observations by more than a decade (cf. [Edwin & Roberts (1983)]) and their potential as seismic tools is described in [Roberts et al. (1984)]. For the details of the derivation of the dispersion relation the reader is referred to these papers. In these first models

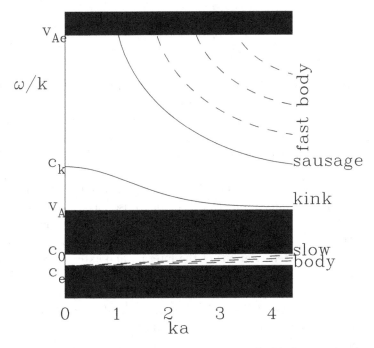

Fig. 4.13 Diagram of the phase velocity for waves in a cylindrical magnetic tube under coronal conditions, as a function of longitudinal wavenumber. c_0 and c_e are the sound speed inside and outside the magnetic cylinder respectively. v_A and $v_{A\,e}$ are the Alfvén speed inside and outside the cylinder. c_k is as defined in the text. (figure based on [Edwin & Roberts (1983)]).

magnetic loops are described as a straight cylinder embedded in a uniform external medium, where inside the cylinder the magnetic field strength is higher than outside, and therefore the plasma density is lower, in order to maintain pressure balance across the discontinuity. In this configuration there is a wealth of distinct oscillation modes possible, which is graphically depicted in Fig. 4.13. In the notation of [Edwin & Roberts (1983)] the characteristic speeds c_0 and c_e are the sound speed c_s internal and external to the magnetic flux tube. v_A and $v_{A\ e}$ are the corresponding values for the Alfvén speed (cf. Eq. (4.116)). Two further characteristic speeds correspond to pinching modes of the flux tube c_T and transverse deformations c_k: 'sausage modes' and 'kink modes' respectively. These speeds are related to the sound speed and Alfvén speed as:

$$c_T \equiv \frac{c_s v_A}{\sqrt{c_s^2 + v_A^2}}$$

$$c_k \equiv \sqrt{\frac{\rho_e v_{A\ e}^2 + \rho_i v_{A\ i}^2}{\rho_e + \rho_i}} . \tag{4.155}$$

There are two regions of propagating modes, separated by a hatched area in Fig. 4.13. In each area there are a number of solid and dashed lines corresponding to overtones of standing slow and fast magneto-acoustic waves: compressive longitudinal modes. The kink mode is a transverse deformation of the entire tube which therefore bears more resemblance to an Alfvén wave.

If the equilibrium conditions of the loop and the external medium are sufficiently well determined, one can attempt to use measurements of frequencies and wavelengths of loop oscillations, to determine which branch the oscillations belong to, i.e. solve the mode identification problem, and from that improve determinations of the internal and external conditions. If, as is probable, the waves are excited by impulsive events in the loop and are therefore decaying propagating waves rather than standing waves, this can also be used as a tool for diagnostics of the density of material in the loop, since each event can then be treated as an 'experimental' Green's function of the system (cf. [Roberts *et al.* (1984)]). A more detailed and comprehensive discussion of recent research can be found in the review [Nakariakov & Verwichte (2005)].

The field is at this stage evolving rapidly. For the purposes of the discussion here it is of interest to point out several paths that are likely to be explored to improve the value of these observations as tools for probing the upper atmosphere of the Sun. Firstly there is the combined effect of stratification and curvature. The loops for which it is easiest to obtain time-resolved observations tend to be large ie. they extend over a considerable range in radii and their curvature can not be neglected over the wavelength of the oscillations. One expects the curvature to affect the dispersion relations significantly. Therefore interpretations on the basis of the straight flux-tube analysis shown above can be misleading. Initial steps to improve the analysis, correcting for curvature and stratification are reported in [Verwichte et al. (2006)], which builds on models of twisted and curved current-carrying loops in [Cargill et al. (1994)]. One important feature of such loops is that the combined effect of twist and curvature introduces an additional restoring force, referred to as 'hoop force', with associated oscillatory modes. The difference between curved and straight flux tubes is therefore qualitative as well as quantitative. Finally one should also mention that the curved loops as seen in the images of TRACE are rarely isolated features. For well resolved images one often sees a striated fan of material, in which individual strands might be identifiable as the flux loops discussed above. The oscillations of these individual strands appear to show some phase correlation among them, which could be due to all being excited by a single external excitation event, but it could also imply coupling between loops. As is pointed out in [Murawski (1993)] for an idealised case of magnetised sheets, such coupling also affects dispersion relations. This also is a reason for caution in interpreting observational results in terms of the current generation of models, and a direction in which further developments are likely.

Chapter 5

Asteroseismology

The purpose of asteroseismology is to determine parameters of stars that are either inaccessible to classical observation or can not be determined with an accuracy that is sufficient for many purposes. The foremost example of these parameters is the mass of stars. Currently, accurate masses of stars can only be determined if they are members of close binary systems. In this case the dynamical information obtained by measuring the radial velocity variations of the member stars, due to their motion around a common centre of mass, combined with measurements of their motion on the sky can yield individual stellar masses. For single stars the mass is not known to better than 20 %, even for the best spectroscopic data and stellar atmospheres modelling currently available. If one wishes to determine the mass function of stars and investigate its universality, for instance in order to investigate star formation processes, the current data are heavily biased towards stars that have formed in binary systems, which may not reflect the variety of conditions under which stars can form and thus provide an incorrect mass function. Investigations of any other process which involves stellar masses as a parameter is similarly limited in accuracy because of this problem. Since stellar oscillation frequencies scale in a very direct way with the stellar mass, they can be used, together with classical observations, to measure stellar masses of individual stars to far higher accuracy. A review of asteroseismic results up to 1994 can be found in [Brown & Gilliland (1994)], but with the advent of new observing facilities, significant advances have already been made after that.

5.1 The Data

The techniques available for observing stellar variations are the same techniques that are also used in global helioseismology: photometry and spectroscopy. It has been suggested that in compact high mass objects such as neutron stars, oscillations could give rise to gravitational waves which, if detected, could be used to measure oscillation frequencies in the same way as modes detected through more traditional means (cf. [Andersson & Kokkotas (1998)]). This has not yet been realised, and will in any case remain applicable only to extreme states of matter, since a large amount of material needs to accelerate rapidly to produce an appreciable signal. A major difference between seismology applied to the Sun and to stars is that stars are only in exceptional circumstances spatially resolved, and never in the detail that is possible for the Sun. This has consequences in particular for identifying the modes that are detected, which is discussed in section 5.1.2.

Another difference between oscillations in the Sun and stars is that in stars other than the Sun, other excitation mechanisms than stochastic excitation by convection play a role. These other excitation mechanisms can be much more specific in the sense that only one or a few modes are excited. Also, with these excitation mechanisms it is much more likely that some modes are excited to such high amplitudes that non-linear and non-adiabatic effects can no longer be ignored in the modelling. It is necessary to have some knowledge of the type of variability displayed by a star, before one can decide what observational technique is likely to be most productive.

In Fig. 5.1 a Hertzsprung-Russel diagram is shown for stars for which the parallax has been accurately determined with the Hipparcos satellite so that both the luminosity and the effective temperature are known with high precision. Schematically indicated on the diagram is the location of various types of pulsating variable stars. Also indicated are some evolutionary tracks for selected stellar masses, and the zero-age main sequence (ZAMS) which is the location of stars as they commence with fusion of Hydrogen as the main source of their luminosity. The horizontal branch (HB) indicated by dash-dotted lines is the locus of stars that have exhausted Hydrogen in the core, have passed through a giant phase, and now are undergoing fusion of He in the core. After also exhausting He in their core, stars with sufficiently high mass undergo a second giant phase: the asymptotic giant branch or AGB. During this phase a considerable part of the envelope is shed in the form of a stellar wind. The core, consisting of an electron degenerate mixture of C and O overlaid by a thin layer of He and H, is eventually laid bare which means that the effective temperature rises considerably. Their trajectory in Fig. 5.1 is indicated by the dotted line. Pressures and temperatures in these configurations never reach sufficiently high regimes for subsequent fusion processes to produce appreciable amounts of energy, and so these stars cool down, decreasing in luminosity and temperature as indicated, becoming white dwarfs.

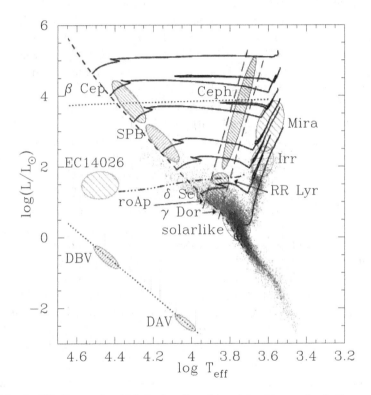

Fig. 5.1 An HR diagram in which the location of a variety of types of pulsating variable stars is indicated, overplotted on the locations of stars from the Hipparcos database for which accurate luminosities can be determined. The short dashed curve indicates the ZAMS and the solid lines are evolution tracks of stars of 1, 2, 3, 4, 7, 12 and 20 $M_\odot$. The band indicated by long dashed lines, enclosing the Cepheid variables, is the classical instability strip. The dot-dashed line is the HB and the dotted curve indicates the white dwarf cooling track. (modified version of figure from [Christensen-Dalsgaard (1999)])

The solar-like oscillators are those stars for which the most important excitation mechanism is stochastic excitation by convection, as in the Sun.

It is expected that this mechanism operates in all stars with a surface convection zone. However, the amplitude to which the oscillations are excited, and therefore whether they can be detected or not, is not well known because of the uncertainties surrounding the excitation mechanism and its relationship with turbulent convection (cf. section 2.10). Certainly there are positive detections of oscillation power for stars on the main sequence with masses both slightly above and below that of the Sun. Also, there is some evidence that giant stars, occupying the region of the HR diagram of the Irregular and Mira type variables, may well be excited at least in part by this mechanism ([Christensen-Dalsgaard et al. (2001)]). Another mechanism which operates in Mira variables is an opacity mechanism: when gas in the atmosphere is compressed the opacity increases and heat from the radiation field is added to the gas, which is then released upon expansion. In Mira variables the zone where H and He change from being pre-dominantly in neutral form to being predominantly in (singly) ionised form is spatially relatively thin and nearly coincident with that layer in the photosphere where the radial pulsation modes change from being nearly adiabatic to strongly non-adiabatic. This combination of conditions is favourable for a heat engine to operate such that the amplitudes of oscillations grow to a finite amplitude before non-linear amplitude limiting mechanisms set in, stopping further increases.

Going up in effective temperature, this ionisation zone moves outward to very shallow layers of the star, where the pulsation becomes almost isothermal since the layer is nearly transparent to radiation, so that this driving mechanism becomes ineffective. However, at somewhat higher effective temperatures the second ionisation zone of He meets the same conditions as described above, since it has emerged from adiabatic layers and has moved into the transition region. This gives rise to the classical instability strip occupied by RR Lyrae and Cepheid variables. From the location in the HR diagram it can be seen that the RR Lyrae type variables, at lower luminosity, are likely to be a heterogeneous group of stars. A minority has just exhausted the core H and is on their way to become giant stars and therefore traverses the instability strip quite rapidly from left to right in the diagram. The majority is stars that are now in the core He fusion stage on the HB. Note that a substantial fraction of the luminosity of these stars is generated by a shell of H fusion just outside the He core, so that the designation as core He fusion stars is something of a misnomer. The Cepheid variables are rather higher mass stars that in some cases traverse the instability strip several times during the course of their evolution, in particular at the high mass / high luminosity end. Where this instability strip crosses the main sequence, one finds δ Sct type variables. These stars oscillate in radial and non-radial p- and g- modes, and are probably all rotating sufficiently rapidly that rotation cannot be treated as a perturbation effect as is the case in the Sun. In the γ Dor stars both the opacity mechanism and stochastic excitation could be playing a role, since these stars are among the highest mass main sequence stars that have some convection in the outer layers. However the main excitation mechanism appears to be due to convective blocking: the overturning time scale of the convection is of the same order or larger than the pulsation time scale so that the convection cannot adapt sufficiently quickly to transport the increase in luminosity during a pulsation cycle. This blocking then plays the role in the heat engine that the opacity plays in classical variables. This mechanism is originally described in [Pesnell (1987)] and [Li (1992)], and applied to the γ Dor stars in [Guzik et al. (2000)]. roAp stars distinguish themselves by showing evidence of radial separation of chemical elements producing very unusual atmospheric compositions. These are also stars with strong magnetic fields, which affects the oscillation properties. The instability strip that all of these stars occupy can be extended down in luminosity to where it crosses the white dwarf cooling track in the region indicated by DAV in Fig 5.1, where the coolest variable white dwarfs are found.

At still higher effective temperatures an 'opacity bump' due to iron peak elements can act as driver for oscillations. The same mechanism can be responsible for driving mostly g-modes in the slowly pulsating B stars (SPB) and also in the DBV type white dwarfs, and p-modes in the β Ceph stars. In the EC14026 or sub-dwarf B (sdB) stars, which are on the extreme hot end of the HB, this mechanism also operates. There are theoretical indications that for these stars this mechanism can only give rise to pulsational instability if the iron peak elements are sufficiently enhanced in abundance compared to the normal range present in stars, which can occur through radiative levitation. At higher effective temperatures, falling just outside the diagram are the nuclei of planetary nebulae, which can be variable. These are located where the dotted line bends around to become the white dwarf cooling track. Here also are variable hot white dwarfs for which the planetary nebula if it is there at all, has too low a surface brightness to be detected (cf. [O'Brien (2000)]).

A recent discussion of all of the above types of variables can also be found in [Thompson, Cunha, & Monteiro (Eds.) (2003)]. With this many types of pulsational variability and a variety of driving mechanisms, there is clearly sufficient potential for using the oscillations as tools for diagnostics of the structure and relevant physical processes in these stars. As mentioned the data in itself is no different from that for the Sun, and leads to measurements of frequencies and amplitudes. However, the diagnostic tools that are applied to these can not necessarily be directly carried over from helioseismology, because of this richness in types of oscillatory variability. Also, because stars are not resolved and provide considerably fewer photons than the Sun does, collecting the primary data poses some challenges worth discussing.

5.1.1 Signal and noise in the observations

Quite apart from the inability to spatially resolve stars, their large distances also imply that the photon flux received is much lower compared to the Sun, by factors easily in excess of 10^{10}. This means that there is an associated loss in signal-to-noise ratio for the same observing conditions. To alleviate this problem one must use larger telescopes, collect photons for longer periods of time to the extent that the expected pulsation periods allow, and use wider photometric bands and/or more spectral lines. Another option is to obtain longer time series, because in the Fourier domain the noise level at any given temporal frequency is reduced as the time series is longer. This can be particularly effective for those stars that are pulsationally unstable and therefore coherent in their pulsation over the entire length of the time series. However, for the stochastically excited stars the pulsation is normally damped, and life-time of a mode can be short. The result of this is that acoustic power is spread out over a Lorentzian profile (cf. section 1.5) and a higher signal-to-noise ratio is needed to determine frequencies. Another problem is that stars can have substantially different dynamical timescales which means that obtaining a long and well-sampled time series can itself be a challenge. An extreme example of this are the Mira variables which have pulsational periods that can be several years. Even stars with periods of several hours to a day are problematic however, because they need to be observed during several day-night cycles. In order to prevent regular gaps in the time series with the associated problems, one must then organise multi-site observing campaigns, or observe from space. There are no networks as yet of telescopes dedicated to asteroseismology, as there are for helioseismology, although there are attempts in this direction (e.g. SONG: http://astro.phys.au.dk/SONG/ and an older initiative by the same name http://www.noao.edu/noao/song/).

Informal ad-hoc collaborations do exist and have successfully obtained long time series for selected stars: examples are the Whole Earth telescope (WET) (http://wet.physics.iastate.edu/), and the delta Scuti Network (DSN) (http://www.univie.ac.at/tops/dsn/intro.html). A dedicated single site project is STARE (http://www.hao.ucar.edu/public/research/stare/stare.html) which is now expanding to multi-site operations. Observing from space has been done with telescopes that are auxiliary instruments on large space telescopes: examples are the fine guidance sensors of the Hubble Space Telescope, WIRE, and also MOST (http://www.astro.ubc.ca/MOST/). A telescope dedicated to planet finding and asteroseismology is COROT (http://corot.oamp.fr/) to be launched late 2006.

Photometry is used as an observing technique for the large amplitude (radial) pulsating stars such as Cepheids and Mira variables, and also for stars such as δ Sct stars which have amplitudes of up to some tens of mmag. For pulsating stars that typically have lower amplitudes of the oscillations than these, i.e. mmag-level or below, ground-based photometry is ruled out for the same reasons as outlined in section 1.1. Even to reach the mmag level precision it is necessary to do differential photometry, preferably with two or more reference stars. A detailed description of the technique can be found in [Young et al. (1991)], which also discusses in detail the various sources of noise and bias. The need for more than one reference star in doing the differential photometry is in part to correct for systematic effects over the duration of the observations, both in the atmosphere and in the telescope, and in part to account for the fact that it is not unusual for a reference star to also be variable at the mmag level. The use of a second reference star makes it possible to ascribe any detected oscillation uniquely to one of the observed stars. It should be noted that the Earth's atmosphere is not the only source of noise. Even satellite based observations, such as will be taken by COROT, suffer from noise due to pointing jitter. Since the photon sensitivity of CCD type devices is not uniform from pixel to pixel, pointing jitter will result in a noise term in photometry. Apart from maintaining high pointing accuracy, a way to combat this noise is to defocus the star, so that the light is spread out over many pixels and sensitivity differences are suppressed by averaging. Various noise and data reduction aspects of CCD photometry are discussed in detail in [Kjeldsen & Frandsen (1992)].

Spectroscopy can provide additional information even for those stars observed using photometry. One reason for this is that the weighting with the cosine of the angle between a normal to the surface and the line of sight is different for measurements in velocity and in intensity (cf. section 1.4). In the main the spectroscopic techniques rely on determining Doppler shifts of spectra, on which the rest of this section concentrates. It should be mentioned however that a 'hybrid' technique, which relies on equivalent width variations, has also been used (cf. [Kjeldsen et al. (1999)]). Since equivalent width variations are somewhat similar to intensity in their relationship with stellar atmospheric conditions, they are more photometric in character but suffer less from some of the noise sources that standard photometric methods do.

Another feature of spectroscopic observations is that for stars with broadened line profiles, due to moderate to rapid rotation, oscillations for modes with l values higher than $l = 3$ can still be detected as 'bumps' travelling through the line profile which can be used to identify such modes (cf. [Smith (1977)], [Vogt & Penrod (1983)]). Perhaps the most important reason for spectroscopic observations, whether in addition to photometric methods or not, is that the signal-to-noise ratio for spectroscopic observations tends to work out in favour of this technique. Currently the foremost instruments used for spectroscopic asteroseismology are the UCLES instrument on the Anglo-Australian Telescope, HIRES at the Keck telescope, and the HARPS instrument at ESO. It should be mentioned that in particular the HARPS instrument is the latest in a line of spectrographs, and its predecessors CORAVEL, CORALIE and ELODIE have been of great significance for spectroscopic asteroseismology as well. All of these have been constructed primarily as tools for on-going long-term campaigns directed at finding extrasolar planets, which have requirements similar to what is needed for asteroseismology. The principles along which such instruments are built is laid out in [Connes (1985)].

The spectroscopic instruments all rely on taking time series of high-resolution spectra over as broad a spectral range as practicable, in order to maximise the number of spectral lines measured simultaneously. The technique for extracting Doppler shift velocities from spectra is not very different from what is done for the Sun, using a single line, but now using as many lines as possible. Some streamlining of the procedure is therefore useful. The most straightforward solution is to cross-correlate the spectra in the time series with a reference. The reference can be a template (cf. [Baranne et al. (1996)]) or a spectrum from the star itself (cf [Chelli (2000)]). In the template method the reference is a set of top-hat functions centered on wavelengths in the spectrum where relevant lines are for a given spectral type. This 'model spectrum' is shifted by the Earths' barycentric velocity, to eliminate its effect from the cross-correlation. The computation of the cross-correlation is carried out in the wavelength domain.

The problem with the template method is that it is sub-optimal in the use made of the information in the spectral lines, as pointed out in [Chelli (2000)] and [Bouchy et al. (2001)]. In these papers it is proposed instead to use as reference spectrum one taken from the star itself. A difference between the two methods is that the latter defines optimal weights in the wavelength domain, along the lines shown in section 1.2 (cf. Eq. (1.11)), whereas the former works instead on a Fourier transform of the spectrum. The expression (1.11) requires slight modification for the case described in [Bouchy et al. (2001)] because the data is now not the difference between the intensity on left- and right-hand sides of a single line profile. Instead one considers the difference over a large wavelength range, between a shifted profile and a reference profile, which are assumed to have the same overall intensity level. The single profile shape P and its derivative with respect to the variable u must therefore be replaced by the shape S of the reference spectrum so that:

$$v_j - v_{\text{ref}} = \sum_i w_i (I_{i,j} - I_{i,\text{ref}})$$

$$w_i = c \left[\frac{1}{\sigma_i^2} \lambda \frac{dS}{d\lambda} \bigg|_{\frac{i}{Rf_{\text{over}}}} \right] \left[\sum_k \frac{1}{\sigma_k^2} \lambda \frac{dS}{d\lambda} \bigg|_{\frac{k}{Rf_{\text{over}}}}^2 \right]^{-1} . \quad (5.1)$$

As is pointed out in [Bouchy et al. (2001)] the noise is a combination of photon noise for which $\sigma^2 \propto S$ and detector noise σ_D so that:

$$\sigma_i^2 = S_i + \sigma_D^2 . \quad (5.2)$$

In [Chelli (2000)] instead the spectra are Fourier transformed so that the cross-correlation of a reference spectrum and a Doppler shifted spectrum is done by multiplication of their Fourier transforms:

$$C(k) = S(k) S_{\text{ref}}^*(k) . \quad (5.3)$$

It is argued in [Chelli (2000)] that if S and S_{ref} only differ by a constant shift, then to find this shift it is best to minimise for v, in the least squares sense, the imaginary part of the quantity:

$$\widetilde{C}(k) \equiv e^{-2i\pi k \lambda_0 v/c} C(k) . \quad (5.4)$$

Writing this out, a solution is needed for:

$$\frac{\partial}{\partial v}\left\{\sum_j \frac{1}{\sigma_j^2}\left[\frac{\tilde{C}_j - \tilde{C}_j^*}{2i}\right]^2\right\} = 0 .\qquad(5.5)$$

Since this is non-linear in v the procedure must be carried out iteratively using a Newton-Raphson type procedure where the increment Δv at each step is:

$$\begin{aligned}\Delta v &= \frac{\sum_j \frac{1}{\sigma_j^2}\left(\frac{\tilde{C}_j - \tilde{C}_j^*}{2i}\right) 2\pi k_j \frac{\lambda_0}{c}\left(\frac{\tilde{C}_j + \tilde{C}_j^*}{2}\right)}{\sum_j \frac{1}{\sigma_j^2} 4\pi^2 k_j^2 \frac{\lambda_0^2}{c^2}\left[\left(\frac{\tilde{C}_j + \tilde{C}_j^*}{2}\right)^2 + \left(\frac{\tilde{C}_j - \tilde{C}_j^*}{2i}\right)^2\right]}\\
&= \frac{c\Delta\lambda}{2\pi\lambda_0}\frac{\sum_j \frac{1}{\sigma_j^2} j\left(\frac{\tilde{C}_j - \tilde{C}_j^*}{2i}\right)\left(\frac{\tilde{C}_j + \tilde{C}_j^*}{2}\right)}{\sum_j \frac{1}{\sigma_j^2} j^2\left[\left(\frac{\tilde{C}_j + \tilde{C}_j^*}{2}\right)^2 + \left(\frac{\tilde{C}_j - \tilde{C}_j^*}{2i}\right)^2\right]}\end{aligned}\qquad(5.6)$$

in which it is used that in the Fourier domain the spectrum is regularly sampled with a spacing $1/\Delta\lambda$ in which $\Delta\lambda$ is the full width of the window over which the spectrum is recorded. Note that as the minimum is approached, the second term in the denominator becomes very small, and this expression reduces to that given in [Chelli (2000)]. In the noise analyses, the reference spectrum is usually set to be noise free, which is of no great consequence for the overall error budget. Evidently one can construct a reference by appropriate averaging or co-adding of spectra ([Galland et al. (2005)]), which also shows that this method can be applied to more rapidly rotating stars which were previously considered to be inaccessible for these high precision velocity measurements due to the rotational line broadening.

It should be pointed out that these methods assume that the entire spectrum shifts in phase and with the same velocity amplitude for the entire spectrum. For exoplanet searches this is appropriate, since the entire body of the star moves. For asteroseismology however, one needs to be more cautious. Different lines in the spectrum are formed at different depths in the atmosphere. As long as all lines originate from a region enclosed between the same two nodes, or outside the outermost node, of any mode expected to be detected they should all be in phase, even if not at the same amplitude. However, if the star has a chromosphere it may have emission lines as well as absorption lines, or emission cores within the absorption line profile. These chromospheric features may well not show any shift at all, or one that is not in phase with other lines. These lines must therefore be excluded from the cross-correlation. In the template method this is straightforward, since one can simply omit a top hat, but in the method where the star provides its own reference this is not an option. The best way to implement suppression of such lines from the analysis is not to replace such lines, using a mask of some sort, but instead to modify the σ_i in Eq. (5.2):

$$\sigma_i^2 = \frac{S_i}{1-q_i} + \sigma_D^2 . \qquad (5.7)$$

For most of the spectrum the factor q_i is then set to 0, but for those regions of the spectrum which are to be suppressed q_i could be increased smoothly to values close to unity. For the Fourier domain analysis of [Chelli (2000)] this does not apply. Here instead one is forced to use a mask, replacing extraneous features by interpolating continuum flux, before carrying out the Fourier transform.

One strength of these two methods for extracting velocities from spectra is that they work on a single spectrum at a time, which means that in principle they can be applied in real time as the spectra are obtained. However, it is important to realise that in contrast to exoplanet research, in asteroseismology the primary quantities of interest are the frequencies of oscillations present in the signal, rather than their amplitude or phase. This suggests a quite different approach to the ones described above. One first performs a temporal Fourier transform of every channel in the spectrum independently. For this it is necessary to collect and store spectra over a certain length of time, i.e. an observing run, and do the analysis afterwards, which is already typically done for asteroseismology. Once in the temporal Fourier domain, the next step is to co-add the Fourier amplitude of the spectra along the spectral direction, using a standard optimal extraction algorithm (cf. [Horne (1986)]). The problems outlined above concerning amplitude differences and possible phase differences for particular spectral features, is circumvented by working with the Fourier amplitude or power. Phase differences between spectral features appear as different phases in the Fourier domain and are therefore ignored in the algorithm, velocity amplitude differences simply result in that the Fourier amplitude is smeared out over more channels in the spectral direction, which is afterwards collected in the co-adding using optimal extraction. Potentially there are some problems with this method as well. In regions where the spectrum has high curvature as a function of λ the signal as a function of time in each channel can be anharmonic, which means that after Fourier transformation the signal is distributed over all frequencies that are integer multiples of the true oscillation frequency. The quantitative criterion for this not to cause significant problems is:

$$\frac{1}{2} \frac{\Delta\lambda}{\lambda} \frac{\lambda \frac{\partial^2 S}{\partial \lambda^2}}{\frac{\partial S}{\partial \lambda}} < \frac{\sigma_I}{I} \qquad (5.8)$$

where to lowest order $\Delta\lambda/\lambda = v_{\rm rad}/c$. From this one can see two ways of combating problems arising from curvature. If the velocity is dominated by a component which is well separated in frequency, or well constrained, such as the Earth's barycentric motion, the individual spectra can be shifted and re-sampled appropriately, to eliminate this component. In this way the $\Delta\lambda/\lambda$ is reduced. The second option is to modify the weights in the optimal extraction procedure. Normally one would weight the data with the inverse of the variance. In the case at hand one uses the zero-frequency component of the Fourier transform of S to determine the appropriate derivatives with respect to wavelength λ, and set the weights instead by taking:

$$w \propto \left[\eta \left(\frac{\lambda \frac{\partial^2 S}{\partial \lambda^2}}{2 \frac{\partial S}{\partial \lambda}} \right)^2 + \sigma^2 \right]^{-1}. \qquad (5.9)$$

The factor η is adjusted $\propto S^2 \langle v_{\rm rad}^2 \rangle / c^2$ in order to ensure this suppression of features is only done where necessary.

Evidently an approach such as this suffers if there are spectra missing from the time series, regular gaps, and the like. All the usual problems that time-series analysis suffers from are not addressed by this method in itself. However, any other linear method that can be applied to time series that are not regularly sampled can also be applied here. Further discussion of methods that address this problem can be found in 1.3 and is also taken up again in sections 5.1.3 and 5.3.1.

It is of interest to note that in the temporal Fourier domain the zero-frequency component is the time-averaged spectrum for the time-series, which is smeared out compared to the underlying stellar spectrum because of the effect of the oscillations. To reverse such smearing one needs to co-add the spectra taken, after shifting each spectrum or subsection thereof by the appropriate amount. For this, the amplitude and phase of each of the spectra are necessary. To carry this out, one can combine the above technique with that of [Chelli (2000)] or [Bouchy et al. (2001)]. In this the starting assumption is that the entire spectrum shifts in phase with the same amplitude. This phase and amplitude are determined by either of the methods of [Bouchy et al. (2001)] or [Chelli (2000)]. The spectra are shifted and resampled accordingly, after which the method described above is applied to determine residual signals. Any features which show significant residuals can then be corrected individually, by selecting appropriate spectral windows and re-calculating the residual shifts for these windows.

A spectrum that is built up in this way, is limited by the response function of the instrument, telescope optics, and atmospheric seeing. Given that the shifts of the spectra are determined to an accuracy that is a fraction of this resolution, it is in fact possible to improve on the resolution of the averaged spectrum. The flux $I_{i,j}$ in channel i of the jth spectrum is related to the underlying spectrum $S(\lambda)$ by:

$$
\begin{aligned}
I_{i,j} &= \int_0^\infty d\lambda \; S(\lambda) K_{ij}(\lambda) \\
&= \int_{-\infty}^\infty d\ln\lambda \; \lambda S(\lambda) K_{ij}(\lambda) \\
&\equiv \int_{-\infty}^\infty dx \; \widetilde{S}(x) \widetilde{K}_{ij}(x)
\end{aligned}
\tag{5.10}
$$

where K is the response function and $x \equiv \ln\lambda$. Since the wavelength satisfies:

$$\lambda = \lambda_{\rm rest} \sqrt{\frac{1 + v_{\rm rad}/c}{1 - v_{\rm rad}/c}} \tag{5.11}$$

the assumption that the radial velocity has a single value for each shifted spectrum implies that in the variable x each spectrum j is shifted by a constant:

$$\Delta_j = \frac{1}{2}\left[\ln\left(1 + v_{\rm rad}(j)/c\right) - \ln\left(1 - v_{\rm rad}(j)/c\right)\right] \tag{5.12}$$

which does not make use of a Taylor expansion. If furthermore the response function is constant over the spectrum and does not change during the observing run, then:

$$\widetilde{K}_{ij}(x) = \widetilde{K}(x - x_i - \Delta_j) \ . \quad (5.13)$$

In either case the measurements I_{ij} are integral constraints on the true spectrum S, which is evident from Eq. (5.10). As long as the response functions K_{ij} are known, or the corresponding $\widetilde{K}_{ij}$ or $\widetilde{K}$ and Δ_j, then determining S is a deconvolution problem. This can be approached using the inverse techniques discussed in detail in sections 3.5 or 3.7. It is thus clear that a spectrum can be obtained at higher resolution by trading off against the signal-to-noise ratio in that spectrum. Such a technique is an adaptation of general image reconstruction techniques (cf. [Starck et al. (2002)] for a review). If furthermore Eq. (5.13) is satisfied, the inversion can be carried out with less computational overhead than is otherwise the case (cf. [Pijpers (1999)] for the corresponding technique in image reconstruction).

5.1.2 The mode identification problem

In section 1.4 the problem of isolating modes and assigning appropriate l and m values to each oscillation frequency is discussed, in the case of resolved observations of the Solar disc. There are very few stars for which it is possible to resolve the stellar disc. Even for the nearest and largest stars the angle that the disc subtends on the sky is so small that even with the largest telescopes the disc remains essentially pointlike. It requires special techniques such as interferometry to measure stellar radii. Although interferometry as a technique has been widely used to do astronomical observations at radio wavelengths since the 1970s it is only very recently that the technique can be applied routinely at optical wavelengths.

To illustrate the techniques for building up images it is convenient to start by describing the image reconstruction for the simplest possible array, consisting of only two elements. This set-up is shown in figure 5.2. For a point source at infinity the incoming electro-magnetic waves are plane waves. The electric vector measured at the focus is:

$$V_1 \propto E \cos(\omega t)$$
$$V_2 \propto E \cos(\omega t - \phi)$$
$$= E \cos\left(\omega t - \frac{2\pi |B|}{\lambda} \cos\theta\right) \ . \quad (5.14)$$

It is normal to record only the correlations i.e. the product of V_1 and V_2

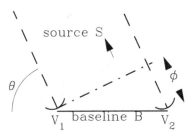

Fig. 5.2 A simple two-element correlating interferometer. The base line is a vector of length B. The direction towards the source is the unit vector s. ϕ is difference in phase measured at the two elements due to the difference in path length, for an incoming plane wave.

using hardware multiplication techniques. A high frequency term is filtered out immediately and the response R then is:

$$R \propto S \cos\left(\frac{2\pi|B|}{\lambda}\cos\theta\right) . \tag{5.15}$$

Here the flux density or power S of the source has replaced the E^2 term. The $\cos\theta$ clearly comes from an inner product of the base line vector $\mathbf{B}$ and the vector towards the source $\mathbf{s}$. The above discussion applies directly to interferometry at radio-wavelengths where the influence of the atmosphere is negligible. At optical wavelengths additional photometric corrections need to be applied, because otherwise transparency variations of the atmosphere dominate over the true interferometric signal. A thorough discussion can be found in [Coudé du Foresto et al. (1997)]. A more general formulation of Eq. (5.15) in terms of this inner product of $\mathbf{B}$ (now measured in units of the wavelength λ) and $\mathbf{s}$ is then:

$$R \propto S \cos(2\pi \mathbf{B} \cdot \mathbf{s}) . \tag{5.16}$$

It is clear that displacing the source along a direction perpendicular to $\mathbf{B}$ does not change the inner product $\mathbf{B} \cdot \mathbf{s}$ and therefore the response R is invariant to such a displacement. This is the mathematical expression of having spatial resolution in only one direction for a linear array. For a source that is extended on the sky rather than a point source the simple product of S and the phase delay term in equation (5.16) becomes a convolution integral:

$$R = \int d\boldsymbol{\sigma} I(\boldsymbol{\sigma}) \cos\left[2\pi \mathbf{B} \cdot (\mathbf{s} + \boldsymbol{\sigma})\right] \ . \tag{5.17}$$

Here I is the brightness distribution of the source modified by the response (beam pattern) $\boldsymbol{\sigma}$ of an individual telescope, so I is itself a convolution of the intrinsic distribution and A. Now $\mathbf{s}$ is merely a convenient reference point near the source. Using the projected spacing $\mathbf{b}$ on the sky defined by

$$\mathbf{b} \equiv \mathbf{B} - (\mathbf{s} \cdot \mathbf{B})\mathbf{s} \tag{5.18}$$

and also the approximation that the angular extent of the region observed at any given time is small so that $\boldsymbol{\sigma}$ is nearly perpendicular to $\mathbf{s}$ the response reduces to:

$$R = V \exp\left[i 2\pi \mathbf{B} \cdot \mathbf{s}\right] \tag{5.19}$$

with the complex visibility V defined by:

$$V \equiv \int d\boldsymbol{\sigma} I(\boldsymbol{\sigma}) \exp\left[i 2\pi \mathbf{b} \cdot \boldsymbol{\sigma}\right] \ . \tag{5.20}$$

Since $\mathbf{B} \cdot \mathbf{s}$ is known, the complex visibility V can be treated as the observable quantity. The phase in the complex plane of V in particular is of interest because it is particularly sensitive to displacements of an otherwise invariant distribution.

Even with the baseline lengths and numbers of simultaneous baselines currently available from interferometers there are still quite severe limitations to the number of resolution elements over stellar surfaces. In [Jankov et al. (2001)] it is proposed to combine Doppler imaging with interferometry to combine the strengths of each method to provide surface images. This technique could well be applied to image the surface structures associated with non-radial pulsations, but would be limited to stars the rotate sufficiently rapidly. It has been suggested that rather than attempting to image the stellar surface one could measure the displacement of the photocentre of the stellar disk using the closure phase, which requires a minimum of three telescopes in an interferometric array, where fringes are collected for all three baselines. In a set-up with three telescopes the sum of the three complex phases of the visibility V can be shown to be 0 for intensity distributions that are axially symmetric. The closure phase is therefore particularly sensitive to departures from axial symmetry. An application of this method to determine the gravitational darkening of a rapidly rotating star can be found in [Domiciano de Souza et al. (2005)]. The closure phase measurement is sufficiently sensitive to discriminate between the lower l-values which are relevant for asteroseismology, which means that for stars in which a single oscillation mode is strongly dominant an l value can be determined. However, it is generally the case that stars in which several, or even very many, modes are simultaneously excited provide much more information on their internal structure and for that reason are highly desirable asteroseismic targets. In this case the photocentre displacement is 'washed out' by averaging over many l values. A solution to this problem is to do 'stroboscopic interferometry', which is to say that one combines time-series analysis techniques with interferometric measurements.

In stroboscopic interferometry it is assumed that a set of oscillation frequencies is already measured using standard asteroseismic techniques. The aim is to pursue an identification specifically for these frequencies. When carrying out the interferometric measurements one uses a selected oscillation frequency and phase-locks the interferometric measurements to this frequency. In this way, all surface structure that is **not** associated with this frequency is removed, which should greatly improve the phase closure signal strength. Since the asteroseismic measurements are, under normal circumstances, not sensitive to l values higher than ~ 5, the stroboscopic measurement should provide a very strong constraint on the l value, and hence also on the n value of the mode. It is important to note that the phase-locking does not need to be carried out during the observations, which would create insurmountable problems in its practical implementation. Rather, one collects the interferometric fringes at the cadence one would normally do, which is sufficiently high compared to the oscillation frequencies of stars. The frequency filtering can be done as a post-processing step by an appropriate weighting procedure. The design of such weights is a standard time-series analysis/inverse problem. A further advantage of doing the latter is that one can design the appropriate weights for each of the measured oscillation frequencies separately, and use the same set of interferometric observations to constrain the identification of all oscillation modes.

In the future it may well be possible to resolve stellar discs interferometrically into a few elements and obtain some details. It will be some time before such techniques can be used routinely to identify node-line patterns over stellar discs for nearby stars. Even then the resolution will never reach the same level as the spatial resolution possible for the Sun. It is therefore necessary to find other means of identifying the node line pattern that corresponds to each measured oscillation frequency in a star. There are several different and complementary ways to do this, at least for low degrees l. Whatever the method, it is clear that as l becomes larger, the averaging over patches of visible surface that are in opposite phases to each other will produce a progressively smaller signal. Given a finite signal-to-noise ratio of measurements, there is always an upper limit to the l value which one can hope to detect.

One can quantify the expected signal as a function of l. First, consider a star that is rotating so slowly, that it can be treated as spherically symmetric. In this case there is no preferred coordinate system within which to describe the oscillations and one can make any convenient choice. The most convenient choice is to take a coordinate system in which the polar axis is pointing towards the observer. One can see that in this case any oscillation mode with $m \neq 0$ has an even number of node lines pointing radially away from the centre of the disc. In unresolved observations there is therefore perfect cancellation of all oscillation signal with $m \neq 0$ and one need only consider $m = 0$. The signal for unresolved observations in a spectral band X of an oscillation with degree l and for $m = 0$ is:

$$S_X(l) = 2\sqrt{2l+1} \int_X d\lambda \int_0^1 du P_l(u) I(\lambda, u) u \qquad (5.21)$$

where $u \equiv \cos\theta$, and $I(\lambda, u)$ is the centre-to-limb variation of intensity for band X: the limb darkening profile. For a uniform brightness disc I is independent of u but realistic brightness profiles are brighter at the centre than at the limb at optical wavelengths. The limb-darkening effectively functions as a non-optimal spatial filter for l (cf. [Heynderickx et al. (1994)], [Bedding et al. (1996)]). For the Sun the brightness profile is known and for some nearby stars the brightness profile has been measured interferometrically. For most stars one will have to rely on the accuracy of model atmospheres to provide $I(\lambda, u)$. With the knowledge of $I(\lambda, u)$, the calculation of $S_X(l)$ is straightforward. The reverse is not necessarily the case. The problem lies in that the physical amplitude of the modes is not known a priori. S_X only measures by what factor any given signal from surface motion is reduced because of the averaging. a value, or even a ratio of values for two modes, therefore does not necessarily reveal the l value appropriate to either mode.

Observing in several different bands, preferably simultaneously, provides a way around this problem. Observationally this idea has been applied to δ Scuti star mode identification (cf. [Viskum, et al. (1998)]) using ratios of amplitudes in several bands. A formal χ^2-minimisation technique for the value of the harmonic degree l, that uses these ratios as data and also the phase differences between bands, is shown in [Balona & Evers (1999)]. A more theoretical approach can be found e.g. in [Pijpers (2001)]. Ideally one would combine the observed relative amplitudes $S_X = \sum_l S_X(l)$ with weights $g_X(l_0)$ so that:

$$\sum_X g_X(l_0) \left(\sum_l S_X(l) \right) = N_{l_0} \delta_{l\, l_0} \tag{5.22}$$

where N_{l_0} is a normalisation factor, and $\delta_{l\, l_0}$ is the Kronecker delta symbol. In practice only a finite amount of data is available, with a finite signal-to-noise ratio, so this ideal cannot be attained. Solving this problem using SOLA (cf. section 3.5.2) could be achieved by minimising for the $g_X(l_0)$ the following:

$$\sum_X \left[g_X(l_0) \sum_l 2\sqrt{2l+1} \int_0^1 du \int d\lambda\, P_l(u) u I_X(\lambda, u) - \mathcal{T}_l(|l - l_0|) \right]^2$$
$$+ \mu \sum_{X\, X'} g_X(l_0) g_{X'}(l_0) E_{X\, X'} \tag{5.23}$$

where μ is the usual SOLA error weighting term, and $E_{X\, X'}$ is the (co-)variance matrix of the errors in the observables. In implementations of the SOLA method for other inverse problems it is common to use a Gaussian form for the filter $\mathcal{T}_l$ around l_0 in which the width is a free parameter to be set as small as feasible given the measurement errors and the amount of data available. In the case at hand a Gaussian might be inappropriate. To illustrate this consider a set of weights $g_X(l_0)$ such that:

$$\sum_X g_X(l_0) \int d\lambda\, u I_X(\lambda, u) = P_{l_0}(u) \ . \tag{5.24}$$

If it were possible to do this and if the entire stellar surface were to have been observed, the orthogonality of Legendre polynomials would ensure that Eq. (5.22) would be satisfied exactly. This can be seen by multiplying in Eq. (5.21) left- and right-hand sides with the $g_X(l_0)$, summing over X, and integrating over u from -1 to 1. For real observations covering only half the stellar surface the integration is from 0 to 1, as written in Eq. (5.21), and Eq. (5.22) is replaced by (cf. [Gradshteyn & Ryzhik (1994)]):

$$\sum_X g_X(l_0) \left(\sum_l S_X(l) \right) = \frac{2}{\sqrt{2l_0+1}} \quad l = l_0$$

$$= 0 \quad |l - l_0| = \text{even}$$

$$= \frac{2(-1/4)^{(l_e+l_o-1)/2} l_e! l_o! \sqrt{2l+1}}{(l_o - l_e)(l_e + l_o + 1)\left[(l_e/2)!((l_o-1)/2)!\right]^2}$$

$$l_e = \text{even}(l, l_0), \ l_o = \text{odd}(l, l_0) \quad (5.25)$$

where the function $\text{even}(l, l_0)$ selects the even element of the pair l, l_0 and $\text{odd}(l, l_0)$ selects the odd element. Although it is possible that Eq. (5.24) does not correspond to the best attainable filter, it is not unlikely that it comes close. In particular it has the desirable feature of producing 0 cross-talk between modes differing in l by an even number, because the linear combination of Eq. (5.25) is 0 for $|l - l_0| = $ even. From Eq. (2.150) it can be seen that modes differing in l by an odd number are separated in frequency by an odd multiple of $\sim \Delta\nu/2$, and therefore well separated. Modes (n_1, l_1) and (n_2, l_2) differing in l by an even number can nearly coincide in frequency if $(n_1 - n_2) = (l_2 - l_1)/2$. It is therefore exactly modes with $|l - l_0| = $ even for which one wishes to achieve a minimal cross-talk with the mode identification filter. Therefore minimisation of Eq. (5.23) can be replaced with minimisation for the $g_X(l_0)$ of:

$$\int_0^1 du \left[\sum_X g_X(l_0) \int d\lambda\, u I_X(\lambda, u) - P_{l_0}(u) \right]^2$$

$$+ \mu \sum_{X X'} g_X(l_0) g_{X'}(l_0) E_{X X'}. \quad (5.26)$$

In practice this minimisation procedure usually will not achieve the exact equality of Eq. (5.24) and thus not achieve an exactly 0 cross-talk between modes for which $|l - l_0| = $ even.

Although proposed in [Pijpers (2001)] the method has not yet been extensively tested, and in particular it is not clear that in practice a sufficient amount of data with sufficiently high signal-to-noise can be obtained to make the technique workable. As in any inverse problem, better results can be achieved for increasing numbers of bands (or other independent observables) X, and for increasing signal-to-noise ratio in each observable. The advantage compared to minimisation of Eq. (5.23) is that an infinite summation over l, which would have to be truncated in practice, is avoided altogether. In either case it is essential to have accurate knowledge of limb darkening profiles as a function of wavelength. Also, it is necessary to note that this method assumes slow rotation. Stars such as δ Sct stars rotate sufficiently rapidly that it is necessary to take into account that there is a preferred axis, the rotation axis, and that one should therefore expect to see cross-talk between modes that is dependent on the angle between the rotation axis and the line of sight. As the work reported in [Viskum, et al. (1998)] shows, this does not preclude identification of modes from colour amplitude ratios but it does introduce additional scatter. Variations in equivalent width of for instance the H Balmer lines provides some additional information. These lines show a strong limb darkening for delta Scuti stars, which is quite different from the broader band photometric colours. This raises the expectation that this line provides strong additional information. However, as pointed out in [Balona (2000)], these lines are sensitive to the temperature variations in the oscillations and not just the velocity, with the result that the information gained compared to broader band photometric colours is not as dramatic as initially expected.

The equivalent width of a line is an integral over the difference between the full flux as a function of wavelength and the (interpolated) continuum: $S - S_{\rm cont}$. This is the lowest order velocity moment M_0 of the line profile where the moments M_n are defined by:

$$M_n = \int {\rm d}v \, v^n S\left(\lambda_c + \frac{v}{c}\right) \qquad (5.27)$$

in which the λ_c is an appropriately chosen reference wavelength for any particular line. Higher-order moments of any given line should in principle be more sensitive to surface velocity fields than M_0 itself. This is the basis of the moment method presented in [Balona (1986a)], [Balona (1986b)] and further developed in [Aerts et al. (1992)], [Aerts (1996)], and [Briquet & Aerts (2003)], which use the ratios M_n/M_0 for $n = 1, 2, 3$. One reason for using these few lowest order moments of lines rather than the full line profile is computational expedience since there are less quantities entering a χ^2-minimisation when fitting pulsation modes. Another reason lies in the signal-to-noise ratio which decreases rapidly with the order of the moment. The information content in the line profile shape is therefore captured well with just the lowest order moments.

In [Balona (1986a)] expressions are derived for the first three moments, where the velocity field is a combination of two components. The first component is rotation around an axis which is inclined with angle i with respect to the line of sight, and the second is pulsation in a mode with spherical harmonic degree and order l and m using the appropriate expressions Eqs. (2.89) and (2.92) for the associated velocity field. The rotation rate is assumed to be small in the sense that the associated deformation of the star or the modification of the oscillation eigenfunctions due to the rotation are neglected. For a slowly rotating, monoperiodic star the first three moment ratios are:

$$\frac{M_1}{M_0} = v_p A_1(l,m,i) \sin\left[(\omega - m\Omega)t + \psi\right]$$

$$\frac{M_2}{M_0} = v_p^2 \left\{ A_{20}(l,m,i) + A_{22}(l,m,i) \sin\left[2(\omega - m\Omega)t + 2\psi + \frac{3\pi}{2}\right] \right\}$$

$$+ v_p v_\Omega A_{21}(l,m,i) \sin\left[(\omega - m\Omega)t + \psi + \frac{3\pi}{2}\right]$$

$$+ \sigma^2 + b_2 v_\Omega^2$$

(5.28)

$$\frac{M_3}{M_0} = v_p^3 \left\{ A_{31}(l,m,i) \sin\left[(\omega - m\Omega)t + \psi\right] \right.$$

$$+ A_{33}(l,m,i) \sin\left[3(\omega - m\Omega)t + 3\psi\right] \}$$

$$+ v_p^2 v_\Omega A_{32}(l,m,i) \sin\left[2(\omega - m\Omega)t + 2\psi + \frac{3\pi}{2}\right]$$

$$+ v_p \left[v_\Omega^2 B_\Omega(l,m,i) + \sigma^2 B_\sigma(l,m,i)\right] \sin\left[(\omega - m\Omega)t + \psi\right]$$

where v_p and v_Ω are the velocity amplitude of the pulsation and the velocity of the rotation respectively. ψ is a reference phase which is constant. b_2 is a constant arising from the limb darkening function chosen, and σ is the intrinsic width of the line. the amplitude functions $A_1, A_{20}, A_{21}, A_{22}, A_{31}, A_{32}, A_{33}$, as well as the B_Ω and B_σ, depend on the spherical harmonic degree and order l and m and the inclination angle i. Expressions for these in terms of the spherical harmonic functions can be found in e.g. [Aerts et al. (1992)]. In [Balona (1986b)] a least-squares algorithm is presented which uses these moments as data in order to determine the l and m and also the (uniform) rotation rate Ω and its axis inclination i. The method proceeds by finding one best fitting singlet l, m at a time. In [Aerts et al. (1992)] a slightly different χ^2 measure, called discriminant, is presented which uses a combination of the moments to decouple the determination of the pulsational parameters from the rotational ones. The method is then extended in [Aerts (1996)] to use more of the information of these moments than envisaged in [Aerts et al. (1992)]. In particular all of the above amplitude functions are utilised rather than the subset of three used in [Aerts et al. (1992)].

Of particular interest is that Eq. (5.28) shows that the second and third order moment contain a mixture of periodic signals with frequencies that are multiples of $\omega - m\Omega$, even for a star that is purely monoperiodic. Clearly, higher order moments for multiperiodic stars will very quickly produce very complicated time signatures. For multiperiodic stars where several modes have comparable velocity amplitudes, the velocity moments will have cross product terms between the modes which affect the fitting procedure. As is noted in these papers, parameters that should have consistent values when fitting the various modes that are present in multiperiodic stars, such as v_Ω and i, in fact vary considerably from singlet to singlet. In [Briquet & Aerts (2003)] the moment method is therefore re-derived, taking into account the fact that the stars that this method is applied to, such as β Cep stars, are multiperiodic and that all detected modes ought to be fitted for simultaneously. Since the method must attempt to fit many singlets simultaneously it is then straightforward to also account for the fact that the rotation of the star affects the eigenfunctions. In a rotating star the spherical harmonic functions are no longer orthogonal. For moderate to rapid rotation the perturbation approach described in section 3.3 breaks down. This is of particular concern in g-modes which are very closely spaced in frequency. These are exactly the modes observed in higher mass non-radial pulsators such as the β Cep stars, so this problem must be addressed. In [Lee & Saio (1987a)] it is shown that each oscillation eigenfunction in a rotating star can be described in terms of a linear combination of the spherical harmonic functions. Even fitting for a single eigenfunction in a moderately to rapidly rotating star therefore requires multiple (l, m) combinations. Since the method described in [Briquet & Aerts (2003)] implements the fitting of multiple singlets simultaneously the properties of moderately rapidly rotating stars as described in [Lee & Saio (1987a)] can be accounted for immediately. The effect of moderate and rapid rotation on the structure and oscillation of stars is discussed further in section 5.2.5.

Another problem of the methods discussed up to this point is that line profiles and even broad band colours are sensitive to temperature and pressure variations as well as velocity fields. An observed variation in any of these observables is therefore not trivially translated into a velocity field. It is necessary to model photometric and spectroscopic observables from a stellar atmosphere of a rotating star, that is perturbed by one or more oscillations, before being able to use these observables for mode identification.

In [Cugier et al. (1994)] photometric amplitude ratios are calculated for several photometric systems that are in widespread use, for $l = 0, 1, 2$ modes in β Cep models. In this work the stellar rotation is ignored. Radiative transfer is calculated in the diffusion approximation without convection. Although convection occurs in thin outer layers of β Cep stars, very little of the energy flux is carried by convection, so that this approximation is justified. The first step in the calculation is to use a non-adiabatic pulsation code to obtain the function $f(r)$ which describes the radial variation of the energy flux perturbation:

$$\frac{\delta 4\pi r^2 F_{\rm rad}}{L} \equiv f(r) Y_l^m(\theta, \phi) e^{i\omega t} \ . \tag{5.29}$$

Using this, the next step is to approximate the optically thin layers of the atmosphere as locally plane parallel layers characterised by an effective temperature and gravity which are functions of θ and ϕ, since they are perturbed from equilibrium by the oscillation:

$$\begin{aligned}
\frac{\delta R}{R} &\equiv \epsilon {\rm Re}\left[Y_l^m(\theta, \phi) e^{i\omega t}\right] \\
\frac{\delta T_{\rm eff}}{T_{\rm eff}} &= \epsilon \frac{1}{4} {\rm Re}\left[f(r) Y_l^m(\theta, \phi) e^{i\omega t}\right] - \frac{1}{2}\frac{\delta R}{R} \\
\frac{\delta g}{g} &= -\left[\frac{R^3 \omega^2}{GM} + 4 - \frac{GM}{R^3 \omega^2} l(l+1)\right] \frac{\delta R}{R} \ .
\end{aligned} \tag{5.30}$$

The stellar monochromatic fluxes are obtained by integrating the contributions of each surface element, with the appropriate local value of g and $T_{\rm eff}$, over the stellar disc. Integrating over wavelength for relevant photometric passbands then produces the observables, which are both the amplitude and relative phase of variations in photometric passbands. The non-adiabatic calculation is essential in particular to obtain the correct relative phases for the variations in the different photometric bands since these are very sensitive to non-adiabatic effects. In [Cugier et al. (1994)] it is shown that modes with different l are sufficiently well separated that for stars with a single dominant mode this mode can be identified. In [Townsend (1997)] the treatment of [Lee & Saio (1987a)] is also the basis of the oscillation calculations. The goal in [Townsend (1997)] is to model line profiles, and therefore particular attention is paid to the intrinsic line shape for each surface element, taking into account the same variations in g and $T_{\rm eff}$ as outlined above. The code described in [Townsend (1997)] also implements limb darkening and the lowest order effect of flattening of the star due to rotation, with which is associated a T_{eff} and g that are higher at the poles than at the equator. Interpolation on a grid of pre-calculated non-LTE synthetic spectra is then used to produce a spectrum integrated over the disc of the star. A useful analysis of the importance of the various effects on the observational diagnostics of oscillations can be found in [Schrijvers & Telting (1999)]. Instead of the detailed spectral synthesis used in [Townsend (1997)], in [Schrijvers & Telting (1999)] a much simpler parametrisation is used, which reproduces qualitatively the behaviour of the atmosphere, but allows for a clear separation of the influence of the effects of local temperature changes, and brightness and equivalent width variations. For instance in [Schrijvers & Telting (1999)] it is shown very clearly how non-adiabatic effects can produce asymmetries in the line profiles, which would be interpreted as a velocity field if one were to rely on adiabatic behaviour of the pulsation in the stellar atmosphere. In this context it is also important to point out that these various effects on the line profile compete and can even cancel, so that in some spectral ranges or spectral lines no variation is seen despite the fact that a star is oscillating. This inter term cancellation is described in detail in [Townsend (2002)].

On the basis of the above discussion one can argue that, as has been done for the Sun, one should take care to select appropriate spectral lines for asteroseismology on stars for any given spectral type. An analysis of the type reported in [Townsend (1997)] and [Townsend (2002)] would be needed in order to investigate whether any given line is temperature sensitive or not. Those lines for which the velocity field is clearly the dominant component for line profile variations, are then the appropriate lines to use as diagnostics of oscillations. This is the approach taken in [De Ridder et al. (2002)] which models the SiII doublet around 413 nm, and the SiIII triplet around 456 nm. The latter are of particular use for β Cep variables while the former are prominent in slowly pulsating B stars (SPBs). In [De Ridder et al. (2002)] it is found that profiles of these lines are relatively insensitive to temperature variations, but that for instance the phase difference between the lowest moments of the line profile are sensitive to temperature and gravity variations.

In the light of these discussions on the influence of velocity, temperature, and other effects, on line profiles, it is useful to re-consider the moment method. The second and third order moments are non-linear diagnostics for what is a linear problem, despite the complications introduced by all of the effects discussed in this section. The problem with non-linear diagnostics is generally their sensitivity to noise. Even though values of the second and third order moment may appear significant on the basis of the assumption that the noise is Gaussian, even small deviations from Gaussianity can significantly bias such moments. The central limit theorem states that for a sufficiently large number of samples, the combined distribution will approach a Gaussian distribution, even if the individual samples do not obey Gaussian statistics. However, the rate of convergence with sample number can unfortunately be quite slow, and is slower the higher the order of the moment being considered. In addition there is the problem when applying these higher order moments to stars for which there are several modes with comparable amplitudes, that a host of cross-terms with combination frequencies appear. If, as is commonly the case, the sampling rate is not perfectly regular the higher order moments will generally display a forest of peaks. In this case it can be argued that if it is necessary on computational grounds to parameterise observed line profiles in terms of a small number of components, it is preferable to do this in terms of a linear combination orthogonal functions of v (or $\Delta\lambda$). The theoretically predicted responses of line profiles to oscillations, from an analysis of the type of [De Ridder et al. (2002)] or [Townsend (1997)], should then be used to determine what basis set of functions would minimise the cross-talk between signals from different singlets. This is the equivalent of the approach shown in Eq. (5.21) and (5.22) but with the correct response functions rather than simple spherical harmonics, which correspond to the case that the line profile is sensitive only to the velocity field, and that the star is not rotating. The appropriate basis set can be constructed numerically. The first step is to calculate the response functions for a relevant range of l and m values, and on a grid of Ω and i values. Next, the surface integral over the visible hemisphere of the star, of the cross product of every pair of these response functions is calculated. These cross-product integrals fill a symmetric matrix. A singular value decomposition of this matrix then provides functions corresponding to the largest singular values, which constitute the optimal linear discriminant filter for this problem. Effectively this describes a wavelet transform of the line profile, where the choice of wavelets is optimised for the problem at hand, in particular including the properties of the noise.

5.1.3 Irregular sampling

In asteroseismology even more than in helioseismology, there are many factors that prevent the acquisition of regularly sampled time series. The problems associated with this are evidently the same as in helioseismology (cf. section 1.3). It is nevertheless worthwhile returning to this subject here, because the sampling may well be quite far from regular, rather than merely having 'missing data' in a series that is otherwise regularly sampled. Also, for those stars which are undergoing pulsations with large amplitudes, such as the 'classical pulsators' (RR Lyraes, Cepheids, Mira variables) nonlinear effects play a role and the light variations, although periodic, are not necessarily sinusoidal. Techniques based on Fourier transforms or on fitting sinusoids are then not necessarily the best approach for determining frequencies. Recent discussions of time series analysis for irregularly sampled data in astronomy can be found in [Schwarzenberg-Czerny (1998a)] and [Vio et al. (2000)]. This problem is also discussed in the context of inverse problems in section 5.3.1.

Perhaps the most straightforward technique is to repeatedly subtract the highest amplitude periodic component from the observed data, which also eliminates its aliases and recursively extract all periodic components, each with their appropriate amplitude and phase as well as the frequency. This philosophy is similar to the CLEAN method, from the field of radio-interferometric imaging [Högbom (1974)], and a recent discussion can be found in [Wilcox & Wilcox (1995)]. A problem is that for irregularly sampled data, with a finite signal to noise, of multimode pulsating stars, the strongest peak in power as a function of frequency could well be an alias rather than the true frequency. There is therefore a need for more robust approaches. In [Scargle (1989)] discrete Fourier transform evaluation is discussed for unevenly sampled data. This method is referred to as the Lomb-Scargle periodogram technique (cf. [Lomb (1976)], [Scargle (1982)]). The method has the advantage that it can be carried out relatively easily and with little loss of speed compared to standard discrete/fast Fourier transform techniques. An implementation can be found in [Press et al. (1992)]. Normally periodograms are interpreted in terms of the confidence level or probability that a given peak as a function of frequency is not due to noise. In order to do this one therefore needs to use the appropriate distribution function for this probability. It is pointed out in [Schwarzenberg-Czerny (1998b)] that there is a difference between the distributions depending on whether one assumes that the variance known or whether it is determined from the data itself and therefore also a stochastic variable, which has implications for confidence levels of detection. This is important in particular for those signals that are at low signal-to-noise ratios.

In [Koen (1999)] sinusoids are fitted to data, as with standard periodogram techniques, but particular attention is paid to the role of correlated noise in time series that are nearly regularly sampled. This paper responds in particular to problems pointed out in [Schwarzenberg-Czerny (1991)] with the standard least squares fitting procedures for sinusoids such as the Lomb-Scargle periodogram, in the presence of correlated noise. In [Koen (1999)] the starting point of the discussion is that the signal y is assumed to be obtained in K blocks of N_k samples per block, which are separated by periods during which no observations are available. The ith measurement of the kth block y_{ik} is given by:

$$y_{ik} = \sum_j (A_j \cos\omega_j t_{ik} + B_j \sin\omega_j t_{ik}) + n_{ik} + \mu_k \qquad (5.31)$$

in which n is the noise and μ represents zero-point shifts between the blocks of observations. The noise n is assumed not to change its statistical properties during any given block of observations, but can change between blocks. μ is a zero-point offset which can also change between blocks. When carrying out unweighted least-squares estimation one can show that for uncorrelated noise the expectation value of the least-squares estimates for these parameters are the true values, and their covariance matrix is given by:

$$\text{cov}(\mathbf{z}_j \mathbf{z}_j^t) = \frac{2\sigma_n^2}{N(A_j^2 + B_j^2)} \begin{pmatrix} A_j^2 + 4B_j^2 & -3A_j B_j & -12B_j/N \\ -3A_j B_j & 4A_j^2 + B_j^2 & 12A_j/N \\ -12B_j/N & 12A_j/N & 12/N^2 \end{pmatrix} \qquad (5.32)$$

where the parameters A_j B_j and ω_j to be fitted for are arranged in vector form $\mathbf{z}_j = (A_j, B_j, \omega_j)$. These equations are derived analytically under assumptions that N is large, and that there are no gaps in the data: they become exact for $K = 1$ and $N_1 = N \to \infty$. For finite samples, explicit Monte Carlo type simulations show similar behaviour with N but with a different numerical factor. In [Schwarzenberg-Czerny (1991)] it is shown that for correlated noise with a typical correlation length of D times the sampling interval the variance estimate for the frequency (i.e. the bottom right corner element of Eq. (5.32) needs to be multiplied with the factor D, as long as $D \ll 2\pi/\omega_j$ for all j. In [Koen (1999)] one can also find the result that for stationary noise processes, i.e. noise processes for which the mean value and autocorrelation are time-independent, the same equation holds, but the factor σ_n^2 is replaced by the power spectrum of the noise process $P_n(\omega)$. In [Koen (1999)] it is pointed out that for correlated noise, the ordinary least-squares estimates are not optimal in the sense that the variances are larger than necessary and the expectation values for the parameters may be biased. If Eq. (5.31) is linearised with respect to ω and the parameter μ is incorporated into the vector **z** it can be written formally as:

$$\mathbf{y} = \mathbf{X} \cdot \mathbf{z} + \mathbf{n} \tag{5.33}$$

where **X** is the design matrix, of which the elements are known. The ordinary least-squares estimate for **z** is:

$$\mathbf{z} = \left(\mathbf{X}^t\mathbf{X}\right)^{-1}\mathbf{X}^t\mathbf{y} \tag{5.34}$$

whereas the correct procedure for unbiased estimates with lowest variance, is the weighted least-squares estimate:

$$\mathbf{z} = \left(\mathbf{X}^t\mathbf{\Sigma}^{-1}\mathbf{X}\right)^{-1}\mathbf{X}^t\mathbf{\Sigma}^{-1}\mathbf{y} \tag{5.35}$$

in which $\mathbf{\Sigma} \equiv \mathrm{cov}(\mathbf{n}, \mathbf{n}^t)$ is the covariance matrix of the noise process. The problem is that often the statistics of the noise process, and therefore $\mathbf{\Sigma}$, are unknown. In [Koen (1999)] simulations are done using a noise process which is a combination of an auto-regressive (AR) part with a history length p, and a moving average (MA) part over q samples:

$$\begin{aligned} n_t &= \xi_t + u_t \\ \xi_t &= \xi_{t-1} + \eta_t \\ u_t &= \sum_{i=1}^{p} \alpha_i u_{t-i} + \sum_{j=1}^{q} \beta_j \epsilon_{t-j} + \epsilon_t \end{aligned} \tag{5.36}$$

in which ϵ_t and η_t are both white noise. ξ_t is a slowly varying mean function, and u_t is the stationary noise process normally referred to in the literature as ARMA(p, q). Data obtained under realistic conditions but without significant periodicities are then used to determine typical parameters for an ARMA process reproducing their behaviour. It is shown that an ARMA(1,1) process is appropriate, but that there are indications that the mean is not stationary. A variant of ARMA models known as ARIMA (autoregressive integrated moving average), works with $n_{t+1} - n_t$ or higher order differences instead of working directly with n_t. The designation is ARIMA(p, d, q) in which the three parameters refer to the order of the autoregression, the differencing and the moving average part of the process. The observational data used in [Koen (1999)] is well represented by an ARIMA(0, 1, 1) model. The paper then proceeds by applying this to data that does have periodic signals in it.

The fitting procedure is iterative and therefore needs starting values, which can be obtained by a standard periodogram technique. The fitting then obtains obtains improved values for all z_j through an iterative least-squares procedure. To obtain the parameters for any one component, the data is pre-whitened by subtracting all other components. This is done for all components and the process is repeated until convergence is achieved. All the periodic signal is then subtracted from the observed time series, and an ARIMA(0, 1, 1) model is fit to the residuals. With this in hand the covariance matrix Σ is obtained. This is then used to re-determine the z_j but now with the weighted least-squares method of Eq. (5.35) rather than Eq. (5.34). The standard errors of the parameters of the periodic signals are then established through Monte Carlo simulation, using the observational residuals as samples of the error process, which is referred to as bootstrapping in [Koen (1999)]. The algorithm does not use regularity of sampling anywhere in the procedure, and therefore can be applied to irregularly sampled data as well, although the covariance matrix Σ from the ARIMA process differs slightly between regularly and irregularly sampled data.

One issue with this method is that it needs starting values for the parameter vector **z**. This is not an insurmountable obstacle but there is evidently scope for alternative methods. Also, if the periodic signals are not sinusoidal, as is the case in particular for larger amplitude variables, it is useful to consider a different approach to extracting the periodic signals from observed time series which may also be irregularly sampled. In [Vaniček (1971)] it is discussed that instead of sine functions one can use other functions, or the sum of Fourier components to represent the periodic but not sinusoidal signal. An appropriate choice of base functions can reduce considerably the number of components necessary for a satisfactory fit, which is computationally advantageous and also more robust against errors. This method is applied to time series for double mode Cepheids in [Andreasen (1987)]. Part of the problem with irregularly sampled data is that standard sine and cosine functions are only orthogonal for regularly sampled data. In [Grison (1994)] a specific method is presented to construct polynomials that are orthogonal for the sampling available. In this way a generalisation of the Lomb-Scargle periodogram is obtained. The construction scheme for the even and odd base functions $\Phi^{e,o}$ is recursive:

$$\begin{pmatrix} \Phi_k^e \\ \Phi_k^o \end{pmatrix} = \begin{pmatrix} \cos(k\omega t + \phi_k) \\ \sin(k\omega t + \phi_k) \end{pmatrix} - \sum_{i=1}^{k-1} \mathbf{R}_{ik} \cdot \begin{pmatrix} \Phi_i^e \\ \Phi_i^o \end{pmatrix} \qquad (5.37)$$

in which the recursion is started with the matrix $\mathbf{R}_{i1} \equiv 0$. For subsequent iterations the four elements of the recursion matrix **R** and the phase ϕ_k must be established by applying the orthogonality constraints:

$$\sum_{j=1}^{N} \Phi_i^{e,o}(t_j)\Phi_k^{e,o}(t_j) = \delta_{ik}\delta_{eo} \qquad (5.38)$$

in which on the right-hand side the first Kronecker delta refers to the index of the polynomial and the second Kronecker delta is 1 if both Φ are even polynomials, or both are odd, and 0 otherwise. In [Schwarzenberg-Czerny (1996)] the same idea is pursued, but it is noted that for stars that are multiperiodic, such as double mode RR Lyrae stars, problems do remain for realistic sampling and signal-to-noise ratio of the data. A somewhat different viewpoint is represented by [Foster (1996)] in which the functions proposed to expand the observations onto are similar to Morlet wavelets: sine functions modulated by a Gaussian envelope. This finds application in particular for stars in which the period changes with epoch, as happens in long-period variables. A multiresolution approach is then obtained by using Gaussian envelopes with a succession of widths (cf. [Otazu et al. (2004)]). Methods that use base-functions that are relatively localised both in the time domain and in the Fourier domain are also referred to as time-frequency methods. A different time frequency approach to the above methods is described in section 5.3 on inverse methods since it addresses the problem of power spectral estimation by applying an inverse method.

A different class of techniques is known collectively as phase dispersion minimisation, which is useful in particular if the data is sparsely sampled and there is a single dominant periodic signal which is not necessarily sinusoidal. Techniques of this type are discussed in [Stellingwerf (1978)], [Bossi & La Franceschina (1995)], [Cincotta et al. (1995)], [Jetsu & Pelt (1999)] and [Cincotta et al. (1999)]. In this technique the data is folded at a number K of trial periods P_k. For every trial period chosen, all measurement times are binned in M bins of equal relative phase ϕ_{ik} which is calculated by:

$$\phi_{ik} = \frac{t_i}{P_k} - \text{int}\left(\frac{t_i}{P_k}\right) . \qquad (5.39)$$

From the observations an overall variance of the entire time series, and a variance per bin can be calculated as:

$$\sigma^2 = \frac{1}{N-1} \sum_i (y_i - \bar{y})$$

$$s_j^2 = \frac{1}{n_j - 1} \sum_{i_j} (y_{i_j} - \bar{y}_j) \tag{5.40}$$

$$s^2 = \frac{1}{\sum_j n_j - M} \sum_j (n_j - 1) s_j^2$$

in which n_j is the number of members in bin j, $\bar{y}_j$ is the mean value in bin j and i_j refers to those times t_i for which the relative phase is such that the measurement is a member of bin j. If there is no periodic signal present with trial period P_k then the expectation value of s^2 is equal to σ^2. However, if there is such a period present, then s^2 is reduced. If there is a single dominant periodic signal and high signal-to-noise ratio of the data, this is evidently a fast method for finding periodicities. For multiperiodic and/or noisy time series, the method becomes more problematic. In [Bossi & La Franceschina (1995)] an extension is proposed which essentially pre-whitens data iteratively by subtracting the periodic signal of the previous step. In [Cincotta et al. (1995)] and [Cincotta et al. (1999)] an entropy-based statistical measure for significance is used, rather than the original ratio of s^2/σ^2 in [Stellingwerf (1978)]. In [Jetsu & Pelt (1999)] the phase dispersion method is applied to disjunct sections of the time series, which effectively reduces the spectral resolution. Co-adding the spectra for all sections then produces a low resolution but higher signal-to-noise estimate of the period of periodic signals present in the data. A more refined search, concentrating on intervals around these crude estimates then provides the full analysis. This technique can represent a considerable reduction in computing time in automated period searches, if the spectral resolution and spectral range corresponding to the time series is large, and only a few weak periodic signals are present.

This section so far concentrates on stars for which the sampling, even if irregular, is relatively dense so that all relevant scales of variability are reasonably well resolved. Although one might expect that it becomes easier to obtain well sampled time series, the longer the time scale of variation, this is not necessarily the case. Giant stars such as Miras have periods which tend to lie upwards of 100 d. Observational data on such stars are generally not obtained at professional observatories, but instead privately, by amateurs or professionals, and data is available through organisations such as the American Association of Variable Star Observers (AAVSO) (cf. http://www.aavso.org/). The sampling of the light variations recorded in databases maintained by such organisations is irregular, can be very sparse as well, and is often quite variable in quality which means that the noise of the time series is not at all stationary. In such cases there are particular challenges to determining even quite basic properties of the pulsation. In a series of papers basic analysis techniques are discussed ([Koen & Lombard (1993)]), as well as the detection of intrinsic 'period jitter' ([Koen & Lombard (1995)]) and period changes ([Koen (1996)], [Koen & Lombard (2001)]) in such variables. An analysis of a sample of Mira variables using the methods developed in [Koen & Lombard (2001)] can be found in [Koen & Lombard (2004)]. Intrinsic irregularity is particularly of interest for giant stars because the pulsation amplitudes can be very large, and non-linear interaction between modes, as well as feedback into the mean structure of the star could produce this behaviour. However a combination of stochastic and κ-mechanism driven oscillation produces similar effects (cf. [Bedding et al. (2005)] for an example). It has also been suggested that pulsation of some AGB stars could be described in terms of a dynamical system that is in a state of low-dimensional chaos (cf. [Buchler et al. (2004)]). In this case it is not just the periods but the statistical properties of the quasi-periodic behaviour that might be used for asteroseismic inference.

5.1.4 Non-seismic constraints

Quite apart from their use in mode identification and in determinations of the abundance of chemical elements, the spectra of stars also hold information on surface rotation rates and magnetic fields of stars. Rotation of the stellar surface produces a characteristic broadening of the line profile. The effect of rotation can be expressed as a convolution of a local line profile and a broadening function which takes into account the rotational velocity field over the surface. In essence this is a simplified case of what is described in section 5.1.2 for mode identification through line profile variations. In the right-hand panel of Fig. 5.3 the effect in the Fourier domain of the broadening function is shown. Fourier transforms of line profiles can yield precise rotation rates (cf. [Gray (1992)]), even if the rotation rates are low. In [Gray (1989)] one can find an application of this method to giants, and in [Reiners (2006)] the method is used to investigate surface differential rotation of F-type and cooler stars. A description of the Fourier transform method for surface rotation that depends on latitude is given in [Reiners & Schmitt (2002)]. Differential rotation produces a somewhat more compli-

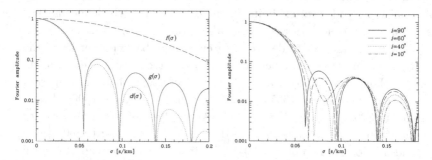

Fig. 5.3 Left panel: The Fourier domain rotational broadening function $g(\sigma)$ for a uniform rotation rate $v \sin i = 12$ kms together with the (Gaussian) Fourier transform of the un-broadened line (dash-dotted), and the product of the two (dotted). Right panel: the function $g(\sigma)$ for a limb darkened differentially rotating star, for several inclination angles of the axis with the line of sight (figure from [Reiners & Schmitt (2002)] reproduced by permission of A&A).

cated behaviour of the rotational broadening function, and the result also becomes dependent on the inclination angle between the rotation axis and the line of sight. The limb-darkening profile of the star must also be taken into account. In the left-hand panel of Fig. 5.3 the calculation of a line profile is done using a parameterised form for the limb-darkening function and for the surface rotation rate. The resulting line profile, averaged over the surface of the star is then Fourier transformed and the result is the function $g(\sigma)$ in which σ is the inverse wavelength shift or equivalently inverse Doppler shift. The difference in the first side-lobe of the function g between the two panels is very clear. The subsequent side-lobes are less dependent on the details of the rotation law and limb-darkening. Although these different process modify $g(\sigma)$ in different ways, there is some ambiguity between the inferred inclination angle and the differential rotation, as is discussed in detail in [Reiners & Schmitt (2002)].

The method described in [Reiners & Schmitt (2002)] does not as yet take the presence of surface inhomogeneities and magnetic fields into account. Sufficiently strong magnetic fields can be detected either through Zeeman splitting or polarisation. An alternative technique is Doppler mapping for sufficiently rapidly rotating stars, already discussed in the context of mode identification in section 5.1.2. This technique can also be used to map the photospheric foot points of magnetic field structures, and even chemical inhomogeneities. A description of the (inverse) method for constructing such maps using MEM can be found in [Donati & Brown (1997)], and an application to a magnetically active star in [Donati & Collier-Cameron (1997)]. Even for stars in which the spots do not cover quite such large fractions of the stellar surface, this technique can still be used to infer differential rotation as is done in [Barnes et al. (2005)].

Stars which are members of binary systems also provide valuable information, since from orbital motions the total mass can be determined, and in the case of eclipsing binary systems even the individual masses of the stars. Although stellar masses are never known as accurately as the solar mass is (see appendix C), a dynamical mass estimate provides a strong constraint in carrying out asteroseismology. The problem with stars that are part of such binary systems is that the stars tidally interact, and if they are close this has implications for their evolution. Here a comparison of the seismic properties of stars with similar spectral types, that are members of binary or multiple systems with a range of separations, and stars that are not members of binaries, will help in assessing the extent of the influence and the severity of the biases introduced by extrapolating the properties of stars in binary systems to populations of stars on a galactic scale. Quite apart from the intrinsic interest that planetary systems around stars other than the Sun have, the orbital motion of these planets can also be used to determine a dynamical mass of the host star, or the system as a whole. Assuming that planets, because of their lower mass, have less tidal influence on their host stars than a stellar companion would have had, this increases the population of stars for which there is a dynamical mass constraint. Such stars are then also good 'standard stars' to test stellar evolution theory on, although there are some indications that there are systematic differences in chemical surface abundances between planet-hosting stars and the general stellar population which can evidently introduce biases.

Apart from its potential use in mode identification, interferometry already provides stellar radii. This information may appear redundant if an effective temperature and a luminosity for the star are measured. However, both of these quantities are in most cases determined through a transformation involving the calculation of the emergent spectrum of a star. Using this, spectra and broad-band colours can be transformed into luminosities and effective temperatures. An interferometric measurement also requires some care because of the influence of limb-darkening but this bias is smaller, so that the interferometric measurement of R is generally more precise than the inference from L and T_{eff} and likely to be more accurate as well. Interferometrically determined angular diameters of giant stars are available through catalogues (cf. the SIMBAD catalogue database: http://simbad.u-strasbg.fr/Simbad). For stars on the main sequence there is less data available, although with the current generation of interferometric optical arrays, it starts to become possible to obtain precise physical radii for nearby stars, even with smaller physical dimensions than the Sun has.

5.2 Modelling

Modelling of stars other than the Sun is not in essence any different. For the purposes of helio- and asteroseismology differences do arise because of the non-seismic constraints that can be placed on parameters controlling the evolution of stars or limiting the allowed range of such parameters. There are also physical effects that for the Sun are of such minor importance that one can ignore their impact on the structure and normal mode oscillation frequencies, but that are very important for other stars, such as rotation.

Although stellar modelling requires solving the same problem as outlined in section 2.2 and 2.3 it is useful to briefly discuss the scaling behaviour of the dependent quantities: density, pressure, temperature etc. with the overall stellar parameters. Normally the equations for stellar structure are solved in dimensionless analogues of these parameters, both because of numerical robustness of the algorithms and because of the fact that this shows more clearly what to expect for the behaviour of stars as a function of their mass and radius. A very similar discussion can also be found in [Christensen-Dalsgaard et al. (2005)] in the context of the limitations in using helioseismic data to measure the gravitational constant G. The radial coordinate r is usually scaled by the stellar radius R. For stars that are rotationally flattened an equatorial radius can be used instead, or some combination of polar and equatorial radii. This scaling also affects the gradient operator:

$$r = R\widetilde{r}$$
$$\nabla = \frac{1}{R}\widetilde{\nabla} \qquad (5.41)$$
$$\nabla^2 = \frac{1}{R^2}\widetilde{\nabla}^2 \ .$$

The natural way to create a dimensionless density $\widetilde{\rho}$ is to scale by the mean density of the star:

$$\rho = \bar{\rho}\widetilde{\rho} = \frac{3M}{4\pi R^3}\widetilde{\rho} \,. \tag{5.42}$$

The Poisson equation (2.4) then provides the best way to scale the gravitational potential:

$$\psi = 4\pi G \bar{\rho} R^2 \widetilde{\psi} = \frac{3GM}{R}\widetilde{\psi} \,. \tag{5.43}$$

Introducing these in the equation of hydrostatic equilibrium (2.3) then provides the scaling for the pressure:

$$p = 4\pi G \bar{\rho}^2 R^2 \widetilde{p} = \frac{9GM^2}{4\pi R^4}\widetilde{p} \,. \tag{5.44}$$

Although the plasma in stars is not perfectly ideal, a good scaling factor for the temperature is nevertheless obtained by using the ideal gas law:

$$T = 4\pi G \bar{\rho} R^2 \frac{m_H}{k}\widetilde{T} = \frac{3Gm_H}{k}\frac{M}{R}\widetilde{T} \,. \tag{5.45}$$

The natural scaling for the energy flux makes use of the luminosity of the star:

$$F = \frac{L}{4\pi R^2}\widetilde{F} \,. \tag{5.46}$$

Using these dimensionless quantities Eqs. (2.3) and (2.4) become:

$$\widetilde{\nabla}\widetilde{p} = -\widetilde{\rho}\widetilde{\nabla}\widetilde{\psi} \tag{5.47}$$

$$\widetilde{\nabla}^2\widetilde{\psi} = \widetilde{\rho} \,. \tag{5.48}$$

It is of interest to note here that neither of these equations have any parameter at all. If there were a direct relationship between p and ρ, then solving this equation would provide the structure of all stars. Historically a polytropic equation of state has been used to construct stellar models $p = K\rho^{1+1/n}$ in which n is a constant polytropic index. This single parameter family of stellar models is referred to as polytropes. Polytropes certainly still have their uses, and rightly find their way into stellar structure textbooks (eg. [Kippenhahn & Weigert (1994)]), which the reader is referred to for details. For realistic equations of state the temperature is required as well however which means that the temperature stratification must be determined through either:

$$\widetilde{\nabla}\widetilde{T} = \frac{\Gamma_2 - 1}{\Gamma_2} \frac{\widetilde{T}}{\widetilde{p}} \widetilde{\nabla}\widetilde{p} \qquad (5.49)$$

or

$$\widetilde{\nabla}\widetilde{T} = -\frac{k^4}{(24\pi)^2 acG^4 m_H^4} \kappa_{\text{eff}} \frac{L}{M^3} \frac{\widetilde{\rho}\widetilde{F}}{\widetilde{T}^3} \qquad (5.50)$$

depending on the dominant mechanism of heat transport. Generally the two are combined in a form of Eq. (5.50), and then the total heat flux $\widetilde{F}$ must be corrected so that only the radiative part of it enters Eq. (5.50) (see the discussion in section 2.2). For stars that are fully convective, the Eq. (5.49) applies throughout, except in very near surface layers where the optical depth becomes of order unity or lower. The dependence of Γ_2 on temperature and density then controls the detailed structure, but this dependence is weak. The stratification of fully convective stars is therefore almost independent of mass or composition. For stars in which radiative transport dominates in substantial regions of the star, the scaling of the second of these equations becomes important. In this case the details of the dependence of opacity κ_{eff} on temperature, density and composition determine the stratification of the star. If one for instance supposes a functional dependence of the form $\kappa_{\text{eff}} = \kappa_0 \rho^\lambda T^\nu$, which is known as a Kramers opacity, and introduces this into Eq. (5.50) with the appropriate dimensionless quantities, one finds that this equation contains a factor f_{rad}:

$$f_{\text{rad}} = L R^{-3\lambda-\nu} M^{-3+\lambda+\nu} . \qquad (5.51)$$

This implies that two stars that have different masses, radii and luminosities, but for which this factor is identical, will have very similar stratifications. In reality the opacity does not follow a Kramers form exactly, although over limited ranges in temperature and density it can be used as an approximation.

The differential equation for the heat flux becomes:

$$\widetilde{\nabla}\widetilde{F} = \widetilde{\rho} \frac{3M}{L} \epsilon . \qquad (5.52)$$

As with the opacity κ_{eff} the energy generation rate ϵ also can be approximated by power laws over limited ranges in temperature and density $\epsilon \propto \rho^p T^q$. The factor in the global stellar parameters that this implies for

Eq. (5.52) is f_{en}:

$$f_{\text{en}} = L^{-1} R^{-3p-q} M^{1+p+q} . \qquad (5.53)$$

This result implies that for two stars to be homologous, i.e. to have very similar stratifications, the two factors f_{rad} and f_{en} must match. Since there is no reason for the exponents for R and M to be the same in these two factors, the radius (for instance) can be eliminated between these two, leading to a mass-luminosity relationship.

On the other hand, solving the equations of stellar structure with given boundary conditions, normally allows only some values of these two parameters: they are eigenvalues of the system of equations with those boundary conditions. As the approximations for κ_{eff} and ϵ with constant exponents have limited ranges of validity, and also depend on the chemical composition, in practice a single eigenvalue has to be replaced with a range of allowed values. This implies that for a given mass only a limited range, or a few limited ranges, of values for the luminosity are possible. Also, stars are never completely radiative, so that these scaling laws are only approximately valid in all cases. However, this simple analysis does show clearly why the HR diagram cannot be completely uniformly populated, even allowing for the spread due to differences in composition from star to star, and composition gradients within each star.

In particular the departures from the above scaling laws reveal the more interesting physical processes: transitions between convective and radiative energy transport, and the effects of particular features in the true opacity which are not captured by the Kramers parametrisation. Also in the resonant frequencies of stars, it is therefore of interest to take out trivial behaviour with overall stellar parameters, in order to concentrate on those properties that reveal physical processes. It is for this reason that in models normally the frequencies of oscillations are scaled by the most relevant time scale: the dynamical time scale (cf. section 2.1). With this scaling a dimensionless frequency is:

$$\omega^2 = \frac{GM}{R^3} \widetilde{\omega}^2 . \qquad (5.54)$$

This is also referred to as the period mean density relationship for stellar oscillations. It is only after removing the proportionality with the square root of the mean density that non-trivial internal physical differences between stars are probed by oscillation frequencies. Inspection of the LAWE equations (2.95), (2.99), (2.100), and (2.101) shows that the frequencies S_l^2

and N^2 must be scaled by the same factor, together with the appropriate scalings for all equilibrium and perturbed quantities to obtain dimensionless forms of these equations.

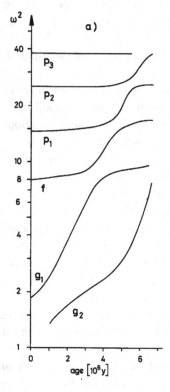

Fig. 5.4 Avoided crossing of modes with time as a star evolves, starting from the ZAMS. The frequency of the lowest order p and g modes and the f-mode are indicated as a function of stellar age. The frequency of the g_1 mode increases due to an increasing gradient in μ with age, so that it interacts with the f mode, and subsequently the frequencies p modes with increasing order are 'bumped' up. (Figure from [Aizenman et al. (1977)] reproduced by permission of A&A.)

One particular way in which frequencies can depart from simple scaling or the asymptotic behaviour of Eq. (2.150) is in avoided crossings. In evolved stars there is normally a sufficiently strong gradient in μ that some modes are propagating with g-mode like behaviour in the interior and p-mode like behaviour in the outer layers, with an evanescent region in between. This tends to displace the frequency of these modes from their asymptotic p-mode location: in models where the frequency is tracked over an evolutionary sequence, it appears to be 'pushed up' in a way that follows an extension of a g-mode branch of oscillation, as shown in Fig. 5.4. The precise n and ν of the mode that is displaced in this way can be very sensitive to details of the internal structure and thus this is a particularly valuable asteroseismic constraint. An example of this behaviour may have been seen in η Boo (cf. [Christensen-Dalsgaard et al. (1995)] and recent re-observation [Carrier et al. (2005)]). This kind of behaviour can also be found due to the effects of rotation (see also section 5.2.5).

Another issue that arises in the modelling of stars concerns the unicity of stellar models. Do the effective temperature, the luminosity, and the surface composition of a star, if they were to be measured with infinitely high precision and accuracy, uniquely constrain parameters such as the mass and the age ? From the discussion above, the answer would appear to be in the negative. If there are several mass-luminosity relationships, then a given luminosity could arise from several different masses. Put another way, it is the behaviour of two quantities, ϵ and κ_{eff}, with density, temperature and composition that determine three global parameters M, R and L. However, this needs to be qualified, since there is a third quantity that plays a more hidden role which is the equation of state, and its dependence on temperature, density and composition breaks that degeneracy. There is a further complication in that stars have a combination of convective and radiative zones, and therefore phenomenological parameters such as a mixing length and an overshoot parameter can be adjusted, which evidently destroys unicity. This problem would be circumvented if convective zones were to be modelled without free parameters, for instance using appropriate hydrodynamical simulations, which would also have to account for the composition gradients in stars, since stars as they evolve have a different chemical composition in their interior as compared to the surface, both due to nuclear reactions and due to settling of heavy elements. Nevertheless, there is no mathematical proof of unicity.

Viewed as an inverse problem it can be argued that the luminosity, effective temperature and surface abundances are integral constraints, since the equations of stellar structure form a set of differential equations that need to be integrated, and the parameters L, T_{eff}, and the surface abundances are surface values for the dependent variables of that set of differential equations, or combinations thereof. It is known that the specification of surface values only in exceptional cases fully specifies the solution of a differential equation, or a set of coupled differential equations. From the modelling of the best measured star, the Sun, it is known that there are many models possible that fit just these surface measurements. Many of these can be excluded by the helioseismic techniques discussed in chapter 3, but not by any other method. Helioseismology and asteroseismology do not provide unicity either: they provide more integral constraints with which more models can be excluded, but the central problem is still ill-posed. The correct manner in which to pose questions is therefore in terms of whether the null-space of solutions to the inverse problem, given all data with the measurement errors, contains solutions that are considered to be qualitatively different. If the answer is affirmative, the next step must be to determine what measurement is needed to discriminate between such solutions, i.e. which helps in eliminating solutions by shrinking the null space. A first step towards a systematic analysis of this type is [Brown et al. (1994)]. Finally it is worth mentioning that the T_{eff} and L of stars are not in fact directly measured quantities: they must be inferred. In the case of the effective temperature, a transformation must be carried out which takes as input measured broad band photometry and spectroscopy. The appropriate transformation is found through solving radiative transfer calculations for the temperature and density stratification through the region where the emergent flux of the star is determined. In the case of the luminosity what is required additionally is a distance determination, or the determination of an angular diameter, for instance through interferometry. It is important to mention this here because these steps not only introduce random errors, but may introduce biases as well.

5.2.1 The ages of stars

For a discussion of the ages of stars, it is important to define what sets the zero-point for the age. Stars form out of the interstellar medium, and become luminous objects, initially through the radiation of gravitational energy released by the collapse of a gas cloud under its own gravity. The luminosity can be substantial, but without any fusion processes yet taking place. The processes controlling the rate of infall of gas are incompletely known. Observationally it is also difficult to asses what exact stage of infall a given young stellar object is in, because it is normally deeply embedded in an interstellar cloud complex for which the extinction is very high, certainly in visible light. A pragmatic choice has therefore been to set the age zero point to that point at which stable H fusion in the core commences: the Zero-Age Main Sequence (ZAMS). During the formation process the star has been fully convective some time in the past, and therefore the model assumption is that stars on the ZAMS are chemically homogeneous. Compared to the Main Sequence life-time of stars, the formation stage is generally thought to be short, since it must scale with the thermal time scale (cf. section 2.1), which means that the full age is only slightly underestimated if it is counted from the ZAMS onward. In order to test the influence of the assumption of chemical homogeneity on the ZAMS, stellar evolution is now also being calculated including some part of the pre-MS evolution. (eg. [Palla & Stahler (1991)], [Palla & Stahler (1993)], [D'Antona & Mazzitelli (1994)], [Siess et al. (2000)]).

Ambiguity in age determinations also arises because of the uncertainty in the abundance of He in stars. There is a lower limit to the He abundance, which is due to the production of He in Big Bang nucleosynthesis. However, generations of stars since the Big Bang have enriched the interstellar medium with heavy elements and have also increased the He abundance. This enrichment is not necessarily homogeneous, and the current stellar content of our galaxy consists of at least two generations of stars with quite different chemical compositions, and with some spread of composition within each of these populations. He is a particular problem for age determinations because its abundance affects quite profoundly the appearance of a star, and the rate at which its evolution progresses. A younger star with a higher abundance of He on the ZAMS cannot be distinguished from an older star with a lower initial He abundance from classical measurements alone. Even for the Sun the reference determination of the age is not from classical observations nor even helioseismically, but from dating of radioactive elements found in meteorites (see appendix C). Helioseismic determinations are model dependent, as is clearly demonstrated in [Dziembowski et al. (1999)], and result in larger uncertainties for the age.

For homogeneous groups of stars it is possible to determine relative ages. Stars can occur as members of groups of stars that, from their radial motions and proper motions on the sky, can be deduced to be dynamically bound or nearly bound. Open and globular clusters are such dynamically bound entities, and the stars that are members of such a group are normally assumed to have formed coevally or very nearly so. Nearby examples of such associations are the Hyades and Pleiades. Certainly the heavy element surface abundance pattern of the confirmed members of such groups are very similar: see e.g. [De Silva et al. (2006)] for a discussion of the abundances of some key elements in the Hyades. When the member stars are plotted in an HR diagram this shows a very distinctive pattern which is consistent with the stars having the same age. By evolving stars using an evolution code, starting with a broad range of stellar masses all on the ZAMS with identical initial abundances, one can construct isochrones for a population of stars. An isochrone is a curve in the HR diagram which connects the locations of stars of different masses at the same age. A group of stars that is formed at the same time must then lie along a single isochrone. A recent compilation of data for galactic open clusters, and uniform age determinations using a single He abundance of $Y = 0.28$ can be found in [Kharchenko et al. (2005)].

The exact shape of isochrones tends to vary with the details of treatment of convection, mass loss and other uncertain factors. Also, the labelling with age of any given isochrone is ambiguous, because of the uncertainty in initial He abundance and other effects. However, He abundance differences between different groups of stars can never be responsible for all of the differences seen in the HR diagrams of all the identified dynamically bound groups. Qualitative judgements on the relative ages of distinct groups of stars can therefore normally be made in this way, although if two groups of stars are inferred to have very similar ages, potential differences in He abundance do give rise to ambiguities. A more detailed discussion of various sources of random and systematic errors in determining cluster ages, with application to three globular clusters, is [Gratton et al. (2003)], and a somewhat older broad review of the subject is [VandenBerg et al. (1996)]. The problems outlined here imply that a determination of the ages of stars is very closely linked to the determination of surface abundances of chemical elements for stars, and in particular the abundance of He.

5.2.2 The composition of stars

Much as the detailed observation of the power as a function of length scale of the cosmic microwave background radiation, the primordial abundances of light elements such as He and Li are strong constraints on theories of the big bang and the associated nucleosynthesis. The ages of the oldest stars have similar cosmological significance, and as is clear from the previous section, this also requires determinations of the abundances of all chemical elements, and in particular He. Also for the luminous classical variables, abundance determinations have importance in the connection between asteroseismology and cosmology. Cepheids, RR Lyrae and Mira variables, can be identified in the Magellanic clouds and in more distant galaxies. Mass-luminosity relationships for variable stars imply period-luminosity relationships. A measurement of the period of luminous variable stars thus allows the determination of an absolute luminosity which together with a brightness in the sky permits determining a distance, which is one of the basic steps in determining the distances to the most distant objects in the universe. However, the period-luminosity relationships are known to have a metallicity dependence which needs to be accounted for. An illustration of this can be found in recent re-determinations for distances to the galactic centre and the Magellanic clouds using Mira variables [Groenewegen & Blommaert (2005)].

Unfortunately there are significant difficulties with determining the abundance of Helium spectroscopically, since it has no permitted transitions in its neutral state. This has the consequence that under the conditions normally encountered in stellar atmospheres for lower mass stars He has no spectral lines that are accessible to observation. This is particularly unfortunate since it is exactly the lowest mass stars that have the longest evolutionary timescales and are of cosmological significance for that reason.

The problem with determining the primordial abundance of Lithium is that there are various processes that affect its abundance in stellar atmospheres. For instance, what is known as the Li gap is a severe depletion of lithium in F type stars that occurs during the main sequence (cf. [Chen et al. (2001)]), in stark contradiction to the predictions of standard (no rotation, no mass loss, no diffusion of chemical elements, no magnetic fields) stellar evolution theory ; thus there are additional processes at work. Mass loss, diffusion, and slow mixing induced by rotation or waves have all been proposed as contributing factors in generating the Li gap (cf. [Schramm et al. (1990)], [Zahn (1992)], [Turcotte et al. (1998a)], [Talon & Charbonnel (1998)]), and evidently these processes must also then influence the Li abundance in other stars.

For all of these processes it would be invaluable to obtain an actual measurement of the depth of surface convection zones in stars because this places constraints on the extent to which each of these can affect the surface Lithium abundance. As is clear from section 3.1.1 it is possible to constrain the helium content of stars by inversion, essentially through the behaviour of the adiabatic exponent Γ_1 in the zone where He changes in its dominant ionisation state from doubly to singly ionised (cf. [Basu et al. (2004)] [Monteiro & Thompson (2005)]). Similarly the change from a radiative zone to a convective zone in a star produces a characteristic change of sound speed gradient at that position, to which the frequencies of oscillations are sensitive (cf. [Monteiro et al. (2000)]). Thus it is possible in principle to constrain both He abundances and convection zone depths using oscillation frequencies of stars. This technique is discussed in more detail in section 5.3.5.

It should be mentioned here that it is not just the extent of surface convection zones which is of importance for stellar evolution and therefore of interest to attempt to determine asteroseismically. Arguably, the convection that occurs on the MS in the cores of stars more massive than the Sun is even more important, since the extent of the convection as a function of age determines the amount of H available for fusion and hence the MS lifetime. Further the molecular weight gradients that this produces can have substantial effects in subsequent evolution. Depending on whether the extent of the convective zone has increased or decreased with time, a more or less clear signature is expected. For instance an initially increasing and then decreasing size produces a discontinuity in the gradient of μ at the location of maximal extent of the convection zone. An increasing relative size produces a discontinuity in the mean molecular weight μ at the location of the convection zone boundary, modified by the convective core overshoot. There is no reason a-priori why the properties of core convection should be the same as those of envelope convection since the physical conditions are quite different. For example, for core convection a constraint exists (cf. [Roxburgh (1989)]) that can be used to put limits to the extent of overshooting, but there is no equivalent that applies to surface convection zones. Overshoot is not captured at all by purely local MLT type prescriptions for convection, since overshoot is a non-local effect. Since overshoot can have a clear impact on stellar structure and also on surface elemental abundances, in particular in advanced stages of stellar evolution, measurement of it is clearly desirable.

Asteroseismic techniques can, at least in principle, help in constraining the abundances of key elements in stars. On the other hand, determinations of the surface abundances of heavy elements are necessary for asteroseismology as well. In particular the opacity of stellar material is sensitive to even minor constituents of stellar matter in specific temperature and density ranges. This is also illustrated by controversies surrounding changes in determinations of solar abundances (cf. [Asplund et al. (2004)], [Asplund et al. (2005a)], [Asplund et al. (2005b)]) and their impact for helioseismology. A number of these changes in the abundances of elements for the Sun are due to improved modelling of the radiative transfer in convective stellar atmospheres, by using 3D hydrodynamical simulations in combination with radiative transfer in spectral lines that are not in local thermal equilibrium (non-LTE). This must also have implications for abundance determinations for other stars, even if these are measured relative to the Sun because the properties of the convection change with spectral type. One example of this is the long standing 'missing opacity' problem: calculations of the emergent spectra in the UV range in the Sun systematically differ from the observed spectrum. Similar problems hold for other stars. It is argued in [Asplund (2004)] that this problem disappears for the Sun with this new generation of models. The implication is that it may do the same when applied to stars, although one should bear in mind that for stars with higher effective temperatures, the surface convection disappears at some point, so it is not obvious that the 3D models are a universal solution.

For helio- and asteroseismology it is important to construct what is known as reference models. These models are not necessarily intended to reproduce exactly all available data. Reference models reproduce L, T_{eff}, and surface abundance measurements, but make use only of the freedom that standard ingredients for stellar modelling allow. Any modern stellar evolution code should therefore in principle be able to reproduce a reference model. The reason to construct reference models is to use these to compare the star with, and to express frequency differences between star and model in terms of structural differences between the star and the model, as discussed in section 3.4. The assumption normally made is that the differences in structural parameters between the true star and the model are sufficiently small, so that the frequency differences can be related in a linear fashion to these structural differences. If the abundances are inaccurate this means the opacity can be sufficiently poorly modelled that this assumption is violated. Evidently as consensus as reached on aspects of ingredients that are considered to be non-standard at some time, these ingredients then become part of reference models. Occasionally reference models are updated.

A particular pattern of abundance anomalies, particularly in more evolved stars, can also indicate processes such as radiative levitation, gravitational settling, or the dredge-up by convection of products of nuclear processes. In AGB stars in particular the detailed analysis of abundance ratios of chemical elements and of various isotopes of elements is an invaluable tool in constraining the evolution (cf. [Renzini & Voli (1981)]), although this is further complicated by the mass loss in these stars, as recent investigations of AGB stars confirm (cf. [Blöcker (1995)] [Ventura & D'Antona (2005)]), as well as other uncertain dynamical processes such as convective overshoot (cf. [Herwig et al. (1997)], [Herwig (2000)]). A recent comprehensive review of AGB star evolution can be found in [Herwig

(2005)]. In each of these cases that presence of abundance anomalies is a tracer of processes occurring at depth. The relative abundances of chemical elements are therefore also integral constraints, in particular for processes controlling the transport of material. One problem here is that the initial conditions are unknown. It is possible, at least in principle, that in some cases the abundance anomaly reflects the mixture of chemical elements in the gas out of which the star was formed, or subsequent accretion processes for instance due to mass transfer in close binaries. If a similar pattern occurs in stars in the same evolutionary stage, but very different locations in the galaxy, this reduces the likelihood of an initial abundance anomaly, and strengthens the case for an effect that is related to particular structural properties of such stars.

Abundance anomalies can also play an important role in allowing or inhibiting the excitation of oscillations. Examples of this are Ap stars and Am stars. Ap stars have overabundances compared to normal A stars of heavy elements, and underabundances of light elements. The same is true for Am stars, but Ap stars also show signs of stronger or more ordered magnetic fields than Am stars do which means that abundances anomalies may be concentrated in patches associated with magnetic features on the surface. Some Ap stars oscillate (roAp), but not all, and the magnetic field is sufficiently strong that it is likely to affect the eigenfunctions, in particular if the axis of the magnetic dipole does not coincide with the rotation axis of the star: oblique rotators. Models for these stars can be found e.g. in [Turcotte et al. (2000)] and [Théado et al. (2005)] and the importance of the magnetic field for modelling the pulsations is explored in [Dziembowski & Goode (1996)] for axisymmetric oscillations, and the generalisation to non-axisymmetric modes in [Bigot et al. (2000)]. Non-axisymmetric and toroidal modes of oscillation are also explored in [Reese et al. (2004)]. In [Cunha & Gough (2000)], as in [Dziembowski & Goode (1996)], a singular perturbation approach is used, following [Campbell & Papaloizou (1986)], and in [Cunha (2006)] this work is extended to magnetic field configurations with dipolar and/or quadrupolar components. One advantage of the treatments of [Cunha & Gough (2000)] is that use is made of the fact that the Lorentz force can only play a significant role for the oscillations where the magnetic pressure is of the same order as the thermal pressure, and is therefore concentrated in a thin boundary layer. This simplifies the analysis compared to [Dziembowski & Goode (1996)]. Singular perturbation theory is necessary, rather than the standard perturbation theory discussed in chapter 3, exactly because in this boundary layer the magnetic field is not a small effect as already pointed out in [Campbell & Papaloizou (1986)]. It is the small extent of the region where the magnetic field dominates, that allows for a solution of the problem, which is then used to predict the oscillation spectrum.

As mentioned in section 5.1 in the modelling of extreme horizontal branch stars abundance anomalies also play a role, but in this case well below the surface. A combination of settling from outer layers, and radiative levitation from below, can produce in these stars a build-up of iron peak elements in a layer which then allows for a κ-mechanism to excite g-modes. Models of the structure of these stars, and calculation of the properties of g-mode oscillations, can be found in [Charpinet et al. (1997)]. A more systematic theoretical survey of the oscillation properties is carried out in [Charpinet et al. (2000)].

5.2.3 Pulsation of stars

Textbooks on the theory of stellar pulsation ([Cox (1980)] and [Unno et. al (1989)]) discuss in considerable detail the LAWE and also the influence of non-linear and non-adiabatic effects on pulsation. In this book the LAWE is presented in chapter 2, and therefore in this chapter only some aspects of non-adiabatic effects and non-linear effects are discussed, where these are particularly relevant for asteroseismology. For further details the reader is referred to these textbooks.

Non-adiabatic effects are important in asteroseismology for two separate reasons. One reason is clear from the discussion in section 5.1.2. The identification of modes often relies on the ability to accurately reproduce stellar spectra and line-profiles. The emergent spectrum is determined in stellar surface layers where the plasma makes a transition from optically thick to optically thin. By definition a region where radiation can escape to be observed at infinity cannot be adiabatic, so non-adiabatic effects must be taken into account in modelling the pulsation for an accurate reproduction of the emergent spectra. The second reason for modelling non-adiabatic effects lies in the excitation mechanism of oscillations, discussed further in section 5.2.4. For stars that are pulsationally unstable there is at least one region in the star where radiative energy is absorbed in the plasma more effectively while the plasma is being compressed and released when the plasma is diluted. It is this that drives the oscillations. Where such regions are located with respect to the nodes of any given oscillation mode, determines whether a mode is excited to finite amplitudes. This implies that stars that are pulsationally unstable are not normally unstable for all possible modes of oscillation. This mode selection is therefore a separate source of information for asteroseismic inference. One example of this is the pulsating HB stars, which models show to be unstable for a limited range of g-modes, but only if iron peak elements are sufficiently enhanced in the driving layer for the oscillations. Here the observation of these modes is direct evidence that radiative levitation and gravitational settling of elements is important in some stars (as well as in the Sun).

Non-linear effects play a role in all stars which are pulsationally unstable, since otherwise the amplitude of unstable oscillations would grow without limit. The amplitude at which the oscillations are observed must therefore provide information on the non-linear mechanisms balancing the energy input from non-adiabatic effects. In some cases however, non-linear effects have a more dramatic effect on the pulsations. An example of this is beat Cepheids, which pulsate simultaneously in two radial modes. The amplitudes of these pulsations are large enough that neither can be considered as a perturbation on a mean state which is in hydrostatic equilibrium. There is an interaction between the modes. Normally in such cases explicit hydrodynamical modelling is necessary. Furthermore, although the convection in these stars is considered to be relatively inefficient, and therefore relatively unimportant for the mean state in hydrodynamical equilibrium, purely radiative hydrodynamical models for the pulsation do not reproduce the pulsational behaviour well. In [Yecko et al. (1998)] one dimensional linear non-adiabatic models are calculated and in [Kolláth et al. (1998)] and [Kolláth et al. (2002)] extended into the non-linear regime, which take turbulent convection into account and also its interaction with the pulsation, albeit with a phenomenological convection model described in [Kuhfuß (1986)]. RR Lyrae variables require much the same treatment, which is discussed in [Feuchtinger (1999a)] and [Feuchtinger (1999b)]. In AGB stars the pulsation amplitude is large enough that in the outer layers of the star the wave profiles deform into radiative shocks. Averaged over many pulsation cycles the mean stratification of these stars produces much larger density scale heights than would be the case if the star were in hydrostatic equilibrium (cf. [Fox & Wood (1985)], [Bowen (1988)], [Struck et al. (2004)]). One would expect that in this case the resonant spectrum, even for low radial overtones of pulsation, is affected by this atmospheric extension and also by the fact that the mean state is not in hydrostatic equilibrium but has a finite outflow velocity (cf. [Pijpers (1997)]).

5.2.4 Excitation

The stochastic excitation mechanism operating in the Sun to excite a broad spectrum of oscillation modes is expected to operate in all stars with surface convection zones, to excite p-modes. The same theory that is developed to model the excitation in the Sun then applies (see section 2.10). The modelling of the mechanisms by which the convection transfers energy to oscillations in most cases still involves free parameters which are calibrated by making the model reproduce the amplitudes and line-widths of the oscillations in the Sun. Hydrodynamical modelling of the convection can circumvent this problem, but has not yet been done systematically over a large range of spectral types (cf. [Stein & Nordlund (2001)] and [Stein et al. (2004)]). For g-modes a step in this direction can be found in [Dintrans et al. (2005)], in which the excitation of g-modes by penetrative convection is investigated through 2D direct simulations of compressible convection. In [Houdek et al. (1999)] one of the excitation models calibrated on the Sun (cf. [Balmforth (1992a)], [Balmforth (1992b)]) is applied to MS stars with masses between 0.9 $M_\odot$ and 2.0 $M_\odot$ to predict the acoustic power and line-widths as a function of frequency. This also must be regarded as a first step, since it is expected that the transfer of energy from convective motions to oscillations depends in detail on atmospheric flow topology and atmospheric opacities which change systematically with spectral type. This is also illustrated by the results reported in [Samadi et al. (2003a)] and [Samadi et al. (2003b)]. There, 3D numerical simulations of convection are used to place constraints on the excitation of p-modes by convection in the Sun, in which it is shown that allowing for departures from Gaussianity of the velocity correlation improves the match between acoustic power observed in the Sun, and that produced in the simulations. Surveys of the acoustic power in a range of spectral types, such as the satellite project COROT are expected to deliver, will provide valuable constraints for modelling such as this and one should expect to see rapid development in the field.

To calculate whether modes are excited or damped in stars, one can follow the same perturbative treatment discussed in sections 3.1 and 3.3, assuming that the non-adiabatic effects are small. The frequency perturbations obtained in such an analysis have a non-zero imaginary component. In this way one obtains what are known as the linear growth rates for modes. If this growth rate is greater than 0 for a mode the star is pulsationally unstable for that mode, otherwise it is stable and no oscillation in that mode is expected to be seen. Starting point is the first of Eq. (2.89). Rather than using the adiabatic Eq. (2.91) the term originally on the left-hand side of (2.77) is restored so that Eq. (2.89) contains the non-adiabatic (perturbation) part of the operator which is:

$$\delta \mathcal{F}(\delta \mathbf{r}) = \frac{i}{\rho \omega} \nabla \left[(\Gamma_3 - 1)(\rho \epsilon - \nabla \cdot \mathbf{F})' \right] . \qquad (5.55)$$

Using this in Eq. (3.23) then produces the frequency perturbation:

$$E_{nl} \delta \omega = \frac{i}{2\omega^2} \int (\Gamma_3 - 1) \frac{\delta \rho^*}{\rho} (\rho \epsilon - \nabla \cdot \mathbf{F})' . \qquad (5.56)$$

In the quasi-adiabatic approximation the adiabatic eigenfunctions are used under the integral sign. If instead the non-adiabatic eigenfunctions are used the full non-adiabatic growth rates are obtained with this same expression. Clearly in the outer layers of stars, the non-adiabatic effects are strong enough that their effect on the pulsations can be substantial. If these outer layers are convective, then the energy transport through convection and its behaviour during the pulsation becomes an important factor in assessing the stability of modes. The implication is that the precise location of the cool (red) edges of the two classical instability strips must depend on details of convective energy transport (cf. [Baker & Gough (1979)]). Recent work on this convection pulsation interaction in stars excited by the κ-mechanism can be found in [Gautschy et al. (1998)] (specifically roAp stars), [Houdek (2000)] (δ Sct stars), [Olivier & Wood (2005)] (AGB stars) and in [Grigahcène (2005)]. In [Mukadam et al. (2006)] it is shown from the pulsation properties of an ensemble of white dwarfs in the DAV instability region that near the red edge of the instability strip pulsation amplitudes are systematically reduced, also indicating that convection increasingly suppresses the mode excitation. In roAp stars the dipole magnetic field interacts with convection as well, suppressing convection at the poles. This means that the κ excitation mechanism is latitude dependent which has consequences for which modes can be excited (cf. [Balmforth et al. (2001)]). The excitation of pulsation in sdB stars is discussed in [Fontaine et al. (2003)].

5.2.5 Rotation

A primary reference for the study of stellar rotation is [Tassoul (2000)], to which the reader is referred for a comprehensive discussion. Only a few basic general results are mentioned here, to set the context for the seismology of rotating stars. Two particular issues are important: how rotation deforms a star and how the deformation and restoring forces affect the pulsation. The study of the deformation of stars has its roots in studies of figures of equilibrium for homogeneous bodies of fluid. There are three different classes:

- The MacLaurin spheroids which are oblate and range from a sphere to an infinitely flat disc for increasing angular momentum.
- The Jacobi ellipsoids which are triaxial in general although some of them are axially symmetric. They are only figures of equilibrium for an angular momentum exceeding a certain value. The latter was proven by Liouville.
- The work on figures of equilibrium was generalised by Lyapunov and Poincaré. In that work it is shown that another type of figure of equilibrium is egg-shaped. This also sometimes referred to as pear-shaped although there is no point of inflection on the surface.

For non-homogeneous, centrally condensed, figures of equilibrium similar studies have been done by Jeans, Milne, and Chandrasekhar. To illustrate the magnitude of the effect ; for the Sun the flattening due to rotation is known from various techniques (cf. [Pijpers (1998)], [Armstrong & Kuhn (1999)]). This is usually expressed quantitatively in terms of the gravitational quadrupole moment of the mass distribution which is $\sim 2 \ 10^{-7}$. This means that the Sun is very nearly a perfect sphere, for which this number would be identical to 0. The Sun is a slowly rotating star as can be seen from the kinetic energy in rotation of the Sun, expressed in units of $GM_\odot^2/R_\odot$, which is is $\sim 7 \ 10^{-7}$ ([Pijpers (1998)]). In higher mass stars in particular it is expected that both these quantities can be much closer to unity. The reason for this is that stars are likely to be formed with initially high angular momentum, much of which is then shed over their MS lifetime through a magnetised wind. More massive stars, without a surface convective layer may not have dynamo mechanisms and therefore smaller magnetic fields. A weaker magnetic field in a wind combined with a shorter MS lifetime implies that massive stars on average have higher angular momentum than low mass stars, which is also found in observational studies.

Among the early theoretical work is also the proof by Laplace that in a uniformly rotating figure of equilibrium, the level surfaces coincide with the surfaces of equal pressure (*isobaric* surfaces) and with the surfaces of equal density (*isopycnic* surfaces). The first of these theorems about surfaces follows from the equation of hydrostatic equilibrium in a co-rotating reference frame, where the centrifugal force and gravity are combined in the effective gravity. Proving the coincidence of isobaric and isopycnic surfaces is more subtle. Before discussing this it is useful to introduce a few definitions. A *barocline* or baroclinic star is a star for which a relation exists with the following form:

$$p = p(\rho, T, \lambda_1, \lambda_2, \ldots) \,. \tag{5.57}$$

Here the λ_i are unspecified parameters than can be functions of the independent coordinates. This is so general that one can safely say every star is a barocline. A *barotrope* is a star for which there is a much less general relation:

$$p = p(\rho) \,. \tag{5.58}$$

From this it is clear that all polytropes are also barotropes. One of the differences between barotropes and baroclines is that in baroclines isobaric and isopycnic surfaces do not necessarily coincide. However if in a barocline the following holds:

$$\frac{\partial \Omega}{\partial z} = 0 \tag{5.59}$$

i.e. the rotation rate Ω is constant over cylinders centered on the axis of rotation, then the isobaric and isopycnic surfaces do coincide. Stars for which this is true are called *pseudo-barotropes*. Note that it is also true that if Eq. (5.59) is **not** satisfied the isobaric and isopycnic surfaces are always inclined to each other by a finite angle. The proof of these theorems is shown in [Tassoul (2000)] and rests on the conditions under which the effective gravity can be derived from a potential. From the same reasoning also follow the Bjerknes-Rosseland rules:

- If $\partial \Omega/\partial z < 0$ at some point then the isopycnic surfaces are **more** oblate than the isobaric surfaces at that point. The reverse is also true.
- If $\partial \Omega/\partial z < 0$ everywhere from equator to pole, and the gas is chemically homogeneous and satisfies a perfect gas equation of state, then isothermal surfaces are **less** oblate than isobaric surfaces. Again the reverse is also true.

The latter rule means that the poles of a rotating star have a lower temperature than the equator if $\partial \Omega/\partial z > 0$ everywhere and vice versa.

These theorems follow from conditions of hydrostatic equilibrium. The next steps are then to investigate the equation of transport of energy, and to investigate the stability of the equilibrium solutions. The investigation of the energy equation yields a 'paradox' first formulated by von Zeipel ([von Zeipel (1924)]):

- Pseudo-barotropic models in a state of permanent rotation cannot be used to describe rotating stars in strict radiative equilibrium.

A rigorous proof of this statement is given in the appendix of [Roxburgh (1966)]. In addition to the equations of hydrostatic equilibrium the effective gravito-centrifugal potential Ψ is defined:

$$\Psi = \Phi - \int_0^\varpi \Omega^2(\varpi')\varpi' d\varpi' \qquad (5.60)$$

in which Φ is the gravitational potential and ϖ the distance from the axis. Furthermore a function $\mathcal{F}$ is defined:

$$\mathcal{F} \equiv -\frac{4ac}{3}\frac{T^3}{\kappa\rho}. \qquad (5.61)$$

For pseudo-barotropes considered here it is clear that all quantities on the right hand side can be expressed as function of the parameter Ψ. This is like a Lagrangian coordinate transform where instead of r or m, Ψ is chosen as independent variable. That this is possible follows from the general theorems discussed above. A little manipulation of these equations and the basic equations of stellar structure then yields:

$$\mathcal{F}'(\Psi)|\nabla\Psi|^2 + \mathcal{F}(\Psi)\left[4\pi G\rho - \frac{1}{\varpi}\frac{d}{d\varpi}\left(\Omega^2\varpi^2\right)\right] = \rho\varepsilon_{nucl}. \qquad (5.62)$$

If Ω is constant the derivative between the square brackets reduces to $2\Omega^2$. On surfaces of constant Ψ the derivative of $\mathcal{F}$ with respect to Ψ disappears and the result is the requirement that:

$$\varepsilon_{nucl} \propto 1 - \frac{\Omega^2}{2\pi G\rho}. \qquad (5.63)$$

Evidently the energy generation rate is given by nuclear physics and cannot have this relationship with Ω. This leads to von Zeipel's paradox. In [Roxburgh (1966)] it is shown that using $\Omega = \Omega(\varpi)$ yields essentially the same result. One possible resolution of this paradox is that Ω is a function of z also, and then the star is no longer a barotrope. This was shown to be an unstable solution. It will evolve to a state where $\partial\Omega/\partial z = 0$. The timescale for this to occur is the Kelvin-Helmholtz timescale. The other possibility is that strict radiative equilibrium breaks down. According to Eddington and Vogt the latter option causes a thermal imbalance along level surfaces (Ψ = constant) and a *meridional circulation* is set up. This is a circulation in meridional planes with an internal down-welling at the equator and an up-welling at the poles. It has also been shown (cf. [Tassoul (2000)]) that in convective layers meridional circulation can occur if the turbulent viscosity is anisotropic. For a consistent treatment of rotation in stars it is essential to introduce a finite viscosity, because otherwise the boundary conditions for the flow that make physical sense cannot be satisfied. Introduction of this viscosity leads to a boundary layer inside the star that allows a continuous transition from a meridional flow with a finite velocity component perpendicular to the boundary of the domain (the star), to a flow with a vanishing radial component **on** the boundary. It has also been argued that this meridional flow must be turbulent even in radiative layers. The cause of the turbulence lies in the fact that the meridional flow is subject to Kelvin-Helmholtz instabilities.

The investigation of (dynamical) stability yields the Solberg-Høiland criteria:

$$\frac{1}{\varpi^3}\frac{\partial j^2}{\partial \varpi} + \frac{1}{c_P}\frac{\gamma-1}{\Gamma_3-1}(-\mathbf{g}\cdot\nabla S) > 0$$
$$-g_z\left(\frac{\partial j^2}{\partial \varpi}\frac{\partial S}{\partial z} - \frac{\partial j^2}{\partial z}\frac{\partial S}{\partial \varpi}\right) > 0. \quad (5.64)$$

In these criteria S is the entropy and j is the angular momentum per unit mass. For homo-entropic stars $\nabla S \equiv 0$ and these criteria both reduce to:

$$\frac{\mathrm{d}}{\mathrm{d}\varpi}\left[(\Omega\varpi^2)^2\right] > 0 \quad (5.65)$$

which is the Rayleigh criterion for stability for an inviscid, incompressible fluid. In the limit of no rotation $j \equiv 0$ and the Solberg-Høiland criteria reduce to:

$$\mathbf{g}\cdot\nabla S < 0. \quad (5.66)$$

This implies that the entropy in a stable spherical system must increase outward.

The direct effect of rotation on the equilibrium structure of stars is particularly important for high mass stars where a decreased total luminosity due to the rapid rotation can lead to longer life times. An important point to realise is that the mixing induced by rotation cannot be very efficient because the existence of a chemical gradient stabilises against such circulation, which follows from the Ledoux stability criterion and also from arguments regarding the energy in rotation that is available for mixing. As a consequence it is probably ruled out that nuclear fuel can be substantially replenished in stellar cores due to the meridional circulation alone. For later stages of evolution where nuclear burning can occur in shells and the internal mass distribution changes, the situation is much less clear.

Even though the main sequence life times of individual low mass stars are not affected much by rotation, the age determination of clusters is affected. The reason for this is that the effective temperature determination for rotating stars depends on the incidence angle of observation. The effect is more severe for higher mass stars that generally have a higher angular momentum per unit mass. This will cause the turn-off point of the main sequence to shift to lower effective temperatures and therefore the cluster will appear older than it is. Meridional circulation also affects the surface abundances of elements such as Li and Be, which has an impact on the inferences of the Li and Be abundances that resulted from primordial nucleosynthesis. In massive stars with very low initial heavy element abundance Z, i.e. the first generation(s) of stars, there is further complex interplay between rotation, mass-loss, circulation and elemental abundances which is explored in [Maeder & Meynet (2006)] and [Meynet et al. (2006)]. Further details on evolution modelling of rotating stars can also be found in the review [Maeder & Meynet (2000)].

In single stars the boundary conditions impose constraints on the amount of angular momentum, if any, that is transferred to the external medium through a stellar wind. When modelling this, generally a thin boundary layer is created over which the external and internal flow is matched. The thickness of this layer is determined by microphysics: viscosity and radiative processes. The layer is referred to as a thermo-viscous boundary layer. In single stars this layer is passive in the sense that it only serves to match the flow in the body of the star to the boundary conditions. In binaries the flow on the surface, i.e. the boundary conditions, are imposed on the star due to the presence of a companion. This forced surface flow must then be matched by the internal flow. There are various mechanisms that force binaries into rotation that is synchronous with the orbital motion. Close binary stars will relatively rapidly synchronise but their internal velocities will take longer to dissipate which again may affect the abundances of some chemical elements in binaries. This is one of the reasons that using binary stars to represent a general population may introduce biases.

In (rapidly) rotating stars a separate set of oscillation modes is possible, which is known as r-modes, where the centrifugal and coriolis forces play a role together with pressure and gravity. The r-modes arise out of the toroidal modes mentioned in section 3.1 (see also [Aizenman & Smeyers (1977)]). For a non-rotating star these modes all have 0 frequency but in a rotating star this degeneracy is lifted and they become similar to what in the Earth's atmosphere is known as Rossby waves. The first detailed analysis of the properties of these modes in rotating stars can be found in [Papaloizou & Pringle (1978)]. To lowest order the frequencies of these r-modes in a radiative star are:

$$\omega_l = m\Omega \left(-1 + \frac{1}{l(l+1)}\right) \qquad (5.67)$$

where l and m are the usual spherical harmonic order and degree. The next higher order terms have an $1/l^2(l+1)^2$ dependence. One problem with the analysis of [Papaloizou & Pringle (1978)] is that their perturbation analysis produces singular terms where $N^2 = 0$ (cf. Eq. (2.96)) i.e. become convectively unstable. If star rotates sufficiently rapidly it becomes deformed (ellipsoidal) to such an extent that g-, p-, and r-modes are all mixed, and the asymptotic relations do not hold. However, even for slow rotation the various types of modes can interact, because of resonances in particular between r and high-order g-modes. Discussions of these effects can be found in e.g. [Smeyers et al. (1981)], [Wolff (1998)] and [Dzhalilov & Staude (2004)]. Since the frequency of r-modes is proportional to the rotation rate, their direct relevance for observations has so far been limited to either very rapidly rotating compact stars, such as neutron stars, or to variations on solar cycle time scales in the Sun (cf. [Wolff (2002)], [Knaack et al. (2005)]). Indirectly they have an important role for stars in which g-modes are observed because of the role they can play in resonant interactions. The treatment of the influence of (uniform) rotation on g-modes is discussed in [Lee & Saio (1987a)] and non-adiabatic effects are considered in [Lee & Saio (1987b)] to determine linear growth rates in order to find the extent of instability strips for g- mode pulsations. This study is extended in [Townsend (2003)] which also considers non-adiabatic effects, but specifically with the purpose of predicting the light variations due to g-modes. In all of these papers the treatment of the oscillations is done in what is called 'the traditional approximation'. In the traditional approximation of the oscillations the deformation of the star is ignored, and the oscillation equations are written in a spherical coordinate system as is done in chapter 2 with the polar axis along the rotation vector $\mathbf{\Omega}$. Written in spherical coordinates the rotation vector for an arbitrary element inside the star then has a radial and horizontal component. The contribution of the horizontal component in the balance of forces is neglected. As mentioned above, this approximation requires that $N \gg |\mathbf{\Omega}|$ and $N \gg \omega$, and therefore breaks down in convectively unstable regions. Treated in this manner the LAWE is still separable in radius, but the force terms that remain can be expressed as a coupling between spherical harmonics with the same m and the same parity about the equator: i.e. only every other l value for any given m. In principle the coupling terms involve an infinite number of terms, but in [Lee & Saio (1987a)] the coupling terms are truncated after only a few terms. An alternative is not to use an expansion at all, but instead to integrate the latitudinal equations numerically (cf. [Bildsten et al. (1996)]). A somewhat different approach, which can be used for r-modes and g-modes is the anelastic approximation (cf. [Dintrans & Rieutord (2000)]). In this approximation the pressure fluctuations are considered negligible compared to the displacement, i.e. the first term in Eq. (2.94) is dropped. An analysis of domains in the star on which the LAWE is a hyperbolic operator (implying propagating modes) or an elliptical operator (implying evanescence of modes) then reveals quite complex domains of propagation for modes. It is known that hyperbolic operators allow for a formulation in which certain quantities are conserved on characteristics. For instance certain hydrodynamical simulation codes (Riemann solvers) are based in the principle that in the flow, some quantities are conserved and propagate locally at a velocity relative to the ambient gas that is the local sound speed, and other quantities are advected with the flow itself. In [Dintrans & Rieutord (2000)] the characteristics are constructed of the LAWE operator in the domains where it is hyperbolic, which shows that these characteristics are either domain filling ('ergodic') for modes corresponding to the r-modes, or periodic for modes that correspond to the g-modes. On the basis of this in particular the r-mode frequency spectrum is expected to become very complex for rapid rotators. The absence of a clear structure will clearly make their asteroseismic use very difficult if not impossible.

An alternative approach to the calculation of frequencies is to treat the problem perturbatively in Ω as is done in section 3.3, but including higher-order terms. A second-order treatment is discussed in [Gough & Thompson

(1990)], which also considers the effect of toroidal or poloidal magnetic fields, and in [Dziembowski & Goode (1992)] and a third-order treatment in [Soufi et al. (1998)]. In all of these treatments the deformation of the star is included to the relevant order in the calculation of the frequencies. At second order, one of the effects of the rotation is that the rotationally split multiplets of oscillations are shifted collectively, compared to the first order treatment, so that the $m = 0$ oscillation frequency is not the same value as it is for a non-rotating star. In [Dziembowski & Goode (1992)] full differential rotation of the star is taken into account. In [Gough & Thompson (1990)] and [Soufi et al. (1998)] the analysis is limited to a radial dependence of Ω. In [Dziembowski & Goode (1992)] and [Soufi et al. (1998)] particular attention is also paid to the fact that rotation can bring the modes for the same m but different n or l differing by 2 into near-resonance, i.e. they become nearly degenerate. In this case the coupling between such modes causes the behaviour of the eigenfunctions to become 'mixed'. If Ω is viewed as a free parameter, an increasing Ω bringing two frequencies into and out of such a degeneracy causes an 'avoided crossing'. This phenomenon is also seen in evolutionary sequences of stars where strong μ-gradients can give rise to near degeneracies between modes of different n (but the same l) (cf. Fig. 5.4 which is taken from [Aizenman et al. (1977)]). When the two frequencies are plotted as a function of Ω, two curves are seen which appear to cross, very similar to what is show in Fig. 5.4. When the crossing point is investigated in detail one finds that the character of the two modes is exchanged so that there is no actual crossing of the curves and the lower curve appears as a continuation of the upper curve and vice versa. Especially for modes which are undergoing such avoided crossings, the location of their frequencies is strongly displaced compared to the regular comb pattern that is expected from first order theory for rotationally split multiplets.

5.3 Inverse Methods

5.3.1 *Time-series analysis as an inverse problem*

In order to obtain the frequency content of irregularly sampled time series, without performing Fourier transforms on unevenly sampled data, one can employ what is known as a Hilbert transform. In radio receiver technology this would be referred to as mixing the signal with a local oscillator, and then applying a low-pass filter. The frequency ω of the local oscillator is 'tunable': the width of a band-pass is covered, sampling regularly in frequency with some sampling step Δf. This sampling distance Δf is chosen in combination with the width of the low-pass filter subsequently applied, so that some oversampling is performed.

One proceeds as follows. For each frequency ω_k define sets of 'local oscillator' weights $q_{i,k}$, $p_{i,k}$:

$$q_{i,k} = 2\cos(\omega_k t_i)$$
$$p_{i,k} = -2\sin(\omega_k t_i) \qquad (5.68)$$

and also define a set of low-pass filter weights $z_{i,l}$:

$$z_{i,l} = \frac{1}{\Delta_{LPF}\sqrt{\pi}} \exp\left[-\left(\frac{t_i - T_l}{\Delta_{LPF}}\right)^2\right]. \qquad (5.69)$$

The central times T_l are spaced by some factor of order unity times Δ_{LPF}. Δ_{LPF} is large in order to only let through signal in a narrow frequency band, and therefore the number of times T_l is small. For a quantity y observed in a time series, one can define a cosine-weighted average $\langle y_{\cos}\rangle$ and a sine-weighted average $\langle y_{\sin}\rangle$ as follows:

$$\langle y_{\cos}\rangle_{lk} = \sum_i z_{i,l} q_{i,k} y_i$$
$$\langle y_{\sin}\rangle_{lk} = \sum_i z_{i,l} p_{i,k} y_i . \qquad (5.70)$$

In order to see what the procedure achieves, consider a time series for which:

$$y_i = A\cos(\omega t_i + \phi) . \qquad (5.71)$$

In this case the following holds:

$$\begin{aligned} q_{i,k} y_i &= 2A\cos(\omega t_i + \phi)\cos(\omega_k t_i) \\ &= A\left[\cos(\omega t_i + \phi - \omega_k t_i) + \cos(\omega t_i + \phi + \omega_k t_i)\right] \\ &= A\left[\cos\left((\omega - \omega_k)t_i + \phi\right) + \cos\left((\omega + \omega_k)t_i + \phi\right)\right] \\ &= A\cos(\phi)\left[\cos\left((\omega - \omega_k)t_i\right) + \cos\left((\omega + \omega_k)t_i\right)\right] \\ &\quad - A\sin(\phi)\left[\sin\left((\omega - \omega_k)t_i\right) + \sin\left((\omega + \omega_k)t_i\right)\right] \qquad (5.72) \end{aligned}$$

$$\begin{aligned}
p_{i,k} y_i &= -2A\cos(\omega t_i + \phi)\sin(\omega_k t_i) \\
&= A\left[\sin(\omega t_i + \phi - \omega_k t_i) - \sin(\omega t_i + \phi + \omega_k t_i)\right] \\
&= A\left[\sin\left((\omega - \omega_k)t_i + \phi\right) - \sin\left((\omega + \omega_k)t_i + \phi\right)\right] \\
&= A\cos(\phi)\left[\sin\left((\omega - \omega_k)t_i\right) - \sin\left((\omega + \omega_k)t_i\right)\right] \\
&\quad + A\sin(\phi)\left[\cos\left((\omega - \omega_k)t_i\right) - \cos\left((\omega + \omega_k)t_i\right)\right] . \quad (5.73)
\end{aligned}$$

The low-pass filtering step using the weights z_i means evaluating the following four summations:

$$\sum_i z_{i,l} \cos\left((\omega - \omega_k)t_i\right) = 1 - \epsilon \quad \text{if } \omega \approx \omega_k$$
$$\approx 0 \quad \text{otherwise}$$
$$\sum_i z_{i,l} \sin\left((\omega - \omega_k)t_i\right) = \epsilon \quad \text{if } \omega \approx \omega_k$$
$$\approx 0 \quad \text{otherwise}$$
$$\sum_i z_{i,l} \cos\left((\omega + \omega_k)t_i\right) \approx 0$$
$$\sum_i z_{i,l} \sin\left((\omega + \omega_k)t_i\right) \approx 0 . \quad (5.74)$$

The final two approximations in (5.74) are valid because $\omega + \omega_k$ is large and therefore the averaging over rapidly oscillating terms makes these disappear quickly. For $\omega = \omega_k$, $\epsilon = 0$ in the first two terms in (5.74). The behaviour of ϵ as $|\omega - \omega_k|$ increases is not uniform, due to the fact that the time series is finite in length, has gaps, and is not perfectly regularly sampled. The combination of multiplying a signal with sine or cosine weights at a fixed reference frequency and subsequently low pass filtering, produces a narrow-band pass filter around that fixed frequency, but with some sideband structure due to these effects. This is the equivalent of the window function. In order to improve on this, one can try to use the detailed profile of the low pass filter weights z. To make this explicit the weights are now ζ_i and in principle one wishes to be free to choose weights differently depending on the frequency ω_k of the local oscillator. The time T_l on which the low-pass filter is centered in Eq. (5.69) now enters the discussion by explicitly setting the reference time for the phase ϕ in Eq. (5.71), which means replacing t_i with $t_i - T_l$. With the weights ζ yet to be determined two functions of ω are defined as follows:

$$\sum_i \zeta_{i,kl} \cos\left((\omega - \omega_k)(t_i - T_l)\right) \equiv \Xi(\omega)$$

$$\sum_i \zeta_{i,kl} \sin\left((\omega - \omega_k)(t_i - T_l)\right) \equiv X(\omega) \ . \quad (5.75)$$

The method has three separate aims to achieve with its choice for the weights. These aims may well not be perfectly compatible and therefore require a form of compromise:

- The first function $\Xi(\omega)$ needs to be a function peaked at $\omega = \omega_k$ and smoothly dropping away, with as little side-lobe structure as possible. For instance:

$$\Xi(\omega) \approx \frac{1}{\sqrt{\pi}\Delta_\omega} \exp\left[-\left(\frac{\omega - \omega_k}{\Delta_\omega}\right)^2\right] \quad (5.76)$$

with as small a width Δ_ω as achievable.
- The second function $X(\omega)$ needs to be as close to 0 as possible everywhere
- The errors in the resulting amplitudes A^2 need to be kept as small as possible.

In the framework of the SOLA method (cf. section 3.5.2) the best compromise between these three aims can be achieved by minimising for $\zeta_{i,kl}$ the quantity $\mathcal{W}(\zeta)$ defined by:

$$\mathcal{W}(\zeta) \equiv (1-\mu) \int_{\omega_{\text{low}}}^{\omega_{\text{high}}} d\omega \ [\Xi(\omega) - \mathcal{T}(\omega)]^2 + \mu \int_{\omega_{\text{low}}}^{\omega_{\text{high}}} d\omega \ [X(\omega)]^2$$

$$+ \lambda \sum_{i,j} \zeta_{i,kl} \zeta_{j,kl} \Sigma_{ij} \quad (5.77)$$

while taking into account the normalisation constraint:

$$\sum_i \zeta_{i,kl} \int d\omega \ \cos\left((\omega - \omega_k)(t_i - T_l)\right) \equiv 1 \ . \quad (5.78)$$

Here the target function $\mathcal{T}(\omega)$ is taken to be a Gaussian as in (5.76). The 'target function' for $X(\omega)$ is 0. The matrix Σ_{ij} is the variance-covariance matrix of the data errors. It is usual to assume that the errors on the data are Gaussian distributed and independent, so that Σ is a diagonal matrix, but this is not essential for the algorithm. The methods described in section 5.1.3 to estimate or model error properties from residuals using ARIMA models can be applied here as well. The parameters $\mu \in [0, 1]$ and λ weight the relative importance of the three terms. Their value needs to be set by trial and error. A very low choice of λ normally leads to unacceptably high errors on the inferred A^2, whereas a very high value generally produces the undesirable effect of increasing width or structure in Ξ. Similar arguments hold for μ. A point that is crucial for this time series analysis version of the SOLA algorithm, is that ω_{low} and ω_{high} must be finite. This is necessary because the integrations in Eq. (5.77) are over products of cosines, which causes problems if they are to be integrated over an infinite range. The algorithm is therefore limited in application to bandwidth limited data. In practice this is not a severe restriction for its application in asteroseismology.

Carrying out the minimisation of (5.77) for the coefficients ζ_i reduces to solving a set of linear equations $\mathbf{A} \cdot \zeta = b$. The symmetric matrix $\mathbf{A}$ is the sum of three components, $\mathbf{A} = (1 - \mu)\mathbf{C} + \mu\mathbf{S} + \lambda\mathbf{\Sigma}$ in which:

$$C_{ij} = \int_{\omega_{\text{low}}}^{\omega_{\text{high}}} d\omega \; \cos\left((\omega - \omega_k)(t_i - T_l)\right) \cos\left((\omega - \omega_k)(t_j - T_l)\right)$$

$$S_{ij} = \int_{\omega_{\text{low}}}^{\omega_{\text{high}}} d\omega \; \sin\left((\omega - \omega_k)(t_i - T_l)\right) \sin\left((\omega - \omega_k)(t_j - T_l)\right) \; . \quad (5.79)$$

The vector b is given by:

$$b_i = (1 - \mu) \int_{\omega_{\text{low}}}^{\omega_{\text{high}}} d\omega \; \cos\left((\omega - \omega_k)(t_i - T_l)\right) \mathcal{T}(\omega) \; . \quad (5.80)$$

The integrals in Eq. (5.79) can be carried out analytically. The limits of integration are set to $\omega_{\text{low}} = \omega_c - \Delta_\omega \text{ band}$ and $\omega_{\text{high}} = \omega_c + \Delta_\omega \text{ band}$. The matrix elements C_{ij} can then be reduced to:

$$C_{ij} = \int_{\omega_{\text{low}}}^{\omega_{\text{high}}} d\omega \; \cos\left((\omega - \omega_k)(t_i - T_l)\right) \cos\left((\omega - \omega_k)(t_j - T_l)\right)$$

$$= \frac{1}{2} \int_{\omega_{\text{low}}}^{\omega_{\text{high}}} d\omega \cos\left((\omega - \omega_k)(t_i + t_j - 2T_l)\right) + \cos\left((\omega - \omega_k)(t_i - t_j)\right)$$

for $i \neq j$
$$= \frac{[\sin((\omega_{\text{high}} - \omega_k)(t_i + t_j - 2T_l)) + \sin((\omega_k - \omega_{\text{low}})(t_i + t_j - 2T_l))]}{2(t_i + t_j - 2T_l)}$$
$$+ \frac{[\sin((\omega_{\text{high}} - \omega_k)(t_i - t_j)) + \sin((\omega_k - \omega_{\text{low}})(t_i - t_j))]}{2(t_i - t_j)}$$
$$= \frac{\sin(\Delta_{\omega\,\text{band}}(t_i + t_j - 2T_l))}{(t_i + t_j - 2T_l)} \cos((\omega_c - \omega_k)(t_i + t_j - 2T_l))$$
$$+ \frac{\sin(\Delta_{\omega\,\text{band}}(t_i - t_j))}{(t_i - t_j)} \cos((\omega_c - \omega_k)(t_i - t_j))$$

or if $i = j$
$$= \frac{\sin(2\Delta_{\omega\,\text{band}}(t_i - T_l))}{2(t_i - T_l)} \cos(2(\omega_c - \omega_k)(t_i - T_l)) + \Delta_{\omega\,\text{band}}\ . \quad (5.81)$$

Similarly the matrix elements S_{ij} can be reduced to:

$$S_{ij} = \int_{\omega_{\text{low}}}^{\omega_{\text{high}}} d\omega \sin((\omega - \omega_k)(t_i - T_l))\sin((\omega - \omega_k)(t_j - T_l))$$

for $i \neq j$
$$= -\frac{[\sin((\omega_{\text{high}} - \omega_k)(t_i + t_j - 2T_l)) + \sin((\omega_k - \omega_{\text{low}})(t_i + t_j - 2T_l))]}{2(t_i + t_j - 2T_l)}$$
$$+ \frac{[\sin((\omega_{\text{high}} - \omega_k)(t_i - t_j)) + \sin((\omega_k - \omega_{\text{low}})(t_i - t_j))]}{2(t_i - t_j)}$$
$$= -\frac{\sin(\Delta_{\omega\,\text{band}}(t_i + t_j - 2T_l))}{(t_i + t_j - 2T_l)} \cos((\omega_c - \omega_k)(t_i + t_j - 2T_l))$$
$$+ \frac{\sin(\Delta_{\omega\,\text{band}}(t_i - t_j))}{(t_i - t_j)} \cos((\omega_c - \omega_k)(t_i - t_j))$$

or if $i = j$
$$= -\frac{\sin(2\Delta_{\omega\,\text{band}}(t_i - T_l))}{2(t_i - T_l)} \cos(2(\omega_c - \omega_k)(t_i - T_l)) + \Delta_{\omega\,\text{band}}\ . \quad (5.82)$$

Combining Eqs. (5.81) and (5.82) gives expressions for the diagonal terms of **A**:

$$A_{ii} = (1 - 2\mu)\frac{\sin(2\Delta_\omega(t_i - T_l))}{2(t_i - T_l)} \cos(2(\omega_c - \omega_k)(t_i - T_l))$$
$$+ \Delta_{\omega\,\text{band}} + \lambda \Sigma_{ii} \quad (5.83)$$

and for the off-diagonal terms $i \neq j$:

$$\begin{aligned}A_{ij} &= A_{ji} \\ &= (1-2\mu)\frac{\sin\left(\Delta_{\omega\,\text{band}}(t_i + t_j - 2T_l)\right)}{(t_i + t_j - 2T_l)}\cos\left((\omega_c - \omega_k)(t_i + t_j - 2T_l)\right) \\ &\quad + \sin\left(\Delta_{\omega\,\text{band}}(t_i - t_j)\right)(t_i - t_j)\cos\left((\omega_c - \omega_k)(t_i - t_j)\right) + \lambda\Sigma_{ij}\,.\end{aligned} \quad (5.84)$$

From Eqs. (5.83) and (5.84) it is clear that there is a dependence of **A** on the choice of reference time T_l. One can eliminate any dependence of **A** on T_l by setting the weighting factor $\mu = 1/2$, i.e. by placing equal importance on reproducing the Ξ and X functions defined in (5.75) through the minimisation of (5.77). A priori there is no reason to consider the reproduction of one of the two target functions more important than the other, so that setting $\mu = 1/2$ is not unreasonable. In the limit of infinitely high frequency resolution $\Delta_\omega \downarrow 0$ the vector b has no dependence on reference time T_l. For a finite number N of samples in the time series, this is not realistic however. One would expect the best achievable frequency resolution to be proportional to the total duration of the sampling: $\Delta_f \propto 1/(t_N - t_1)$ where the proportionality constant is of order unity and depends on the noise level as well, as is also suggested by the discussion in section 5.1.3.

Once the vector and matrix elements are computed, it is a straightforward numerical matter to determine the weights $\zeta_{i,kl}$. Replacing in Eq. (5.70) the weights z with the weights ζ, but otherwise following the same procedure as before then yields oscillation spectra at a resolution that is the best compromise between reducing sidelobes, limiting data error propagation, and resolving fine structure. An example of this method as applied to a realistic set of sampling times is shown in Fig. 5.5. Note that the method is linear, and that therefore the propagation of the measurement errors is straightforward. It does carry a significant computational burden however, since for every trial frequency ω_k a matrix factorisation is to be carried out. It is therefore worthwhile combining this method with a crude period search, so that one can reduce the number of trial frequencies.

Finally it is useful to point out that this method has elements in common with the method described in [Grison (1994)] (see section 5.1.3) which constructs an orthogonal base set taking the, irregularly distributed, sampling times into account. If an SVD is carried out on the matrix **A** above, the eigenvectors associated with the singular values describe an orthogonal set of functions. In this case the orthogonal functions are constructed such that not only the irregular sampling, but also the behaviour of the errors is considered.

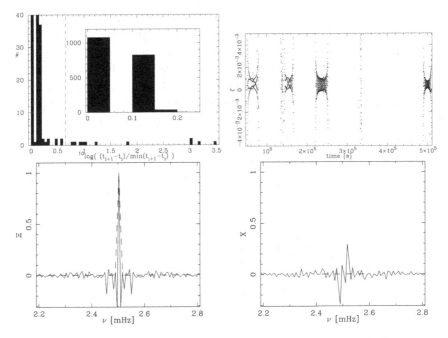

Fig. 5.5 An example of the SOLA spectral reconstruction method, applied to a sampling, which is almost regular over blocks of time, with large gaps in between. Upper right-hand panel: a histogram of the time intervals between successive measurements, normalised to the minimum value with the inset panel a blow-up of the left-most bins. Top right-hand panel, the coefficients ζ determined for $\mu = 1/2$, $T_l \sim 3.23$ d, $\omega_k/2\pi = 2.5$ mHz, $\Delta_\omega/2\pi = 0.69$ μHz. Lower left and right hand panels are the window functions Ξ and X as defined in the text.

5.3.2 Linear and global methods

Once a usable estimate of the stellar mass has been obtained one can construct a model for the star using the best available atomic data required to construct stellar models: the equation of state of the plasma, the nuclear reaction rates, and the opacity (all as a function of composition, density and temperature). Usually a model that has the correct effective temperature and luminosity can be obtained by evolving it from the zero-age main sequence, where the star is assumed to be fully mixed with cosmic abundances of the elements. In this case one may start with the assumption that the oscillation frequencies predicted for that model are quite close to the observed values and that the difference between observed and model frequencies is a good **linear** measure of differences in the sound speed: a reference model is obtained.

In practice it may not always be possible to uniquely identify modes from spectroscopy or photometry. In particular for the stochastically excited pulsations the amplitudes are small, and the signal-to-noise ratio of individual line profiles does not allow for the techniques discussed in section 5.1.2. The asymptotic theory of stellar oscillations may help in such cases to identify modes with sufficient precision to then construct appropriate detailed stellar models. From chapter 2 the asymptotic prediction for frequencies of p-modes Eq. (2.150) is repeated:

$$\nu_{nl} \approx (n + \frac{l}{2} + \frac{1}{4} + \alpha)\Delta\nu - (Al(l+1) - \beta)\frac{\Delta\nu^2}{\nu_{nl}} \quad (5.85)$$

where:

$$\Delta\nu = \left[2\int_0^R \frac{\mathrm{d}r}{c}\right]^{-1}$$

$$A = \frac{1}{4\pi^2 \Delta\nu}\left[\frac{c(R)}{R} - \int_0^R \frac{\mathrm{d}c}{\mathrm{d}r}\frac{\mathrm{d}r}{r}\right]. \quad (5.86)$$

The phases α and β are constants which express various inaccuracies in the approximations that go into deriving this asymptotic expression, in particular at the outer boundaries or the stellar centre (cf. section 2.9). From this follow the standard definitions of the large and small separations:

$$\Delta_l(n) \equiv \nu_{n\,l} - \nu_{n-1\,l}$$
$$d_{l\,l+2}(n) \equiv \nu_{n\,l} - \nu_{n-1\,l+2} \quad (5.87)$$

which have a (weak) dependence on n, so that in practice sometimes averages, indicated by $\langle\rangle$, over a range of n are used. If a sufficiently large number of frequencies are detected, the large and small separations can be determined even without having a definitive identification of all the modes. The two parameters are plotted for models over a range of masses and evolutionary stage on the main sequence. A diagram such as this was first presented in [Christensen-Dalsgaard (1993)]. From this it can be seen that a measurement of these two parameters can constrain the mass and central H abundance, as long as the heavy element abundances are known. For this reason such a diagram is sometimes referred to as a seismic HR diagram. It seems reasonable therefore to include at least the large separation as an additional constraint that a reference model has to satisfy, in addition

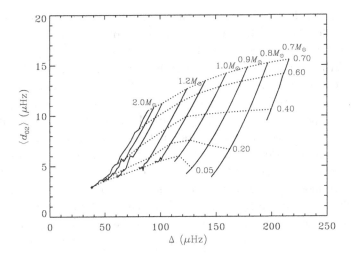

Fig. 5.6 The large and small separations between oscillation frequencies for a range of stellar masses (solid lines) and evolution stages, as measured by the central H abundance (dashed lines) (figure courtesy J. Christensen-Dalsgaard).

to L, $T_{\rm eff}$ and/or R. In order to avoid biases, one should use the same operational definition of a large separation both for the data and the models. This is to say that if a large separation is determined from the data by averaging $2(\nu_{n\,l+1} - \nu_{n\,l})$ or $\nu_{n+1,l} - \nu_{n,l}$ over a range of frequencies, it should be determined by averaging over the same range in frequency from the models.

Much of the discussion of this section concentrates on p-modes, which is most appropriate for the solar-like stars which excite large numbers of modes due to stochastic excitation by surface convection zones, However, a similar treatment for the frequencies of g-modes it is also possible, as long as care is taken in using using oscillation periods rather than frequencies, since instead of being nearly uniformly spaced in frequency, the frequencies of g-modes are nearly uniformly spaced in period.

The task of obtaining appropriate reference models, i.e. global model fitting, can in itself carry a considerable computational burden. Traditional use of stellar evolution codes is to run models with a fixed mass and composition from the ZAMS forward in time. A grid can be built up varying the input parameters, which can take up substantial amounts of computing time and memory since the number of separate parameters to vary is large. In order to obtain an appropriate reference model for a star with measured L and $T_{\rm eff}$ one then conducts a grid search. The grid search in itself is quite fast, but it does not guarantee that a model thus obtained is globally the best model that can be found. An example of this is pulsating white dwarf stars, where the grid search based models turned out to be too limited. A method based on a genetic algorithm provided different solutions for the internal structure of these white dwarfs which had not been considered in creating the original grid, but which were physically acceptable solutions (cf. [Metcalfe et al. (2003)] [Metcalfe et al. (2003)] [Metcalfe et al. (2004)]). In genetic algorithms a population of models is evolved. From one generation of models to the next, those that match observations most closely are given a higher likelihood to be present in the next generation, with slightly modified parameters, and the more poorly matching models are less likely to be retained. Also, a certain amount of mutation is allowed for from one generation to the next: i.e. a substantial change in one or more parameters. The evolution part of the algorithm is to provide convergence of parameters to an optimal set of values. The mutation part is to maximise the likelihood of finding the globally best fitting set, rather than a local minimum in terms of model misfit. Although clearly successful in this case, some cautionary remarks are in order. A genetic algorithm is a brute force method in the sense that it requires calculating/evolving large numbers of models, several 10^6 is a typical number used. For the white dwarf models in question the computational burden per model is very small since their stratification is relatively simple and well captured even by polytropes. For more sophisticated models this is an important issue. There are gains compared to a comprehensive grid-based approach, but it does remain worthwhile to restrict the search space as much as possible before starting computing. The second problem with genetic algorithms is more fundamental. One can demonstrate that close to minima of model misfit the mutation part of genetic algorithms drives the population away from the minimum faster than the evolution drives the population towards the minimum. Full convergence is therefore either slow or absent. More traditional minimisation algorithms such as the Levenberg-Marquardt method (cf. [Press et al. (1992)]) have much better convergence properties. These methods converge very fast and therefore require orders of magnitude fewer model calculations. The problem with these methods is that they are not guaranteed to find a global minimum. Only by restarting the method for a number of substantially different parameter choices can one be reasonably confident that a global minimum has been found. Ideally therefore one would envisage using a combination of methods. An initial genetic algorithm based search to identify promising parameter subspaces, and a

Levenberg-Marquardt type method to find the true minimum in each of these subspaces.

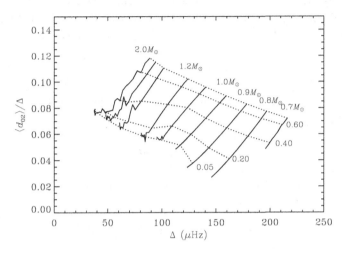

Fig. 5.7 The equivalent of the seismic HR diagram, using the scaled small separation (figure courtesy J. Christensen-Dalsgaard).

Once a reference model is found an asteroseismic analysis of the frequencies can be focussed on departures from this uniform spacing. One way to do this, is to divide frequencies, and frequency differences, by a reference large separation. There is another reason to do this which is related to the scaling of frequencies by the mean density, discussed in section 5.2. The large separation scales with the mean density in the same way that the frequencies do. By taking the ratio of these two quantities, to first order the mean density dependence is removed. It is shown in [Roxburgh & Vorontsov (2003)], [Roxburgh (2005)] and [Otí Floranes et al. (2005)] that this can greatly facilitate asteroseismic inference of frequency differences. In Fig. 5.6 the ratio of the mean small separation to the mean large separation $\langle d_{02}\rangle/\Delta$ is shown, which corresponds to Fig. 5.5. The most important reason for the scaled differences to be better measures of interior structure differences between stars and models, is that the influence of surface effects is eliminated to lowest order. This is demonstrated by an appropriate asymptotic treatment of the LAWE which can be found in [Roxburgh & Vorontsov (2000a)], [Roxburgh & Vorontsov (2000b)] and [Roxburgh & Vorontsov (2001)], which lay the theoretical foundation for methods that use the ratio of (linear combinations of) frequencies with the large separation.

From the variational kernels for frequency differences the same reduction of sensitivity to surface effects can be seen, that the asymptotic theory shows. In the notation of [Roxburgh & Vorontsov (2003)] and [Otí Floranes et al. (2005)] the scaled versions r of the small separations d are:

$$r_{l\,l+2}(n) \equiv \frac{d_{l\,l+2}(n)}{\Delta_{l+1}(n)} \; . \tag{5.88}$$

The frequency differences are related to structural differences $\delta \mathbf{Q}$ through (cf. Eq. (3.2)):

$$\delta \nu_{n\,l} = \int \mathbf{K}_{nl} \cdot \delta \mathbf{Q} \tag{5.89}$$

and therefore differences in the small separation between a star and a reference model are:

$$\delta d_{l\,l+2}(n) = \int [\mathbf{K}_{n\,l} - \mathbf{K}_{n-1\,l+2}] \cdot \delta \mathbf{Q} \; . \tag{5.90}$$

In the near-surface layers the eigenfunctions, and therefore also the kernels, are functions of ν only, and independent of l (cf. section 3.1.2) so that:

$$\mathbf{K}_d \equiv \mathbf{K}_{n\,l} - \mathbf{K}_{n-1\,l+2} \approx d_{l\,l+2}(n)\frac{\partial \mathbf{K}_{n\,l}}{\partial \nu} \tag{5.91}$$

which follows if the kernels $\mathbf{K}$ are written as functions of frequency and a first-order Taylor expansion in frequency is used. Similarly a kernel for the large separation satisfies in the near-surface layers:

$$\mathbf{K}_\Delta \equiv \mathbf{K}_{n\,l} - \mathbf{K}_{n-1\,l} \approx \Delta_l(n)\frac{\partial \mathbf{K}_{n\,l}}{\partial \nu} \; . \tag{5.92}$$

A small difference between star and model in the ratio r is related to small differences in d and Δ, using linearised perturbations:

$$\delta r_{l\,l+2}(n) = \frac{1}{\Delta_{l+1}(n)} \left[\delta d_{l\,l+2}(n) - r_{l\,l+2}(n)\delta\Delta_{l+1}(n) \right] \tag{5.93}$$

and the corresponding expression for the kernel is then:

$$\mathbf{K}_r \equiv \frac{1}{\Delta_{l+1}(n)} \left[\mathbf{K}_d - r_{l\,l+2}(n)\mathbf{K}_\Delta \right] \; . \tag{5.94}$$

As is shown in [Otí Floranes et al. (2005)], introduction of the Taylor series expansions valid in the near-surface layers, of Eqs. (5.91) and (5.92), clearly leads to cancellation of the leading order term. A plot of the sound speed

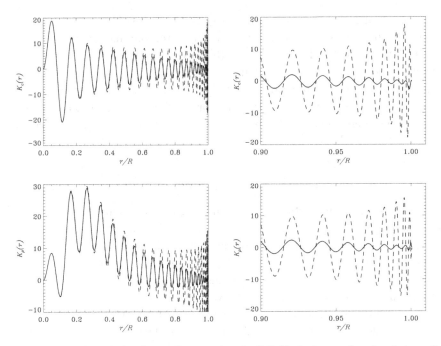

Fig. 5.8 The kernels for the small separation $d_{02}(23)$ (dashed curves) and scaled small separation $r_{02}(23)$ (solid curves). Top panels are the sound speed kernels, bottom panels the density kernels. The right-hand panels are blown up to show the near-surface structure in more detail (figure courtesy M. Thompson).

and density kernels using the full variational kernels shown in Fig. 5.8 also confirms this behaviour. Compare also with the variational kernel of Fig. 3.1 for the frequency differences, which is even more strongly weighted towards the surface.

Apart from eliminating surface effects from inversions the scaling can also be necessary to compare models more appropriately. One example of this is discussed in [Christensen-Dalsgaard & Thompson (1999)]. When comparing models with different physics one can compare models that have, for instance, the same values of dependent parameters at $r = 0$, or one can compare different models with the same surface values. Arguably the latter is more justified observationally, for instance in the case considered in [Christensen-Dalsgaard & Thompson (1999)] when comparing models and frequencies for rapidly rotating stars with those of non-rotating stars. Also worth mentioning in this context is that the linearisation inherent in using the above kernels, derived in section 3.1.1, can be circumvented. If one uses the variational principle in the form of Eq. (3.16) for two different models, one can also express the difference between corresponding frequencies in terms of the right-hand sides of Eq. (3.16) calculated for each of those models. In particular if the frequencies differences are scaled and surface terms are neglected, such expressions can be used to explore how 'close' a reference model needs to be in structure, for the linearised kernels to be an accurate representation of the sensitivity to structural differences between a star and a model.

5.3.3 Detecting discontinuities and kinks

In section 5.2.2 the importance of determining the He abundance in stars is discussed. The asteroseismic measurement of this quantity therefore deserves particular attention. One method makes use of the fact that the behaviour of the adiabatic exponent Γ_1 is sensitive to the He content of the plasma in the regions where He changes its dominant ionisation stage. This occurs in two regions that are relatively thin in spatial extent, certainly compared to the length scale of the oscillation modes that are accessible to asteroseismology. The region in which He changes from being primarily in its neutral stage to being singly ionised coincides with the ionisation region for H, which is inconvenient for determining the He abundance. One therefore concentrates on the layer in which He changes from singly to doubly ionised. Because the region is a thin shell, it is useful to use an inverse technique that makes use of this fact. The technique is an adaptation of one that is originally developed to locate the bottom of the solar convection zone, and detect the presence of overshoot, if any.

Discontinuities in c or in its derivatives, or more generally any phenomenon that gives rise to discontinuities in the acoustic potential of the wave equation, have a characteristic effect on the oscillation frequencies. Compared to smooth models, models that are discontinuous or that have a discontinuity in a higher derivative, shift frequencies by an amount that is a periodic function of the frequency. The details of this demonstration can be found in [Monteiro et al. (1994)] and [Christensen-Dalsgaard et al. (1995)]. The application of this technique to stars can be found in [Monteiro et al. (2000)]. The variation of this method, aimed at measuring the abundance of Helium in stars, through the effect that this has on the ionisation balance very near the surface, can be found in [Monteiro & Thompson (2005)].

Before discussing that method in detail, it is useful to consider a simple toy model as an illustration of the idea. The toy model is not in itself very realistic but it does show the same oscillatory behaviour of the frequencies that the real models do. The full method is then discussed in the next section.

5.3.3.1 Example: Explicit calculations for a crude model

The LAWE in the Cowling approximation Eq. (2.111) with Eq. (2.129) is the starting point for this analysis. Nevertheless it is still quite difficult to solve, so some further approximations are applied to obtain a tractable problem. It is assumed that the modes under consideration are in a sufficiently high frequency regime that $N/\omega \ll 1$. The acoustic cutoff frequency is also set to 0. Setting $N = 0$ is not a bad approximation for p-modes as long as the model is reasonably homogeneous in composition which implies that it should not be very evolved. Setting $\omega_c = 0$ is not a bad approximation anywhere in a star except near the surface. The assumption is that this can be considered to influence the oscillations only in the sense that it changes the boundary conditions as discussed also in sects. 2.9 and 3.1.2, and that it therefore causes a nearly constant shift in the frequencies. Eq. (2.111) then reduces to:

$$\frac{\partial^2 X}{\partial r^2} + \left[\left(\frac{\omega^2}{c(r)^2}\right) - \frac{l(l+1)}{r^2}\right] X = 0 \qquad (5.95)$$

which is nearly the same as the homogeneous sphere, apart from having a sound speed $c(r)$ that depends on the radius. It is useful to use as independent variable the acoustic radius τ similar to what is used in section 2.10:

$$\tau \equiv \int_0^r dr' \, \frac{1}{c(r')} \, . \qquad (5.96)$$

With this definition the derivatives with respect to r become:

$$\frac{\partial}{\partial r} = \frac{1}{c} \frac{\partial}{\partial \tau}$$
$$\frac{\partial^2}{\partial r^2} = -\frac{1}{c^3} \frac{\partial c}{\partial \tau} \frac{\partial}{\partial \tau} + \frac{1}{c^2} \frac{\partial^2}{\partial \tau^2} \, . \qquad (5.97)$$

Using this in Eq. (5.95) and then multiplying by c^2 produces:

$$\frac{\partial^2}{\partial \tau^2} X - \frac{1}{c} \frac{\partial c}{\partial \tau} \frac{\partial X}{\partial \tau} + \left[\omega^2 - \frac{l(l+1)}{\tau^2} \left(\frac{c\tau}{r}\right)^2\right] X = 0 \, . \qquad (5.98)$$

It is preferable to remove the first derivative term again which can be done by defining $Z \equiv c^{-1/2} X$ and substituting, which leads to:

$$\frac{\partial^2}{\partial \tau^2} Z + Z \left[\omega^2 - \frac{l(l+1)}{\tau^2} \left(\frac{c\tau}{r} \right)^2 + \frac{1}{2} \left(\frac{1}{c} \frac{\partial^2 c}{\partial \tau^2} - \frac{3}{2} \left(\frac{1}{c} \frac{\partial c}{\partial \tau} \right)^2 \right) \right] = 0 \,. \quad (5.99)$$

The next step is to use a power law for the sound speed as a function of radius which is evidently not realistic, for one because it diverges for $r = 0$, but it suffices for the purpose at hand:

$$c(r) = c_1 \left(\frac{r}{R} \right)^\alpha \quad (5.100)$$

with $\alpha < 0$ and c_1 is the sound speed at the surface. A more realistic choice would have been:

$$c(r) = \frac{c_0}{1 + \beta \left(\frac{r}{R} \right)^{-\alpha}} \quad (5.101)$$

with c_0 the sound speed in the centre. This is approximately the same as 5.100 if $\beta = c_0/c_1 - 1$ is large enough, except in the stellar core. The effect of using Eq. (5.100) is not very severe since the higher frequency modes have turning points well outside the core region and therefore behave much the same between these two profiles.

With the choice (5.100) for the sound speed it is possible to calculate the functions entering equation (5.99):

$$\begin{aligned}
\tau &= \frac{R}{c_1} \frac{1}{1 - \alpha} \left(\frac{r}{R} \right)^{1-\alpha} \\
\left(\frac{c\tau}{r} \right)^2 &= \frac{1}{(1-\alpha)^2} \\
\frac{\partial c}{\partial r} &= \frac{\alpha c_1}{R} \left(\frac{r}{R} \right)^{\alpha-1} \\
\frac{\partial^2 c}{\partial r^2} &= \frac{\alpha(\alpha-1) c_1}{R^2} \left(\frac{r}{R} \right)^{\alpha-2} \\
\left(\frac{1}{c} \right) \frac{\partial^2 c}{\partial \tau^2} - \frac{3}{2} \left(\frac{1}{c} \frac{\partial c}{\partial \tau} \right)^2 &= c \frac{\partial^2 c}{\partial r^2} - \frac{1}{2} \left(\frac{\partial c}{\partial r} \right)^2 \\
&= \frac{\alpha \left(\frac{1}{2}\alpha - 1 \right)}{(1-\alpha)^2} \frac{1}{\tau^2}
\end{aligned} \quad (5.102)$$

and that means that after substituting these results, Eq. (5.99) becomes:

$$\frac{\partial^2}{\partial \tau^2} Z + Z \left[\omega^2 - \frac{l(l+1) - \frac{1}{2}\alpha \left(\frac{1}{2}\alpha - 1\right)}{(1-\alpha)^2} \frac{1}{\tau^2} \right] = 0 \, . \tag{5.103}$$

At this point it is useful to define a parameter μ:

$$\mu^2 \equiv \frac{l(l+1) - \frac{1}{2}\alpha \left(\frac{1}{2}\alpha - 1\right)}{(1-\alpha)^2} + \frac{1}{4} = \left(\frac{l + \frac{1}{2}}{1-\alpha}\right)^2 \tag{5.104}$$

because with this definition Eq. (5.103) becomes:

$$\frac{\partial^2}{\partial \tau^2} Z + Z \left[\omega^2 - \frac{\mu^2 - \frac{1}{4}}{\tau^2} \right] = 0 \, . \tag{5.105}$$

The solutions of (5.105) can be expressed in terms of Bessel functions of non-integer order:

$$Z_{nl} = \tau^{1/2} J_\mu(\omega_{nl} \tau) \tag{5.106}$$

where the eigenfrequencies ω_{nl} are obtained by matching the zeros of the $Z_{n,l}$ to $r = R$. Recalling Eq. (2.130) and the definition of Z, setting $Z = 0$ corresponds to:

$$c^{3/2} \rho^{1/2} \nabla \cdot \delta \mathbf{r} = 0 \tag{5.107}$$

which is the appropriate free boundary condition for the displacement $\delta \mathbf{r}$. In a more realistic model the outer boundary condition would be modified because of the term with the cut-off frequency which is ignored here. Evidently it is possible to give asymptotic relations for frequencies using this boundary condition but this is not relevant for the discussion at hand.

In order to find the effect of small departures from Eq. (5.100) of the sound speed on the frequencies it is necessary to explicitly calculate the steps described formally in section 3.1. First of all the operator $\mathcal{F}$ is obtained by rearranging Eq. (5.99):

$$\mathcal{F} Z = \left\{ -\frac{\partial^2}{\partial \tau^2} + \left[\frac{l(l+1)}{\tau^2} \left(\frac{c\tau}{r}\right)^2 - \frac{1}{2}\left(\frac{1}{c}\frac{\partial^2 c}{\partial \tau^2} - \frac{3}{2}\left(\frac{1}{c}\frac{\partial c}{\partial \tau}\right)^2\right) \right] \right\} Z \, . \tag{5.108}$$

If the equilibrium model has a sound speed $c_0 + \delta c$ then that must be substituted in Eq. (5.108) to obtain $\delta \mathcal{F}$.

Consider first the term $c\tau/r$, where one must recall that τ is the independent variable and not r. At a given τ a perturbation of the sound speed

δc causes to first order a proportional perturbation of the radius δr so that:

$$\frac{\delta c}{c_0} = \frac{\delta r}{r_0} \qquad (5.109)$$

which can be seen by substituting $c_0 + \delta c$ and $r_0 + \delta r$ into the definition for τ Eq. (5.96). This means that to first order, the perturbation terms of $c\tau/r$ cancel. The only terms in Eq. (5.108) that produce non-zero contributions in a first order perturbation, are in the derivative of the sound speed. It is useful at this point to rewrite these terms slightly:

$$-\frac{1}{2}\left(\frac{1}{c}\frac{\partial^2 c}{\partial \tau^2} - \frac{3}{2}\left(\frac{1}{c}\frac{\partial c}{\partial \tau}\right)^2\right) = c^{\frac{1}{2}}\frac{\partial^2}{\partial \tau^2}c^{-\frac{1}{2}}. \qquad (5.110)$$

Now it is straightforward to substitute $c_0 + \delta c$ and develop to first order:

$$c_0^{\frac{1}{2}}\left(1 + \frac{1}{2}\frac{\delta c}{c_0}\right)\frac{\partial^2}{\partial \tau^2}c_0^{-\frac{1}{2}}\left(1 - \frac{1}{2}\frac{\delta c}{c_0}\right)$$

$$= c_0^{\frac{1}{2}}\left(1 + \frac{1}{2}\frac{\delta c}{c_0}\right)\left[\left(1 - \frac{1}{2}\frac{\delta c}{c_0}\right)\frac{\partial^2 c_0^{-\frac{1}{2}}}{\partial \tau^2} - \frac{\partial c_0^{-\frac{1}{2}}}{\partial \tau}\frac{\partial}{\partial \tau}\left(\frac{\delta c}{c_0}\right) - \frac{1}{2}c_0^{-\frac{1}{2}}\frac{\partial^2}{\partial \tau^2}\left(\frac{\delta c}{c_0}\right)\right]$$

$$= c_0^{\frac{1}{2}}\frac{\partial^2}{\partial \tau^2}c_0^{-\frac{1}{2}} - c_0^{\frac{1}{2}}\frac{\partial}{\partial \tau}c_0^{-\frac{1}{2}}\frac{\partial}{\partial \tau}\left(\frac{\delta c}{c_0}\right) - \frac{1}{2}\frac{\partial^2}{\partial \tau^2}\left(\frac{\delta c}{c_0}\right). \qquad (5.111)$$

With these results the perturbed part $\delta \mathcal{F}$ of the operator is:

$$\delta \mathcal{F} = -c_0^{\frac{1}{2}}\frac{\partial}{\partial \tau}c_0^{-\frac{1}{2}}\frac{\partial}{\partial \tau}\left(\frac{\delta c}{c_0}\right) - \frac{1}{2}\frac{\partial^2}{\partial \tau^2}\left(\frac{\delta c}{c_0}\right)$$

$$= \frac{1}{2}\frac{\partial c_0}{\partial r}\frac{\partial}{\partial \tau}\left(\frac{\delta c}{c_0}\right) - \frac{1}{2}\frac{\partial^2}{\partial \tau^2}\left(\frac{\delta c}{c_0}\right)$$

$$= \frac{\alpha}{2(1-\alpha)}\frac{1}{\tau}\frac{\partial}{\partial \tau}\left(\frac{\delta c}{c_0}\right) - \frac{1}{2}\frac{\partial^2}{\partial \tau^2}\left(\frac{\delta c}{c_0}\right). \qquad (5.112)$$

The next step is to evaluate the inner products that occur in Eq. (3.23). A good inner product to use in this case is a simple integration over τ of the product of the solutions, because by making use of the orthogonality properties of Bessel functions it can be shown (cf. [Gradshteyn & Ryzhik (1994)]) that:

$$\int_0^{\tau(R)} d\tau\, \tau J_\mu(\omega_{nl}\tau) J_\mu(\omega_{n'l}\tau) = \delta_{nn'}\frac{\tau(R)^2}{2}J_{\mu+1}^2(\omega_{nl}\tau(R)). \qquad (5.113)$$

Making use of this the denominator in Eq. (3.23) yields:

$$\langle Z_{n,l} \cdot Z_{n,l} \rangle = \frac{1}{2}\tau(R)^2 J_{\mu+1}^2(\omega_{nl}\tau(R)) \,. \tag{5.114}$$

The two terms in the numerator arising from the two terms of $\delta\mathcal{F}$ can be evaluated separately. The first term becomes:

$$\frac{\alpha}{2(1-\alpha)} \int_0^{\tau(R)} d\tau \, J_\mu^2(\omega_{nl}\tau)\frac{\partial}{\partial \tau}\left(\frac{\delta c}{c_0}\right) = \frac{\alpha}{2(1-\alpha)} \int_0^1 dx \, J_\mu^2(\Lambda_{nl}x)\frac{\partial}{\partial x}\left(\frac{\delta c}{c_0}\right) \tag{5.115}$$

in which $x \equiv \tau/\tau(R)$ and $\Lambda_{nl} \equiv \omega_{nl}\tau(R)$. The second term becomes:

$$-\frac{1}{2} \int_0^{\tau(R)} d\tau \, \tau J_\mu^2(\omega_{nl}\tau)\frac{\partial^2}{\partial \tau^2}\left(\frac{\delta c}{c_0}\right) = -\frac{1}{2} \int_0^1 dx \, xJ_\mu^2(\Lambda_{nl}x)\frac{\partial^2}{\partial x^2}\left(\frac{\delta c}{c_0}\right) \,. \tag{5.116}$$

From this it appears that the frequencies can also be thought of as sensitive to the first and second second derivative of the sound speed with respect to the acoustic depth τ. In realistic stellar models at the transition boundaries between convective and radiative regions the sound speed exhibits a 'kink': the first derivative with respect to r changes over a very short length scale. How abrupt the kink is depends on models/parameters of convection prescriptions but there has to be a kink in all cases. At He ionisation zones the same happens although the transition is not quite as abrupt. This implies that the second derivative of c with respect to r τ shows a peak approaching the behaviour of a delta-function. In the toy model the behaviour is therefore set to:

$$\frac{\partial^2}{\partial x^2}\left(\frac{\delta c}{c}\right) = A\delta(x - x_b) \,. \tag{5.117}$$

Integrating this once yields for the first derivative:

$$\frac{\partial}{\partial x}\left(\frac{\delta c}{c}\right) = AH(x - x_b) + B \tag{5.118}$$

where B is a constant of integration and H is the Heaviside step function. The first term Eq. (5.115) can then be integrated to yield:

$$\frac{\alpha B}{2(1-\alpha)} \int_0^1 dx \, J_\mu^2(\Lambda_{nl}x) + \frac{\alpha A}{2(1-\alpha)} \int_{x_b}^1 dx \, J_\mu^2(\Lambda_{nl}x) \,. \tag{5.119}$$

Substituting the behaviour of the sound speed into Eq. (5.116), then produces the result that the frequencies behave as:

$$\left(\frac{\delta\omega}{\omega}\right)_{nl} = \frac{\alpha}{2(1-\alpha)\omega_{nl}^2 J_{\mu+1}^2(\Lambda_{nl})} \left[B \int_0^1 dx \, J_\mu^2(\Lambda_{nl} x) \right.$$

$$\left. + A \int_{x_b}^1 dx \, J_\mu^2(\Lambda_{nl} x) \right]$$

$$- \frac{A}{\Lambda_{nl}^2} x_b \frac{J_\mu^2(\Lambda_{nl} x_b)}{J_{\mu+1}^2(\Lambda_{nl})} . \tag{5.120}$$

The functions J_μ are oscillatory and for large values of their argument behave as cosines. J_μ^2 or integrals of it therefore also have oscillatory behaviour, but with an offset. This means that if there is a single position x_b at which there is a 'kink' in the sound speed, the frequency shift $\delta\omega$ for a given l alternately increases and decreases as the frequencies ω_{nl} increase for increasing n.

5.3.3.2 Derivative discontinuities in realistic models

When considering the detection of discontinuities in first or higher order derivatives of structural variables or rotation in real stars, it makes sense to do so in the context of the physical effects that can create such discontinuities. In this way more specific use is made of which quantity is discontinuous to optimise the detection. It is this approach which is followed in [Monteiro et al. (1994)], [Christensen-Dalsgaard et al. (1995)], and [Monteiro & Thompson (2005)] in which the technique is described in particular with application to solar data in mind. In [Monteiro et al. (2000)] and in [Basu et al. (2004)] the same idea is generalised for application other low mass stars is for which only modes with low l values are available.

In [Monteiro et al. (1994)] the LAWE is also written with an acoustic depth τ as independent variable as in the previous section, but note that there the definition of τ corresponds to what is $\tau(R) - \tau$ in the notation used in the previous section. In appendix A of [Monteiro et al. (1994)] the variational principle is used to calculated the frequency shift due to the transition between convectively stable and unstable plasma in the Sun, i.e. the base of the convection zone of the Sun. Of interest in particular is the difference in behaviour between models that have overshoot of material into

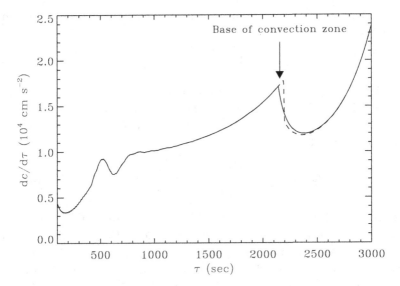

Fig. 5.9 A plot of the derivative of the sound speed with respect to the acoustic depth, as a function of acoustic depth for a model of the present Sun without overshoot, diffusion or settling (solid line) and with an overshoot layer which is adiabatically stratified over a fraction of a pressure scale height (dash-dotted line). The variable τ in the figure is equivalent to $\tau(R) - \tau$ in terms of the variable used in the text, so that the surface is at $\tau = 0$. The feature around $\tau = 400$ is associated with the second ionisation zone of He. (figure courtesy J. Christensen-Dalsgaard)

the radiative layer, and models that do not. In models without overshoot the second derivative of the sound speed with respect to acoustic depth is discontinuous, which implies a δ-function behaviour of the third derivative. The eigenfunctions behave asymptotically as sin or cos functions with a non-uniform argument in the acoustic radius τ (in the notation of this chapter):

$$\widetilde{\delta r}_r(\tau) = \frac{1}{r\sqrt{\rho c_s}} \cos(\Lambda) \tag{5.121}$$

in which Λ is:

$$\Lambda(\tau) \equiv \phi + \omega \int_\tau^{\tau(R)} d\tau \left(1 - \frac{S_l^2}{\omega^2}\right)^{1/2}. \tag{5.122}$$

Using this as eigenfunctions in the variational formulation of the eigenfrequency perturbations then produces a perturbation which has a periodic component, as well as a smooth displacement. The periodic component can be expressed as:

$$\delta\omega_p = A_1(\omega, l)\cos(2\Lambda(\tau_d)) \tag{5.123}$$

where the factor of 2 in the argument arises because the frequency perturbation is related to the square of the eigenfunction. τ_d is the location in acoustic radius of the discontinuity. The amplitude of this periodic component, to leading order, is:

$$A_1(\omega, l) = \frac{1 - 2S_l^2/\omega^2}{(1 - S_l^2/\omega^2)^2} \frac{a_1}{(1 - S_l^2/\omega^2)^{1/2}} \frac{1}{\omega^2}. \tag{5.124}$$

The constant a_1 is the parameter describing the amplitude of the derivative discontinuity, which in a helio- or asteroseismic setting would be fitted for. In models that do have overshoot there is a discontinuity in the first derivative of the sound speed. This case is therefore quite similar to what is discussed in the previous section, with a δ function behaviour for the second derivative. The same behaviour for the periodic component of the frequency perturbation holds as in Eq. (5.123) but now with an amplitude:

$$A_2(\omega, l) = \frac{1 - 2S_l^2/\omega^2}{(1 - S_l^2/\omega^2)^2} a_2 \frac{1}{\omega} \tag{5.125}$$

where the constant a_2 describes the amplitude of the derivative discontinuity. There is also a difference in phase between the two cases so that $\phi_1 - \phi_2 = \pi/4$. In practice then, the frequency differences between a star and a model are fitted for using a smooth component and periodic components with a combination of the A_1 and A_2 amplitudes, and ϕ_1 and ϕ_2 phases. From the period of the periodic shifts the location of a discontinuity would be deduced, and from the amplitude the amount of overshoot. It should be noted here, as is pointed out in [Monteiro et al. (1994)] that this analysis is done assuming the star remains spherically symmetric. In practice overshoot is a localised phenomenon so that one should expect a 'corrugation' of the boundaries of convection zones. The horizontal averaging implicit in the treatment here then smooths out the discontinuity and reduces the signal compared to the expressions given here. In [Christensen-Dalsgaard et al. (1995)] this analysis is extended further in the sense that expressions of the a_1 and a_2 are given in terms of a power series in the small parameter $(\nabla - \nabla_{\rm ad})/\nabla_{\rm ad}$, where $\nabla \equiv \partial \log T/\partial \log p$ and $\nabla_{\rm ad}$ is defined in Eq. (2.6). This makes it possible to relate a frequency signature more directly to a quantity prescribed by or predicted by the models. In [Monteiro et al. (2000)] this method is applied to a range of stellar models, varying masses and ages, in order to investigate the variation of amplitude of this periodic signal with these quantities for low values of l. From this it is concluded that with realistic errors in measured oscillation frequencies the depth of convection zones in stars can be detected, as long as a sufficiently large number of frequencies is detected to establish a periodicity.

In [Basu et al. (2004)] and [Monteiro & Thompson (2005)] a very similar variational technique is used in order to calculate the sensitivity to variations in the adiabatic exponent $\delta\Gamma_1$. In particular the second ionisation zone of He is of interest since the strength of a dip in Γ_1 can be used to determine the He abundance in the Sun and stars. In the Sun this method is complementary to other methods for determining the He abundance which make use of the large number of measured oscillation frequencies, but for stars the technique is important since it does not rely as much on the availability of large amounts of data. The two treatments in [Basu et al. (2004)] and [Monteiro & Thompson (2005)] differ slightly in terms of filtering out the smooth component of the frequency perturbations, but are otherwise very similar. One particular feature worth noting is that the second ionisation zone of He is more extended radially than a convection zone boundary is. If the 'bump' in $\delta\Gamma_1$ is parameterised in terms of a half-width β in acoustic depth, the amplitude A_Γ of the frequency perturbations becomes:

$$A_\Gamma = a_\Gamma \frac{1 - 2S_l^2/3\omega^2}{1 - S_l^2/\omega^2} \frac{\sin^2\left(\omega\beta(1 - S_l^2/\omega^2)^{1/2}\right)}{\omega\beta(1 - S_l^2/\omega^2)} \quad (5.126)$$

where a_Γ is a constant related to the depth of the $\delta\Gamma_1$ dip, and therefore to the He abundance Y. The signal has the same periodicity as described by Eq. (5.123) where Λ must be evaluated at the acoustic radius of the He second ionisation zone. As is pointed out in [Basu et al. (2004)] the signatures of the convection zone transition and the He second ionisation zone are present simultaneously and can therefore be fitted for simultaneously. In [Basu et al. (2004)] the frequencies are scaled by the large separation for the reasons outlined in section 5.3.2, before applying the fitting procedures. It is found that although this scaling does help considerably in pinning down Y an independent measurement of stellar radius and/or mass remains desirable. In [Basu et al. (2004)] it is also found that the optimal range of stellar masses for this method to realistically be applied is near to or slightly larger than the mass of the Sun. For very low mass stars the amplitude of the signal becomes too small compared to the expected observational errors, whereas for stars with higher masses the convection zone and He second ionisation zone signals become too similar to disentangle.

Appendix A

Useful Vector Formulas

a, b, c, and d are arbitrary vectors:

$$a \cdot (b \times c) = b \cdot (c \times a) = c \cdot (a \times b)$$
$$a \times (b \times c) = (a \cdot c)b - (a \cdot b)c$$
$$(a \times b) \cdot (c \times d) = (a \cdot c)(b \cdot d) - (a \cdot d)(b \cdot c) \ . \tag{A.1}$$

A and B are arbitrary vector fields, ψ is a scalar field, and ∇ is the gradient operator.

$$\nabla \times \nabla \psi = 0$$
$$\nabla \cdot (\nabla \times A) = 0 \tag{A.2}$$
$$\nabla \times (\nabla \times A) = \nabla(\nabla \cdot A) - \nabla^2 A$$

$$\nabla \cdot (\psi A) = A \cdot \nabla \psi + \psi \nabla \cdot A$$
$$\nabla \times (\psi A) = \nabla \psi \times A + \psi \nabla \times A \tag{A.3}$$

$$\nabla(A \cdot B) = (A \cdot \nabla)B + (B \cdot \nabla)A + A \times (\nabla \times B) + B \times (\nabla \times A)$$
$$\nabla \cdot (A \times B) = B \cdot (\nabla \times A) - A \cdot (\nabla \times B) \tag{A.4}$$
$$\nabla \times (A \times B) = A(\nabla \cdot B) - B(\nabla \cdot A) + (B \cdot \nabla)A - (A \cdot \nabla)B$$

If x is the coordinate of a point with respect to some origin at a distance $|x| \equiv r$, and defining a unit radial vector $n \equiv x/r$, then:

$$\begin{array}{ll} \nabla \cdot x = 3 & \nabla \times x = 0 \\ \nabla \cdot n = 2/r & \nabla \times n = 0 \end{array} \tag{A.5}$$

$$(A \cdot \nabla)n = \frac{1}{r}[A - n(A \cdot n)] \equiv \frac{A_\perp}{r} \ . \tag{A.6}$$

In the following ϕ and ψ are continuous and differentiable scalar fields, and A is a continuous and differentiable vector field. V is a 3-dimensional volume with volume element d^3x, S is a closed surface bounding V with surface area element $\mathrm{d}a$ which has unit outward pointing normal vector n.

$$\int_V \nabla\psi \mathrm{d}^3x = \int_S \psi n \mathrm{d}a$$

$$\int_V \nabla \cdot A \mathrm{d}^3x = \int_S A \cdot n \mathrm{d}a \qquad (A.7)$$

$$\int_V \nabla \times A \mathrm{d}^3x = \int_S n \times A \mathrm{d}a$$

The following are known as Green's first identity and Green's theorem respectively:

$$\int_V (\phi \nabla^2 \psi + \nabla\phi \cdot \nabla\psi) = \int_S \phi n \cdot \nabla\psi \mathrm{d}a$$

$$\int_V (\phi \nabla^2 \psi - \psi \nabla^2 \phi) = \int_S (\phi \nabla\psi - \psi \nabla\phi) \cdot n \mathrm{d}a \ . \qquad (A.8)$$

In the following S is an open surface with C a contour bounding it with line element $\mathrm{d}l$. The vector n normal to S is found by the right-hand side rule with respect to the direction of the line integral around C. The second identity is known as Stokes' theorem:

$$\int_S n \times \nabla\psi \mathrm{d}a = \oint_C \psi \mathrm{d}l$$

$$\int_S (\nabla \times A) \cdot n \mathrm{d}a = \oint_C A \cdot \mathrm{d}l \ . \qquad (A.9)$$

Appendix B

Explicit Forms of Vector Operations

A and B are arbitrary vector fields, ψ is an arbitrary scalar field. Explicit forms of the gradient operator, the divergence and rotation operators, and the Laplacian are given for, in order:

- A Cartesian coordinate system (x, y, z)
- A cylindrical coordinate system (ρ, ϕ, z)
- A spherical coordinate system (r, θ, ϕ)

$$
\begin{aligned}
\nabla \psi &= \left(\frac{\partial \psi}{\partial x}, \frac{\partial \psi}{\partial y}, \frac{\partial \psi}{\partial z} \right) \\
\nabla \cdot A &= \frac{\partial A_x}{\partial x} + \frac{\partial A_y}{\partial y} + \frac{\partial A_z}{\partial z} \\
\nabla \times A &= \left(\left[\frac{\partial A_z}{\partial y} - \frac{\partial A_y}{\partial z} \right], \left[\frac{\partial A_x}{\partial z} - \frac{\partial A_z}{\partial x} \right], \left[\frac{\partial A_y}{\partial x} - \frac{\partial A_x}{\partial y} \right] \right) \\
\nabla^2 \psi &= \frac{\partial^2 \psi}{\partial x^2} + \frac{\partial^2 \psi}{\partial y^2} + \frac{\partial^2 \psi}{\partial z^2}
\end{aligned}
\quad \text{(B.1)}
$$

$$
\begin{aligned}
\nabla \psi &= \left(\frac{\partial \psi}{\partial \rho}, \frac{1}{\rho}\frac{\partial \psi}{\partial \phi}, \frac{\partial \psi}{\partial z} \right) \\
\nabla \cdot A &= \frac{1}{\rho}\frac{\partial \rho A_\rho}{\partial \rho} + \frac{1}{\rho}\frac{\partial A_\phi}{\partial \phi} + \frac{\partial A_z}{\partial z} \\
\nabla \times A &= \left(\left[\frac{1}{\rho}\frac{\partial A_z}{\partial \phi} - \frac{\partial A_\phi}{\partial z} \right], \left[\frac{\partial A_\rho}{\partial z} - \frac{\partial A_z}{\partial \rho} \right], \frac{1}{\rho}\left[\frac{\partial \rho A_\phi}{\partial \rho} - \frac{\partial A_\rho}{\partial \phi} \right] \right) \\
\nabla^2 \psi &= \frac{1}{\rho}\frac{\partial}{\partial \rho}\left(\rho \frac{\partial \psi}{\partial \rho} \right) + \frac{1}{\rho^2}\frac{\partial^2 \psi}{\partial \phi^2} + \frac{\partial^2 \psi}{\partial z^2}
\end{aligned}
\quad \text{(B.2)}
$$

$$\nabla \psi = \left(\frac{\partial \psi}{\partial r}, \frac{1}{r}\frac{\partial \psi}{\partial \theta}, \frac{1}{r\sin\theta}\frac{\partial \psi}{\partial \phi} \right) \tag{B.3}$$

$$\nabla \cdot A = \frac{1}{r^2}\frac{\partial r^2 A_r}{\partial r} + \frac{1}{r\sin\theta}\frac{\partial \sin\theta A_\theta}{\partial \theta} + \frac{1}{r\sin\theta}\frac{\partial A_\phi}{\partial \phi}$$

$$\nabla \times A = \left(\frac{1}{r\sin\theta}\left[\frac{\partial \sin\theta A_\phi}{\partial \theta} - \frac{\partial A_\theta}{\partial \phi}\right], \left[\frac{1}{r\sin\theta}\frac{\partial A_r}{\partial \phi} - \frac{1}{r}\frac{\partial r A_\phi}{\partial r}\right], \frac{1}{r}\left[\frac{\partial r A_\theta}{\partial r} - \frac{\partial A_r}{\partial \theta}\right] \right)$$

$$\nabla^2 \psi = \frac{1}{r^2}\frac{\partial}{\partial r}\left(r^2 \frac{\partial \psi}{\partial r}\right) + \frac{1}{r^2 \sin\theta}\frac{\partial}{\partial \theta}\left(\sin\theta \frac{\partial \psi}{\partial \theta}\right) + \frac{1}{r^2 \sin^2\theta}\frac{\partial^2 \psi}{\partial \phi^2} \tag{B.4}$$

In the spherical coordinate system the operators ∇ and ∇^2 are sometimes split up into a radial and a tangential component:

$$\nabla_h \psi = \left(0, \frac{\partial \psi}{\partial \theta}, \frac{1}{\sin\theta}\frac{\partial \psi}{\partial \phi} \right) \tag{B.5}$$

$$\nabla_h \cdot A = \frac{1}{\sin\theta}\frac{\partial \sin\theta A_\theta}{\partial \theta} + \frac{1}{\sin\theta}\frac{\partial A_\phi}{\partial \phi}$$

$$\nabla^2 \psi = \nabla_r^2 \psi + \frac{1}{r^2}\nabla_h^2 \psi$$

$$\nabla_r^2 \psi \equiv \frac{1}{r^2}\frac{\partial}{\partial r}\left(r^2 \frac{\partial \psi}{\partial r}\right)$$

$$\nabla_h^2 \psi \equiv \frac{1}{\sin\theta}\frac{\partial}{\partial \theta}\left(\sin\theta \frac{\partial \psi}{\partial \theta}\right) + \frac{1}{\sin^2\theta}\frac{\partial^2 \psi}{\partial \phi^2}. \tag{B.6}$$

One reason for this is that spherical harmonics $Y_l^m(\theta,\phi)$ are eigenfunctions of the operator ∇_h^2:

$$\nabla_h^2 Y_l^m(\theta,\phi) = -l(l+1)Y_l^m(\theta,\phi) \tag{B.7}$$

in which spherical harmonics can be written in terms of Legendre polynomials and harmonic functions:

$$Y_l^m(\theta,\phi) = P_l^m(\cos\theta)e^{im\phi} \tag{B.8}$$

with appropriate normalisation (cf. [Gradshteyn & Ryzhik (1994)]).

Also commonly occurring is the quantity $(A\cdot\nabla)B$ which in each of these three coordinate systems is:

- Cartesian:

$$(A\cdot\nabla)B = \left(A_x\frac{\partial B_x}{\partial x} + A_y\frac{\partial B_x}{\partial y} + A_z\frac{\partial B_x}{\partial z},\ A_x\frac{\partial B_y}{\partial x} + A_y\frac{\partial B_y}{\partial y} + A_z\frac{\partial B_y}{\partial z},\right.$$
$$\left.A_x\frac{\partial B_z}{\partial x} + A_y\frac{\partial B_z}{\partial y} + A_z\frac{\partial B_z}{\partial z}\right). \tag{B.9}$$

- cylindrical:

$$(A\cdot\nabla)B = \left(A_\rho\frac{\partial B_\rho}{\partial \rho} + A_\phi\frac{\partial B_\rho}{\partial \phi} + A_z\frac{\partial B_\rho}{\partial z} - A_\phi B_\phi \rho,\right.$$
$$A_\rho\frac{\partial B_\phi}{\partial \rho} + \frac{A_\phi}{\rho}\frac{\partial B_\phi}{\partial \phi} + A_z\frac{\partial B_\phi}{\partial z} + \frac{A_\phi B_\rho}{\rho},$$
$$\left.A_\rho\frac{\partial B_z}{\partial \rho} + \frac{A_\phi}{\rho}\frac{\partial B_z}{\partial \phi} + A_z\frac{\partial B_z}{\partial z}\right). \tag{B.10}$$

- spherical:

$$(A\cdot\nabla)B = \left(A_r\frac{\partial B_r}{\partial r} + \frac{A_\theta}{r}\frac{\partial B_r}{\partial \theta} + \frac{A_\phi}{r\sin\theta}\frac{\partial B_r}{\partial \phi} - \frac{A_\theta B_\theta + A_\phi B_\phi}{r},\right.$$
$$A_r\frac{\partial B_\theta}{\partial r} + \frac{A_\theta}{r}\frac{\partial B_\theta}{\partial \theta} + \frac{A_\phi}{r\sin\theta}\frac{\partial B_\theta}{\partial \phi} + \frac{A_\theta B_r}{r} - \cot\theta\frac{A_\phi B_\phi}{r},$$
$$\left.A_r\frac{\partial B_\phi}{\partial r} + \frac{A_\theta}{r}\frac{\partial B_\phi}{\partial \theta} + \frac{A_\phi}{r\sin\theta}\frac{\partial B_\phi}{\partial \phi} + \frac{A_\phi B_r}{r} + \cot\theta\frac{A_\phi B_\theta}{r}\right). \tag{B.11}$$

Note that in this formulation the vectors **A** and **B** are in 'physical units'. In the literature sometimes one finds these relationships in a slightly different form, because the components of these vectors are angles in the ϕ direction for a cylindrical coordinate system, and in the θ and ϕ directions for a spherical coordinate system. Conversion is straightforward and for the purposes of this book, the expressions given here are the most useful.

In spherical geometry, it is sometimes also useful the make use of the fact that a divergence-free, or solenoidal, vector field can be decomposed into toroidal **T** and poloidal **P** components, i.e. if $\nabla\cdot\mathbf{B} = 0$ then $\mathbf{B} = \mathbf{T} + \mathbf{P}$

in which:

$$\begin{aligned}\mathbf{T} &\equiv \nabla \times (\Psi \widehat{e}_r) \\ &= \left(0, \frac{1}{r\sin\theta}\frac{\partial \Psi}{\partial \phi}, -\frac{1}{r}\frac{\partial \Psi}{\partial \theta}\right) \\ \mathbf{P} &\equiv \nabla \times \nabla \times (\Phi \widehat{e}_r) \\ &= \left(-\frac{1}{r^2}\nabla_h^2 \Phi, \frac{1}{r}\frac{\partial^2 \Phi}{\partial r \partial \theta}, \frac{1}{r\sin\theta}\frac{\partial^2 \Phi}{\partial r \partial \phi}\right) .\end{aligned} \quad (B.12)$$

If one expands the potential functions Ψ and Φ in terms of spherical harmonics:

$$\begin{aligned}\Psi &= \sum_{lm} t_{lm}(r) Y_l^m \\ \Phi &= \sum_{lm} p_{lm}(r) Y_l^m\end{aligned} \quad (B.13)$$

then the expressions (B.12) reduce to:

$$\begin{aligned}\mathbf{T} &= \left(0, \sum_{lm}\frac{t_{lm}(r)}{r\sin\theta}\frac{\partial Y_l^m}{\partial \phi}, -\sum_{lm}\frac{t_{lm}(r)}{r}\frac{\partial Y_l^m}{\partial \theta}\right) \\ \mathbf{P} &= \left(\sum_{lm}l(l+1)\frac{p_{lm}(r)}{r^2}Y_l^m, \sum_{lm}\frac{1}{r}\frac{\partial p_{lm}(r)}{\partial r}\frac{\partial Y_l^m}{\partial \theta},\right. \\ &\quad \left.\frac{1}{r\sin\theta}\frac{\partial p_{lm}(r)}{\partial r}\frac{\partial Y_l^m}{\partial \phi}\right) .\end{aligned} \quad (B.14)$$

From the definitions (B.12) one can see that the curl of a toroidal field is poloidal. One can also show that the curl of a poloidal field is toroidal (cf [Chandrasekhar (1961)]). Expressions for $\nabla \times \mathbf{T}$ and $\nabla \times \mathbf{P}$ are therefore straightforward to derive:

$$\begin{aligned}\nabla \times \mathbf{T} &= \left(\sum_{lm}l(l+1)\frac{t_{lm}(r)}{r^2}Y_l^m, \sum_{lm}\frac{1}{r}\frac{\partial t_{lm}(r)}{\partial r}\frac{\partial Y_l^m}{\partial \theta},\right. \\ &\quad \left.\frac{1}{r\sin\theta}\frac{\partial t_{lm}(r)}{\partial r}\frac{\partial Y_l^m}{\partial \phi}\right) \\ \nabla \times \mathbf{P} &= \left(0, \sum_{lm}\frac{\widetilde{p}_{lm}(r)}{r\sin\theta}\frac{\partial Y_l^m}{\partial \phi}, -\sum_{lm}\frac{\widetilde{p}_{lm}(r)}{r}\frac{\partial Y_l^m}{\partial \theta}\right)\end{aligned} \quad (B.15)$$

in which the $\widetilde{p}_{lm}$ are defined by:

$$\widetilde{p}_{lm} \equiv l(l+1)\frac{p_{lm}(r)}{r^2} - \frac{\partial^2 p_{lm}(r)}{\partial r^2} \,. \tag{B.16}$$

More details and a demonstration of orthogonality of this decomposition can be found in eg. appendix III of [Chandrasekhar (1961)].

In more general curvilinear coordinate systems it becomes useful to make use of a symbolic notation from differential geometry which is commonly used for instance in the field of general relativity. More details can be found e.g. in chapter 8 of [Misner, Thorne, & Wheeler (1973)]. One first defines the metric of the curvilinear coordinate system by specifying the components of the metric tensor $g_{\alpha\beta}$ in terms of the unit vectors $\mathbf{e}_\alpha$ of the coordinate directions:

$$g_{\alpha\beta} = \mathbf{e}_\alpha \cdot \mathbf{e}_\beta \,. \tag{B.17}$$

For instance for a standard spherical coordinate system $g_{\alpha\beta}$ is:

$$g_{\alpha\beta} \equiv \begin{pmatrix} 1 & 0 & 0 \\ 0 & r^2 & 0 \\ 0 & 0 & r^2\sin^2\theta \end{pmatrix} \tag{B.18}$$

and the inverse of $g_{\alpha\beta}$ is denoted by $g^{\alpha\beta}$. It is common two make use of two shorthand notation conventions:

- Taking a derivative with respect to a coordinate μ is denoted by adding the subscript $,\mu$, so for instance:

$$\partial g_{\beta\gamma}/\partial x^\mu \equiv g_{\beta\gamma,\mu} \,. \tag{B.19}$$

- If a subscript and a superscript are repeated this implies that there is a summation over all coordinate directions:

$$g^{\alpha\beta}g_{\alpha\beta} \equiv \sum_\alpha \sum_\beta g^{\alpha\beta}g_{\alpha\beta} \,. \tag{B.20}$$

The next symbol normally defined to facilitate formulating derivatives in curvi-linear coordinate systems, is the connection coefficient or Christoffel symbol $\Gamma^\alpha_{\beta\gamma}$. The values this takes, for all possible values of the coefficients, is determined by its relation with the metric tensor g:

$$\Gamma_{\mu\beta\gamma} \equiv \frac{1}{2}\left(g_{\mu\beta,\gamma} + g_{\mu\gamma,\beta} - g_{\beta\gamma,\mu}\right)$$

$$\Gamma^\alpha_{\beta\gamma} \equiv g^{\alpha\mu}\Gamma_{\mu\beta\gamma} \,. \tag{B.21}$$

Any transport equation involves the gradient operator ∇. When expressed in a general curvi-linear coordinate system, this needs to be formulated making use of co-variant derivatives, which replace the standard derivatives in a recti-linear system. The co-variant derivative of a variable, with the notation $;\mu$ as subscript, is determined with the help of the Γ. For a scalar quantity S:

$$S_{;\mu} = S_{,\mu} \,. \tag{B.22}$$

For a vector quantity V^α:

$$V^\alpha_{;\mu} = V^\alpha_{,\mu} + \Gamma^\alpha_{\mu\gamma} V^\gamma \,. \tag{B.23}$$

For a tensor quantity T:

$$T^{\alpha\beta}_{;\mu} = T^{\alpha\beta}_{,\mu} + \Gamma^\alpha_{\mu\gamma} T^{\gamma\beta} + \Gamma^\beta_{\mu\gamma} T^{\alpha\gamma} \,. \tag{B.24}$$

The divergence of vectors and tensors would be expressed as:

$$\nabla \cdot V = V^\alpha_{;\alpha} \tag{B.25}$$

$$\nabla \cdot T = T^{\alpha\beta}_{;\alpha} \,. \tag{B.26}$$

In a Cartesian coordinate system all the connection coefficients are 0 so that a covariant derivative is identical to a partial derivative.

Appendix C

Useful Constants

Various physical constants are necessary for the computation of solar and stellar models. These constants must be determined experimentally or are defined for a given system of units such as the SI system. In most cases the current best or recommended values are compiled at dedicated institutes and can by found at their web-sites. For fundamental physical constants such a site is maintained by the National Institute of Standards and Technology (NIST) in the USA: http://physics.nist.gov. The value of the speed of light in vacuum c is a defining constant. The value of the gravitational constant G and its uncertainty is the 2002 CODATA value (see also http://www.codata.org). The Planck constant h, the atomic mass unit m_H and the Boltzmann constant k are from the same source. In the list(s) below the uncertainty in () brackets refers to the last two digits quoted.

$$c = 2.99792458 \times 10^8 \text{ m s}^{-1} \quad \text{exact} \tag{C.1}$$
$$G = [6.6742\ (10)] \times 10^{-11} \text{ m}^3 \text{ kg}^{-1} \text{ s}^{-2} \tag{C.2}$$
$$h = [6.6260693\ (11)] \times 10^{-34} \text{ J s} \tag{C.3}$$
$$k = [1.3806505\ (24)] \times 10^{-23} \text{ J K}^{-1} \tag{C.4}$$
$$m_H = [1.66053886\ (28)] \times 10^{-27} \text{ kg} \tag{C.5}$$

The radiation constant a can be expressed in terms of c, h and k:

$$a = \frac{8\pi^5 k^4}{15 c^3 h^3} = [7.565767\ (51)] \times 10^{-16} \text{ J m}^{-3} \text{ K}^{-4} . \tag{C.6}$$

The product $GM_\odot$ of the gravitational constant and the mass of the Sun can be determined from planetary orbital periods and the semi-major axes of their orbits, or equivalently the standard siderial year and the astronomical unit AU. The average Earth-Sun distance known as astronomical unit (AU) and the siderial year are determined by a consistent multi-parameter fit to solar system dynamical data for a number of major and minor bodies in the solar system. The values quoted here are retrieved from a dedicated solar system dynamics web-site http://ssd.jpl.nasa.gov/ maintained by the Jet Propulsion Laboratory (JPL) in the USA.

$$AU = [1.49597870691\,(03)] \times 10^{11} \text{ m} \qquad (C.7)$$

$$GM_\odot = [1.32712440018\,(08)] \times 10^{20} \text{ m}^3\text{s}^{-2} \qquad (C.8)$$

The mass of the Sun can be determined from (C.2) and (C.8):

$$M_\odot = [1.9884\,(03)]\,10^{30} \times \text{ kg}. \qquad (C.9)$$

A more usual value, for solar physicists, of the solar mass $M_\odot$ is $1.989\,10^{30}$ kg which is essentially obtained by rounding off, when using the older (CODATA 1986) value for the gravitational constant. With that value $G = 6.67259(85)$ m^3 kg^{-1} s^{-2} the solar mass comes out as:

$$M_\odot(\text{CODATA1986}) = [1.98892\,(25)] \times 10^{30} \text{ kg}. \qquad (C.10)$$

Recent, somewhat anomalous, determinations of G have pushed up both the value of G and its uncertainty, which means that the usual value for $M_\odot$ with which most models are calculated is now consistent only at the 2σ level with the current value (C.9) inferred from (C.8) and (C.2).

The value of the solar radius is also somewhat problematic since a recent optical determination [Brown & Christensen-Dalsgaard (1998)] and a helioseismic determination using f-modes [Tripathy & Antia (1999)] differ:

$$R_\odot(optical) = [6.95508\,(26)] \times 10^8 \text{ m} \qquad (C.11)$$

$$R_\odot(seismic) = [6.9577\,(10)] \times 10^8 \text{ m}. \qquad (C.12)$$

Both of these recent determinations are smaller than the "standard" value used for most solar models: $6.9599\,10^8$ m.

What is quoted as the age of the Sun is inferred from the ratios of abundances of a number of radioactive elements found in meteorites. These ratios are set by their production through the nuclear s-process in AGB stars. Through their stellar wind AGB stars enrich the ISM, and the chemical mixture out of which the Sun has formed reflects these ratios. The premise is that as the Sun forms, the rocky planets and meteorites solidify. After solidification the abundance ratio of selected chemical isotopes slowly changes due to their different decay rates, and therefore by measuring a number of isotope abundances the epoch of solidification can be found. Details on the methods and isotopes used can be found in [Göpel et al. (1994)], [Allègre et al. (1995)], and also in the appendix to the review [Bahcall et al. (1995)], written by G.J. Wasserburg. The age that results from this is:

$$t_{\odot\,\text{met}} = 4.57 \pm 0.02 \text{ Gyr} \qquad (C.13)$$

where the uncertainty is due in part to uncertainties surrounding the thermalisation process at solidification. It has been argued that the solidification occurs before the Sun arrives at the ZAMS and that when measured from the ZAMS the current age of the Sun should be set to:

$$t_{\odot\,\text{ZAMS met}} = 4.53 \pm 0.04 \text{ Gyr} \qquad (C.14)$$

where the uncertainty is increased because of uncertainties in the pre-main sequence time scales (cf. [Guenther & Demarque (1997)]). The helioseismically inferred age is model dependent. An important source of uncertainty is the current photospheric He abundance Y_{ph} in the Sun. This abundance can be determined from inversions which depend on the EOS, so that uncertainties in the EOS of plasma in the solar convection zone affect the age estimates. With the seismically determined (cf. [Degl'Innocenti et al. (1997)], [Basu & Antia (2004)]):

$$Y_{\text{ph}} = 0.248 \qquad (C.15)$$

the seismic solar age becomes (cf. [Dziembowski et al. (1999)]):

$$t_{\odot\,\text{ZAMS seis}} = 4.66 \pm 0.11 \text{ Gyr} \qquad (C.16)$$

Special relativistic corrections to the EOS pull the latter age down into agreement with the meteoritic value (cf. [Bonanno et al. (2002)]):

$$t_{\odot\,\text{ZAMS seis}\,'02} = 4.57 \pm 0.11 \text{ Gyr}. \qquad (C.17)$$

Bibliography

Abramowitz, M., Stegun, I.A., (1964) *Handbook of Mathematical Functions*, Dover
Adelberger, E.G., Austin, S.M., Bahcall, J.N., Balantekin, A.B., Bogaert, G., Brown, L.S., Buchmann, L., Cecil, F.E., Champagne, A.E., de Braeckeleer, L., Duba, C.A., Elliott, S.R., Freedman, S.J., Gai, M., Goldring, G., Gould, C.R., Gruzinov, A., Haxton, W.C., Heeger, K.M., Henley, E., Johnson, C.W., Kamionkowski, M., Kavanagh, R.W., Koonin, S.E., Kubodera, K., Langanke, K., Motobayashi, T., Pandharipande, V., Parker, P., Robertson, R.G.H., Rolfs, C., Sawyer, R.F., Shaviv, N., Shoppa, T.D., Snover, K.A., Swanson, E., Tribble, R.E., Turck-Chièze, S., Wilkerson, J.F. (1998) *Rev. Mod. Phys.* **70**, 1265
Adorf, H.-M. (1995) *Ast.Soc.Pac. Conf.Ser.* **77**, 460
Aerts, C., (1996) *Astron.Astroph.* **314**, 115
Aerts, C., De Pauw, M., Waelkens, C., (1992) *Astron.Astroph.* **266**, 294
Aizenman, M.L., Smeyers, P., (1977) *Astroph. Spa.Sc.* **48**, 123
Aizenman, M.L., Smeyers, P., Weigert, A., (1977) *Astron.Astroph.* **58**, 41
Allègre, C.J., Manhès, G., Göpel, C., (1995) *Geoch. Cosmoch. Acta* **59**, 1445
Anderson, E., (1993) *Ast.Soc.Pac. Conf.Ser.* **42**, 11
Andersson, N., Kokkotas, K., (1998) *Mon.Not. Roy.Ast.Soc.* **299**, 1059
Andreasen, G.K., (1987) *Astron.Astroph.* **186**, 159
Angulo, C., (1997) *Nucl. Phys. A* **621**, 591
Appourchaux, T., Andersen, B.N., (1990) *Solar Phys.* **128**, 91
Appourchaux, T., Andersen, B.N., Fröhlich, C., Jiménez, A., Telljohann, U., Wehrli, C., (1997) *Solar Phys.* **170**, 27
Armstrong, J., Kuhn, J.R., (1999) *Astroph.J.* **525**, 533
Aschwanden, M.J., Fletcher, L., Schrijver, C.J., Alexander D., (1999) *Astroph.J.* **520**, 880
Asplund, M., (2004) *Astron.Astroph.* **417**, 769
Asplund, M., Grevesse, N., Sauval, A.J., Allende Prieto, C., Blomme, R., (2005a) *Astron.Astroph.* **431**, 693
Asplund, M., Grevesse, N., Sauval, A.J., Allende Prieto, C., Kiselman, D., (2004) *Astron.Astroph.* **417**, 751

Asplund, M., Grevesse, N., Sauval, A.J., Allende Prieto, C., Kiselman, D., (2005b) *Astron.Astroph.* **435**, 339
Asplund, M., Ludwig, H.-G., Nordlund, Å., Stein, R.F., (2000) *Astron.Astroph.* **359**, 669
Backus, G.E., Gilbert, J.F., (1968) *Geophys.J.,* **16**, 169
Backus, G.E., Gilbert, J.F., (1970) *Phil.Trans.Roy.Soc. London A,* **266**, 123
Badnell, N.R., Bautista, M.A., Butler, K., Delahaye, F., Mendoza, C., Palmeri, P., Zeippen, C.J., Seaton, M.J., (2005) *Mon.Not. Roy.Ast.Soc.* **360**, 458
Bahcall, J.N., Pinsonneault, M.H., Wasserburg, G.J., (1995) *Rev Mod. Phys.* **67**, 781
Baker, N.H., Gough, D.O., (1979) *Astroph.J.* **234**, 232
Balmforth N.J., (1992a) *Mon.Not. Roy.Ast.Soc.* **255**, 603
Balmforth N.J., (1992b) *Mon.Not. Roy.Ast.Soc.* **255**, 639
Balmforth, N.J., Cunha, M.S., Dolez, N., Gough, D.O., Vauclair, S., (2001) *Mon.Not. Roy.Ast.Soc.* **323**, 362
Balona, L.A., (1986a) *Mon.Not. Roy.Ast.Soc.* **219**, 111
Balona, L.A., (1986b) *Mon.Not. Roy.Ast.Soc.* **220**, 647
Balona, L.A., (2000) *Mon.Not. Roy.Ast.Soc.* **319**, 606
Balona, L.A., Evers, E.A., (1999) *Mon.Not. Roy.Ast.Soc.* **302**, 349
Baranne, A., Queloz, D., Mayor, M., Adrianzyk, G., Knispel, G., Kohler, D., Lacroix, D., Meunier, J.-P., Rimbaud, G., Vin, A., (1996) *Astron.Astroph. Supp.* **119**, 373
Barnes, J.R., Collier Cameron, A., Donati, J.-F., James, D.J., Marsden, S.C., Petit, P., (2005) *Mon.Not. Roy.Ast.Soc.* **357**, L1
Basu, S., Antia, H.M., *Astroph.J. Lett.* **606**, L85
Basu, S., Christensen-Dalsgaard, J., (1997) *Astron.Astroph. Lett.* **322**, L5
Basu, S., Mazumdar, A., Antia, H.M., Demarque, P., (2004) *Mon.Not. Roy.Ast.Soc.* **350**, 277
Basu, S., Thompson, M.J., Christensen-Dalsgaard, J., Pérez Hernández, F., (1996) *Mon.Not. Roy.Ast.Soc.* **280**, 651
Beckers, J.M., Brown, T.M., (1980) *Oss. Mem.Oss.Ast. Arcetri* **106**, 189
Bedding, T., Kiss, L.L., Kjeldsen, H., Brewer, B.J., Dind, Z.E., Kawaler, S.D., Zijlstra, A.A., (2005) *Mon.Not. Roy.Ast.Soc.* **361**, 1375
Bedding, T., Kjeldsen, H., Reetz, L., Barbuy, B., (1996) *Mon.Not. Roy.Ast.Soc.* **280**, 1155
Bender, C.M., Orszag, S.A., (1978) *Advanced Mathematical Methods for Scientists and Engineers*, McGraw Hill
Berthomieu, G., Provost, J., (1990) *Astron.Astroph.* **227**, 563
Bildsten, L., Ushomirsky, G., Cutler, C., (1996) *Astroph.J.* **460**, 827
Bigot, L., Provost, J., Berthomieu, G., Dziembowski, W.A., Goode, P.R., (2000) *Astron.Astroph.* **356**, 218
Birch, A.C., Kosovichev, A.G., (2000) *Solar Phys.* **192**, 193
Birch, A.C., Kosovichev, A.G., Price, G.H., Schlottmann, R.B., (2001) *Astroph.J. Lett.* **561**, L229
Blöcker, T., (1995) *Astron.Astroph.* **297**, 727
Bogdan, T.J., (1989) *Astroph.J.* **339**, 1132

Bogdan, T.J., (1997) *Astroph.J.* **477**, 475
Bogdan, T.J., Knölker, M., (1989) *Astroph.J.* **339**, 579
Bogdan, T.J., Knölker, M., (1991) *Astroph.J.* **369**, 219
Bogdan, T.J., Knölker, M., MacGregor, K.B., Kim, E.-J., (1996) *Astroph.J.* **456**, 879
Böhm-Vitense, E., 1958, *Zeits. f. Astroph.* **46**, 108
Bohren, G.F., Huffman, D.R, (1983) *Absorption and Scattering of Light by Small Particles*, Wiley
Bonanno, A., Schlattl, H., Paternò, L., (2002) *Astron.Astroph.* **390**, 1115
Bossi, M., La Franceschina, L., (1995) *Astron.Astroph. Supp.* **113**, 387
Bouchy, F., Pepe, F., Queloz, D., (2001) *Astron.Astroph.* **374**, 733
Bowen, G.H., (1988) *Astroph.J.* **329**, 299
Brandenburg, A., Dobler, W., (2002) *Astron. Nachr.* **323**, 411
Braun, D.C., (1995) *Astroph.J.* **451**, 859
Braun, D.C., Duvall Jr., T.L., LaBonte, B., Jefferies, S.M., Harvey, J.W., Pomerantz, M.A., (1992) *Astroph.J. Lett.* **391**, L113
Briquet, M., Aerts, C., (2003) *Astron.Astroph.* **398**, 687
Brookes, J.R., Isaak, G.R., van der Raay, H.B., (1978) *Mon.Not. Roy.Ast.Soc.* **185**, 1
Brown, T.M., Christensen-Dalsgaard, J., (1990) *Astroph.J.* **349**, 667
Brown, T.M., Christensen-Dalsgaard, J., (1998) *Astroph.J. Lett.* **500**, L195
Brown, T.M., Christensen-Dalsgaard, J., Weibel-Mihalas, B., Gilliland, R.L., (1994) *Astroph.J.* **427**, 1013
Brown, T.M., Gilliland, R.L., (1994) *Ann.Rev. Astron.Astroph.* **32**, 37
Bruls, J.H.M.J., (1993) *Astron.Astroph.* **269**, 509
Bruls, J.H.M.J., Rutten, R.J., (1992) *Astron.Astroph.* **265**, 257
Bruls, J.H.M.J., Rutten, R.J., Shchukina, N.G., (1992) *Astron.Astroph.* **265**, 237
Brummell, N.H., Clune, T.L., Toomre, J., (2002) *Astroph.J.* **570**, 825
Brummell, N.H., Hurlburt, N,E., Toomre, J, (1996) *Astroph.J.* **473**, 494
Brummell, N.H., Hurlburt, N,E., Toomre, J, (1998) *Astroph.J.* **493**, 955
Brun, A.S., Miesch, M.S., Toomre, J., (2004) *Astroph.J.* **614**, 1073
Brun, A.S., Toomre, J., (2002) *Astroph.J.* **570**, 865
Buchler, J.R., Kolláth, Z., Cadmus Jr., R.R., (2004) *Astroph.J.* **613**, 532
Cacciani, A., Varsik, J., Zirin, H., (1990) *Solar Phys.* **125**, 173
Cally, P.S., Bogdan, T.J., (1997) *Astroph.J. Lett.* **486**, L67
Campbell, C.G., Papaloizou, J.C.B., (1986) *Mon.Not. Roy.Ast.Soc.* **220**, 577
Canuto, V.M., Dubovikov, M.S., (1996) *Phys. Fluids* **8**, 571
Canuto, V.M., Dubovikov, M.S., (1998) *Astroph.J.* **493**, 834
Canuto, V.M., Mazzitelli, I., (1991) *Astroph.J.* **370**, 295
Canuto, V.M., Minotti, F., (1991) *Mon.Not. Roy.Ast.Soc.* **328**, 829
Cargill, P.J., Chen, J., Garren, D.A., (1994) *Astroph.J.* **423**, 854
Carlsson, M., Stein, R.S., (1999) *Ast.Soc.Pac. Conf.Ser.* **184**, 206
Carrier, F., Eggenberger, P., Bouchy, F., (2005) *Astron.Astroph.* **434**, 1085
Castor, J.I., (2004) *Radiation Hydrodynamics*, Cambridge Univ. Press
Chandrasekhar, S., (1961) *Hydrodynamic and Hydromagnetic Stability*, Oxford Univ. Press

Chandrasekhar, S., (1964) *Astroph.J.* **139**, 664
Chaplin, W.J., Elsworth, Y., Howe, R., Isaak, G.R., McLeod, C.P., Miller, B.A., van der Raay, H.B., Wheeler, S.J., New, R., (1996) *Solar Phys.* **168**, 1
Charbonneau, P., Christensen-Dalsgaard, J., Henning, R., Larsen, R.M., Schou, J., Thompson, M.J., Tomczyk, S., (1999) *Astroph.J.* **527**, 445
Charpinet, S., Fontaine, G., Brassard, P., Dorman, B., (1997) *Astroph.J.* **489**, 149
Charpinet, S., Fontaine, G., Brassard, P., Dorman, B., (2000) *Astroph.J. Supp.* **131**, 223
Chelli, A., (2000) *Astron.Astroph. Lett.* **358**, L59
Chen, Y.Q., Nissen, P.E., Benoni, T., Zhao, G., (2001) *Astron.Astroph.* **371**, 943
Childress, S., Gilbert, A.D., (1995) *Stretch, Twist, Fold: The Fast Dynamo*, Springer
Christensen-Dalsgaard, J., (1984) in *Solar seismology from space*, JPL Publ. 84-84, 219
Christensen-Dalsgaard, J., (1993) in *Proc. GONG 1992 : Seismic investigation of the Sun and stars*, Ed. T.M. Brown, ASP Conf. Ser. **42**, 347
Christensen-Dalsgaard, J., (1999) in *The non-sleeping universe*, Eds. M.T.V.T.Lago & A.Blanchard, *Astroph. Spa.Sc.* **261**, 1
Christensen-Dalsgaard, J., (2002) *Rev.Mod.Phys.* **74**, 1073
Christensen-Dalsgaard, J., (2003) *Lecture Notes on Stellar Oscillations* 5th Ed., WWW : http://astro.phys.au.dk/~jcd/oscilnotes/
Christensen-Dalsgaard, J., Bedding, T.R., Kjeldsen, H., (1995) *Astroph.J. Lett.* **443**, L29
Christensen-Dalsgaard, J., Däppen, W., Ajukov, S.V., Anderson, E.R., Antia, H.M., Basu, S., Baturin, V.A., Berthomieu, G., Chaboyer, B., Chitre, S.M., Cox, A.N., Demarque, P., Donatowicz, J., Dziembowski, W.A., Gabriel, M., Gough, D.O., Guenther, D.B., Guzik, J.A., Harvey, J.W., Hill, F., Houdek, G., Iglesias, C.A., Kosovichev, A.G., Leibacher, J.W., Morel, P., Proffitt, C.R., Provost, J., Reiter, J., Rhodes Jr., E.J., Rogers, F.J., Roxburgh, I.W., Thompson, M.J., Ulrich, R.K. *Science* **272**, 1286
Christensen-Dalsgaard, J., Di Mauro, M.P., Schlattl, H., Weiss, A., (2005) *Mon.Not. Roy.Ast.Soc.* **356**, 587
Christensen-Dalsgaard, J., Hansen, P.C., Thompson, M.J., (1999) *Mon.Not. Roy.Ast.Soc.* **264**, 541
Christensen-Dalsgaard, J., Kjeldsen, H., Mattei, J., (2001) *Astroph.J. Lett.* **562**, L141
Christensen-Dalsgaard, J., Monteiro, M.J.P.F.G., Thompson, M.J., (1995) *Mon.Not. Roy.Ast.Soc.* **276**, 283
Christensen-Dalsgaard, J., Pérez-Hernández, F., (1992) *Mon.Not. Roy.Ast.Soc.* **257**, 62
Christensen-Dalsgaard, J., Thompson, M.J., (1999) *Astron.Astroph.* **350**, 852
Cincotta, P.M., Méndez, M., Núñez, (1995) *Astroph.J.* **449**, 231
Cincotta, P.M., Helmi, A., Méndez, M., Núñez, J.A., Vucetich, H., (1995) *Mon.Not. Roy.Ast.Soc.* **302**, 582
Clayton, D.D., (1983) *Principles of Stellar Evolution and Nucleosynthesis*, Univ.

Chicago Press
Connes, P. (1985) *Astroph. Spa.Sc.* **110**, 211
Corbard, T., Berthomieu, G., Provost, J., Morel, P., (1999) *Astron.Astroph.* **330**, 1149
Corbard, T., Blanc-Féraud, L., Berthomieu, G., Provost, J., (1999) *Astron.Astroph.* **344**, 696
Coudé du Foresto, V., Ridgway, S., Mariotti, J.-M., (1997) *Astron.Astroph. Supp.* **121**, 379
Courant, R., Hilbert, D., (1953) *Methods of Mathematical Physics Vol. I*, Wiley
Cox, J.P., (1980) *Theory of Stellar Pulsation*, Princeton Univ. Press
Cugier, H., Dziembowski, W.A., Pamyatnykh, A.A., (1994) *Astron.Astroph.* **291**, 143
Cunha, M.S., (2006) *Mon.Not. Roy.Ast.Soc.* **365**, 153
Cunha, M.S., Gough, D.O., (2000) *Mon.Not. Roy.Ast.Soc.* **319**, 1020
D'Antona, F., Mazzitelli, I., (1994) *Astroph.J. Supp.* **90**, 467
Degl'Innocenti, S., Dziembowski, W.A., Fiorentini, G., Ricci, B., (1997) *Astropart. Phys.* **7**, 77
De Ridder, J., Dupret, M.-A., Neuforge, C., Aerts, C., (2002) *Astron.Astroph.* **385**, 572
De Silva, G.M., Sneden, C., Paulson, D.B., Asplund, M., Bland-Hawthorn, J., Bessell, M.S., Freeman, K.C., (2006) *Astron.J.* **131**, 455
Deubner, F.-L., Gough, D.O., (1984) *Ann.Rev. Astron.Astroph.* **22**, 593
Deupree, R.G., (2001) *Astroph.J.* **552**, 268
Dintrans, B., Rieutord, M., (2000) *Astron.Astroph.* **354**, 86
Dintrans, B., Brandenburg, A., Nordlund, Å., Stein, R., (2005) *Astron.Astroph.* **438**, 365
Domiciano de Souza, A., Kervella, P., Jankov, S., Vakili, F., Ohishi, N., Nordgren, T.E., Abe, L., (2005) *Astron.Astroph.* **442**, 567
Donati, J.-F., Brown, S.F., (1997) *Astron.Astroph.* **326**, 1135
Donati, J.-F., Collier-Cameron, A., (1997) *Mon.Not. Roy.Ast.Soc.* **291**, 1
D'Silva, S., (1996) *Astroph.J.* **469**, 964
Duvall Jr., T.L., (1982) *Nature* **300**, 242
Duvall Jr., T.L., Jefferies, S.M., Harvey, J.W., Osaki, Y., Pomerantz, M.A., (1993) *Astroph.J.* **410**, 829
Duvall Jr., T.L., Jefferies, S.M., Harvey, J.W., Pomerantz, M.A., (1993) *Nature* **362**, 430
Duvall Jr., T.L., Kosovichev, A.G., Scherrer, P.H., Bogart, R.S., Bush, R.I., De Forest, C., Hoeksema, J.T., Schou, J., Saba, J.L.R., Tarbell, T.D., Title, A.M., Wolfson, C.J., Milford, P.N., (1997) *Solar Phys.* **170**, 63
Dzhalilov, N.S., Staude, J., (2004) *Astron.Astroph.* **421**, 305
Dzhalilov, N.S., Zhugzhda, Y.D., Staude, J., (1992) *Astron.Astroph.* **257**, 359
Dzhalilov, N.S., Zhugzhda, Y.D., Staude, J., (1994) *Astron.Astroph.* **291**, 1001
Dziembowski, W.A., Fiorentini, G., Ricci, B., Sienkiewicz, R., (1999) *Astron.Astroph.* **343**, 990
Dziembowski, W.A., Goode, P.R., (1989) *Astroph.J.* **347**, 540
Dziembowski, W.A., Goode, P.R., (1992) *Astroph.J.* **394**, 670

Dziembowski, W.A., Goode, P.R., (1996) *Astroph.J.* **458**, 338
Edwin, P.M., Roberts, B., (1983) *Solar Phys.* **88**, 179
Eggenberger, P., Maeder, A., Meynet, G., (2005) *Astron.Astroph. Lett.* **440**, L9
Fahlman, G.G., Ulrych, T.J., (1982) *Mon.Not. Roy.Ast.Soc.* **199**, 53
Fan, Y., Braun, D.C., Chou, D.-Y., (1995) *Astroph.J.* **451**, 877
Feuchtinger, M.U., (1999a) *Astron.Astroph.* **351**, 103
Feuchtinger, M.U., (1999b) *Astron.Astroph. Supp.* **136**, 217
Fierry Fraillon, D., Appourchaux, T., (2001) *Mon.Not. Roy.Ast.Soc.* **324**, 115
Fontaine, G., Brassard, P., Charpinet, S., Green, E.M., Chayer, P., Billres, M., Randall, S.K., (2003) *Astroph.J.* **597**, 518
Fossat E., Kholikov, Sh., Gelly, B., Schmider, F.X., Fierry-Fraillon, D., Grec, G., Palle, P., Cacciani, A., Ehgamberdiev, S., Hoeksema, J.T., Lazek, M., (1999) *Astron.Astroph.* **343**, 608
Foster, G., (1996) *Astron.J.* **112**, 1709
Fox, M.W., Wood, P.R., (1985) *Astroph.J.* **297**, 455
Gabriel, A.H., Bocchia, R., Bonnet, R.M., Cesarsky, C., Christensen-Dalsgaard, J., Damé, L., Delache, P., Deubner, F.-L., Foing, B., Fossat, E., Fröhlich, C., Gorisse, M., Gough, D.O., Grec, G., Hoyng, P., Pallé, P., Paul, J., Robillot, J.-P., Roca-Cortés, T., Stenflo, J.O., Ulrich, R.K., van der Raay, H.B., (1989) *The SOHO Mission — Scientific and Technical aspects of the Instruments*, ESA SP-1104, Paris, 13
Gabriel, M., (2000) *Astron.Astroph.* **353**, 399
Galland, F., Lagrange, A.-M., Udry, S., Chelli, A., Pepe, F., Queloz, D., Beuzit, J.-L., Mayor, M., (2005) *Astron.Astroph.* **443**, 337
Gautschy, A., Saio, H., Harzenmoser, H., (1998) *Mon.Not. Roy.Ast.Soc.* **301**, 31
Georgobiani, D., Stein, R.F., Nordlund, Å., (2003) *Astroph.J.* **596**, 698
Gilman, P.A., Glatzmaier, G.A., (1981) *Astroph.J. Supp.* **45**, 335
Gizon, L., Birch, A.C., (2002) *Astroph.J.* **571**, 966
Gizon, L., Birch, A.C., (2004) *Astroph.J.* **614**, 472
Gizon, L., Birch, A.C., (2005) *Living Rev. Solar Phys.*, cited Feb. 2006 http://www.livingreviews.org/lrsp-2005-6
Goldreich, P., Kumar, P. (1988) *Astroph.J.* **326**, 462
Goldreich, P., Murray, N., Kumar, P. (1994) *Astroph.J.* **424**, 466
González Hernández, I., Komm, R., Hill, F., Howe, R., Corbard, T., Haber, D., (2006) *Astroph.J.* **638**, 576
Goode, P.R., Thompson, M.J., (1992) *Astroph.J.* **395**, 307
Göpel, C., Manhès, G., Allègre, C.-J., (1994) *Earth Planet.Sc.Lett.* **121**, 153
Gough, D.O., (1969) *J. Atmos. Sc.*, **26**, 448
Gough, D.O., (1984) *Phil.Trans.Roy.Soc. London A* **313**, 27
Gough, D.O., Kosovichev, A.G., Toomre, J., Anderson, E., Antia, H.M., Basu, S., Chaboyer, B., Chitre, S.M., Christensen-Dalsgaard, J., Dziembowski, W.A., Eff-Darwich, A., Elliott, J.R., Giles, P.M., Goode, P.R., Guzik, J.A., Harvey, J.W., Hill, F., Leibacher, J.W., Monteiro, M.J.P.F.G., Richard, O., Sekii, T., Shibahashi, H., Takata, M., Thompson, M.J., Vauclair, S., Vorontsov, S.V., (1996a) *Science* **272**, 1296
Gough, D.O., Leibacher, J.W., Scherrer, P.H., Toomre, J., (1996b) *Science* **272**,

1281
Gough, D.O., Thompson, M.J., (1990) *Mon.Not. Roy.Ast.Soc.* **242**, 25
Gough, D.O., Thompson, M.J., (1991) *Solar Interior and Atmosphere*, Eds. A.N.Cox, W.Livingston, M.Matthews, Univ. Arizona Press, 519
Gough, D.O., Toomre, J., (1983) *Solar Phys.* **82**, 401
Gradshteyn, I.S., Ryzhik, I.M., (1994) *Table of Integrals, Series, and Products* 5th Ed., Academic Press.
Gratton, R.G., Bragaglia, A., Carretta, E., Clementini, G., Desidera, S., Grundahl, F., Lucatello, S., (2003) *Astron.Astroph.* **408**, 529
Gray, D.F., (1989) *Astroph.J.* **347**, 1021
Gray, D.F., (1992) *The Observation and Analysis of Stellar Photospheres*, Cambridge Univ. Press.
Grigahcène, A., Dupret, M.-A., Gabriel, M., Garrido, R., Scuflaire, R., (2005) *Astron.Astroph.* **434**, 1055
Grison, P., (1994) *Astron.Astroph.* **289**, 404
Groenewegen, M.A.T., Blommaert, J.A.D.L., (2005) *Astron.Astroph.* **443**, 143
Guenther, D.B., Demarque, P., (1997) *Astroph.J.* **484**, 937
Guzik, J.A., Kaye, A.B., Bradley, P.A., Cox, A.N., Neuforge, C., (2000) *Astroph.J. Lett.* **542**, L57
Haber, D.A., Hindman, B.W., Toomre, J., Bogart, R.S., Larsen, R.M., Hill, F., (2002) *Astroph.J.* **570**, 855
Harvey, J.W., Hill, F., Hubbard, R.P., Kennedy, J.R., Leibacher, J.W., Pintar, J.A., Gilman, P.A., Noyes, R.W., Title, A.M., Toomre, J., Ulrich, R.K., Bhatnagar, A., Kennewell, J.A., Marquette, W., Patron, J., Saa, O., Yasukawa, E., (1996) *Science* **272**, 1284
Herwig, F., (2000) *Astron.Astroph.* **360**, 952
Herwig, F., (2005) *Ann.Rev. Astron.Astroph.* **43**, 435
Herwig, F., Blöcker, T., Schönberner, D., El Eid, M., (1997) *Astron.Astroph. Lett.* **324**, L81
Heynderickx, D., Waelkens, C., Smeyers, P., (1994) *Astron.Astroph. Supp.* **105**, 447
Hill, F., Gough, D., Merryfield, W.J., Toomre, J., (1991) *Astroph.J.* **369**, 237
Hill, F., Stark, P.B., Stebbins, R.T., Anderson, E.R., Antia, H.M., Brown, T.M., Duvall Jr., T.L., Haber, D.A., Harvey, J.W., Hathaway, D.H., Howe, R., Hubbard, R.P., Jones, H.P., Kennedy, J.R., Korzennik, S.G., Kosovichev, A.G., Leibacher, J.W., Libbrecht, K.G., Pintar, J.A., Rhodes Jr., E.J., Schou, J., Thompson, M.J., Tomczyk, S., Toner, C.G., Toussaint, R., Williams, W.E., (1996) *Science* **272**, 1292
Hindman, B.W., Gough, D., Thompson, M.J., Toomre, J., (2005) *Astroph.J.* **621**, 512
Högbom, J.A., (1974) *Astron.Astroph. Supp.* **15**, 417
Horne, K., (1986) *Pub.Ast.Soc.Pac.* **98**, 609
Houdek, G., (2000) in δ *Sct and related stars*, Eds. M. Breger & M.H. Montgomery, ASP Conf. Ser. **210**, 454
Houdek, G., Balmforth, N.J., Christensen-Dalsgaard, J., Gough, D.O., (1999) *Astron.Astroph.* **351**, 582

Hoyng, P., (1990) *Proc. IAU Symp. 138: The Solar Photosphere: Structure, Convection & Magnetic Fields*, Ed. J.O. Stenflo, Kluwer, 359
Hughes, S.J., Pijpers, F.P., Thompson, M.J., (2006) *Astron.Astroph.* submitted
van der Hulst, H.C., (1957) *Light Scattering by Small Particles*, Wiley
Iglesias, C.A., Rogers, F. J., (1996) *Astroph.J.* **464**, 943
Isaak, G.R., McLeod, C.P., Palle, P.L., van der Raay, H.B., Roca-Cortés, T., (1989) *Astron.Astroph.* **208**, 297
Ishimaru, A., (1978) *Wave Propagation in Random Media*, Academic Press (note: re-published 1999, Wiley)
Jackson, D.R., Dowling, D.R., (1991) *J. Acoust.Soc.Am.* **89**, 171
Jackson, J.D., (1999) *Classical Electrodynamics* 3rd Ed., Wiley.
Jankov, S., Vakili, F., Domiciano de Souza, A., Jr., Janot-Pacheco, E., (2001) *Astron.Astroph.* **377**, 721
Jefferies, S.M., Osaki, Y., Shibahashi, H., Duvall Jr., T.L., Harvey, J.W., Pomerantz, M.A., (1994) *Astron.Astroph.* **434**, 795
Jeffrey, W.A., (1988) *Astroph.J.* **327**, 987
Jeffrey, W.A., Rosner, R., (1986a) *Astroph.J.* **310**, 463
Jeffrey, W.A., Rosner, R., (1986b) *Astroph.J.* **310**, 473
Jensen, J.M., Pijpers, F.P., (2003) *Astron.Astroph.* **412**, 257
Jetsu, L., Pelt, J., (1999) *Astron.Astroph. Supp.* **139**, 629
Kalkofen, W. (Ed.), (1988) *Numerical Radiative Transfer*, Cambridge Univ. Press
Kharchenko, N.V., Piskunov, A.E., Röser, S., Schilbach, E., Scholz, R.-D., (2005) *Astron.Astroph.* **438**, 1163
Kippenhahn, R., Weigert, A., (1994) *Stellar Structure and Evolution*, Springer
Kjeldsen, H., Bedding, T.R., Frandsen, S., Dall, T.H., (1999) *Mon.Not. Roy.Ast.Soc.* **303**, 579
Kjeldsen, H., Frandsen, S., (1992) *Pub.Ast.Soc.Pac.* **104**, 413
Knaack, R., Stenflo, J.O., Berdyugina, S.V. *Astron.Astroph.* **438**, 1067
Knobloch, E., (1978) *Astroph.J.* **225**, 1050
Koen, C., (1996) *Mon.Not. Roy.Ast.Soc.* **283**, 471
Koen, C., (1999) *Mon.Not. Roy.Ast.Soc.* **309**, 769
Koen, C., Lombard, F. (1993) *Mon.Not. Roy.Ast.Soc.* **263**, 287
Koen, C., Lombard, F. (1995) *Mon.Not. Roy.Ast.Soc.* **274**, 821
Koen, C., Lombard, F. (2001) *Mon.Not. Roy.Ast.Soc.* **325**, 1124
Koen, C., Lombard, F. (2004) *Mon.Not. Roy.Ast.Soc.* **353**, 98
Kolláth, Z., Beaulieu, J.P., Buchler, J.R., Yecko, P., (1998) *Astroph.J. Lett.* **502**, L55
Kolláth, Z., Buchler, J.R., Szabó, R., Csubry, Z., (2002) *Astron.Astroph.* **385**, 932
Komm, R.W., Gu, Y., Hill, F., Stark, P.B., Fodor, I.K., (1999) *Astroph.J.* **519**, 407
Korzennik, S.G., (2005) *Astroph.J.* **626**, 585
Kosovichev, A.G., Duvall Jr., T.L., (1997) in *SCORe'96: Solar Convection and Oscillations and Their Relationship*, Eds. F. P. Pijpers, J. Christensen-Dalsgaard, C. S. Rosenthal, Kluwer, 241
Kosovichev, A.G., Duvall Jr., T.L., Scherrer, P.H., (2000) *Solar Phys.*, **192**, 159

Kuhfuß, R., (1986) *Astron.Astroph.* **160**, 116
Kumar, P., Lu, E., (1991) *Astroph.J. Lett.* **375**, L35
Landau, L.D., Lifshitz, E.M., (1987) *Course of Theoretical Physics Vol. 6: Fluid Mechanics* 2nd Ed., Pergamon
Lantz, S.R., Fan, Y., (1999) *Astroph.J. Supp.* **121**, 247
Larsen, R.M., Hansen, P.C., (1997) *Astron.Astroph. Supp.* **121**, 587
Lee, U., Saio, H., (1987a) *Mon.Not. Roy.Ast.Soc.* **224**, 513
Lee, U., Saio, H., (1987b) *Mon.Not. Roy.Ast.Soc.* **225**, 643
Li, L.H., Robinson, F.J., Demarque, P., Sofia, S., Guenther, D.B., (2002) *Astroph.J.* **567**, 1192
Li, Y., (1992) *Astron.Astroph.* **257**, 145
Lindsey, C., Braun, D.C., (1990) *Solar Phys.* **126**, 101
Lindsey, C., Braun, D.C., (1997) *Astroph.J.* **485**, 895
Lighthill, J., (1978) *Waves in fluids*, Cambridge Univ. Press.
Lin, C.-H., Däppen, W., (2005) *Astroph.J.* **623**, 556
Lomb, N.R., (1976) *Astroph. Spa.Sc.* **39**, 447
Lynden-Bell, D., Ostriker, J.P., (1967) *Mon.Not. Roy.Ast.Soc.* **136**, 293
Maeder, A., Meynet, G., (2000) *Ann.Rev. Astron.Astroph.* **38**, 143
Maeder, A., Meynet, G., (2003) *Astron.Astroph.* **411**, 543
Maeder, A., Meynet, G., (2004) *Astron.Astroph.* **422**, 225
Maeder, A., Meynet, G., (2005) *Astron.Astroph.* **440**, 1041
Maeder, A., Meynet, G., (2006) *Astron.Astroph. Lett.* **448**, L37
Marchenkov, K., Roxburgh, I.W., Vorontsov, S., (2000) *Mon.Not. Roy.Ast.Soc.* **312**, 39
Marquering, H., Dahlen, F.A., Nolet, G., (1999) *Geophys. J. Int.*, **137**, 805
McComb, W.D., (1990) *The Physics of Fluid Turbulence*, Oxford Univ. Press
Meisner, R.W., Rast, M.P., (2003) *Bull. Am.Ast.Soc.* **34**, 734
Metcalfe, T.S., (2003) *Astroph. Spa.Sc.* **284**, 141
Metcalfe, T.S., Montgomery, M.H., Kanaan, A., (2004) *Astroph.J. Lett.* **605**, L133
Metcalfe, T.S., Montgomery, M.H., Kawaler, S.D., (2003) *Mon.Not. Roy.Ast.Soc.* **344**, L88
Meynet, G., Ekström, S., Maeder, A., (2006) *Astron.Astroph.* **447**, 623
Miesch, M.S., Elliott, J.R., Toomre, J., Clune, T.L., Glatzmaier, G.A., Gilman, P.A., (2000) *Astroph.J.* **532**, 593
Mihalas, D., (1978) *Stellar Atmospheres* 2nd Ed., Freeman
Mihalas, D., Mihalas, B.W., (1984) *Foundations of Radiation Hydrodynamics*, Oxford Univ. Press
Misner, C.W., Thorne, K.S., Wheeler, J.A., (1973) *Gravitation*, Freeman & Co.
Monin, A.S., Yaglom, A.M., (1975) *Statistical Fluid Mechanics Vol. 2*, MIT Press.
Monteiro, M.J.P.F.G., Christensen-Dalsgaard, J., Thompson, M., (1994) *Astron.Astroph.* **283**, 247
Monteiro, M.J.P.F.G., Christensen-Dalsgaard, J., Thompson, M., (2000) *Mon.Not. Roy.Ast.Soc.* **316**, 165
Monteiro, M.J.P.F.G., Thompson, M., (2005) *Mon.Not. Roy.Ast.Soc.* **361**, 1187
Moss, D., Tuominen, I., Brandenburg, A., (1991) *Astron.Astroph.* **245**, 129

Mukadam, A.S., Montgomery, M.H., Winget, D.E., Kepler, S.O., Clemens, J.C., (2006) *Astroph.J.* **640**, 956
Murawski, K., (1993) *Ac. Astron.* **43**, 161
Musielak, Z.E., Rosner, R., Stein, R.F., Ulmschneider, P., (1994) *Astroph.J.* **423**, 474
Nakariakov, V.M., Verwichte, E., (2005) *Living Rev. Solar Phys.*, cited Feb. 2006 http://www.livingreviews.org/lrsp-2005-3
Nayfonov, A., Däppen, W., Hummer, D.G., Mihalas, D., (1999) *Astroph.J.* **526**, 451
Narayan, R., Nityananda, R., (1986) *Ann.Rev. Astron.Astroph.*, **24**, 127
Neckel, H., Labs, D., (1994) *Solar Phys.* **153**, 91
Nigam, R., Kosovichev, A.G., (1998) *Astroph.J. Lett.* **505**, L51
Nigam, R., Kosovichev, A.G., Scherrer, P., Schou, J., (1998) *Astroph.J. Lett.* **495**, L115
O'Brien, M.S. (2000) *Astroph.J.* **532**, 1078
Olivier, E.A., Wood, P.R., (2005) *Mon.Not. Roy.Ast.Soc.* **362**, 1396
Ossendrijver, M., (2003) *Astron. Astroph. Rev.* **11**, 287
Otazu, X., Ribó, M., Paredes, J.M., Peracaula, M., Nuñez, J., (2004) *Mon.Not. Roy.Ast.Soc.* **351**, 215
Otí Floranes, H., Christensen-Dalsgaard, J., Thompson, M.J., (2005) *Mon.Not. Roy.Ast.Soc.* **356**, 671
Palla, F., Stahler, S.W., (1991) *Astroph.J.* **375**, 288
Palla, F., Stahler, S.W., (1993) *Astroph.J.* **418**, 414
Palle, P.L., Régulo, C., Roca-Cortés, T., Sánchez-Duarte, L., Schmider, F.X., (1992) *Astron.Astroph.* **254**, 348
Papaloizou, J.C.B., Pringle, J.E., (1978) *Mon.Not. Roy.Ast.Soc.* **182**, 423
Paquette, C., Pelletier, C., Fontaine, G., Michaud, G., (1986) *Astroph.J. Supp.*, **61**, 177
Parker, E.N., (1955) *Astroph.J.* **122**, 293
Patrón, J., González Hernández, I., Chou, D.-Y., Sun, M.-T., Mu, T.-M., Loudagh, S., Bala, B., Chou, Y.-P., Lin, C.-H., Huang, I.-J., Jiménez, A., Rabello-Soares, M.C., Ai, G., Wang, G.-P., Zirin, H., Marquette, W., Nenow, J., Ehgamberdiev, S., Khalikov, S., TON Team, (1997) *Astroph.J.* **485**, 869
Pérez-Hernández, F., Christensen-Dalsgaard, J. (1994) *Mon.Not. Roy.Ast.Soc.* **267**, 111
Pesnell, W.D., (1987) *Astroph.J.* **314**, 598
Pijpers, F.P., (1993) *Astron.Astroph.* **267**, 471
Pijpers, F.P., (1997) *Astron.Astroph.* **326**, 1235
Pijpers, F.P., (1998) *Mon.Not. Roy.Ast.Soc.* **297**, L76
Pijpers, F.P., (1999) *Mon.Not. Roy.Ast.Soc.* **307**, 659
Pijpers, F.P., (2003) *Astron.Astroph.* **402**, 683
Pijpers, F.P., Thompson, M.J., (1992) *Astron.Astroph. Lett.* **262**, L33
Pijpers, F.P., Thompson, M.J., (1994) *Astron.Astroph.* **281**, 231
Pijpers, F.P., Thompson, M.J., (1996) *Mon.Not. Roy.Ast.Soc.* **279**, 498
Porter, D.H., Woodward, P.R., (2000) *Astroph.J. Supp.* **127**, 159

Press, W.H., Teukolsky, S.A., Vetterling, W.T., Flannery, B.P., (1992) *Numerical Recipes: The Art of Scientific Computing* 2nd Ed., Cambridge Univ. Press
Priest, E.R., Hood, A.W. (Eds.), (1991) *Advances in Solar System Magnetohydrodynamics*, Cambridge Univ. Press
Proffitt, C.R., Michaud, G., (1991) *Astroph.J.* **380**, 238
Rädler, K.-H., Wiedemann, E., Brandenburg, A., Meinel, R., Tuominen, I., (1990) *Astron.Astroph.* **239**, 413
Reiners, A. (2006) *Astron.Astroph.* **446**, 267
Reiners, A., Schmitt, J.H.M.M. (2002) *Astron.Astroph.* **384**, 155
Renzini, A., Voli, M., (1981) *Astron.Astroph.* **94**, 175
Reese, D., Rincon, F., Rieutord, M., (2004) *Astron.Astroph.* **427**, 279
Rhodes, E.J., Cacciani, A., Garneau, G., Misch, T., Progovac, D., Shieber, T., Tomczyk, S., Ulrich, R.K., (1988) *Max 1991: Flare Research at the Next Solar Maximum. Workshop 1 : Scientific Objectives*, NASA-GSFC, 33
Rhodes, E.J., Jr., Kosovichev, A.G., Schou, J., Scherrer, P.H., Reiter, J., (1997) *Solar Phys.* **175**, 287
Rickett, J.E., Claerbout, J.F., (2000) *Solar Phys.* **192**, 203
Ritzwoller, M.H., Lavely, E.M., (1991) *Astroph.J.* **369**, 557
Roberts, B., Edwin, P.M., Benz, A.O., (1984) *Astroph.J.* **279**, 857
Robinson, F.J., Chan, K.L., (2001) *Mon.Not. Roy.Ast.Soc.* **321**, 723
Rosenthal, C.S., (1995) *Astroph.J.* **438**, 434
Rosenthal, C.S., (1998) *Astroph.J.* **508**, 864
Rosenthal, C.S., Christensen-Dalsgaard, J., Nordlund, Å, Stein, R.F., Trampedach, R., (1999) *Astron.Astroph.* **351**, 689
Rosenthal, C.S., Bogdan, T.J., Carlsson, M., Dorch, S.B.F., Hansteen, V., McIntosh, S.W., McMurry, A., Nordlund, Å., Stein, R.F., (2002) *Astroph.J.* **564**, 508
Roxburgh, I.W., (1966) *Mon.Not. Roy.Ast.Soc.* **132**, 201
Roxburgh, I.W., (1989) *Astron.Astroph.* **211**, 361
Roxburgh, I.W., (2005) *Astron.Astroph.* **434**, 665
Roxburgh, I.W., Vorontsov, S., (1995) *Mon.Not. Roy.Ast.Soc.* **272**, 850
Roxburgh, I.W., Vorontsov, S., (1997) *Mon.Not. Roy.Ast.Soc.* **292**, L33
Roxburgh, I.W., Vorontsov, S., (2000) *Mon.Not. Roy.Ast.Soc.* **317**, 141
Roxburgh, I.W., Vorontsov, S., (2000) *Mon.Not. Roy.Ast.Soc.* **317**, 151
Roxburgh, I.W., Vorontsov, S., (2001) *Mon.Not. Roy.Ast.Soc.* **322**, 85
Roxburgh, I.W., Vorontsov, S., (2003) *Astron.Astroph.* **411**, 215
Samadi, R., Nordlund, Å. Stein, R.F., Goupil, M.J., Roxburgh, I., (2003a) *Astron.Astroph.* **403**, 303
Samadi, R., Nordlund, Å. Stein, R.F., Goupil, M.J., Roxburgh, I., (2003b) *Astron.Astroph.* **404**, 1129
Scargle, J.D., (1982) *Astroph.J.* **263**, 835
Scargle, J.D., (1989) *Astroph.J.* **343**, 874
Schekochihin, A.A., Cowley, S.C., (2006) *Magnetohydrodynamics: Historical Evolution and Trends*, Eds. S.Molokov, R.Moreau, H.K.Moffatt, Springer
Scherrer, P.H., Bogart, R.S., Bush, R.I., Hoeksema, J.T., Kosovichev, A.G., Schou, J., Rosenberg, W., Springer, L., Tarbell, T.D., Title, A., Wolfson,

C. J., Zayer, I. & MDI engineering team, (1995) *Solar Phys.* **162**, 129
Schou J., Bogart, R.S., (1998) *Astroph.J. Lett.* **504**, L131
Schramm, D.N., Steigman, G., Dearborn, D.S.P., (1990) *Astroph.J. Lett.* **359**, L55
Schrijvers, C., Telting, J.H., (1999) *Astron.Astroph.* **342**, 453
Schwarzenberg-Czerny, A., (1991) *Mon.Not. Roy.Ast.Soc.* **253**, 189
Schwarzenberg-Czerny, A., (1996) *Astroph.J. Lett.* **460**, L107
Schwarzenberg-Czerny, A., (1998a) *Balt.Ast.* **7**, 43
Schwarzenberg-Czerny, A., (1998b) *Mon.Not. Roy.Ast.Soc.* **301**, 831
Sekii, T. (1993) *Mon.Not. Roy.Ast.Soc.* **264**, 1018
Shergelashvili, B.M., Poedts, S., (2005) *Astron.Astroph.* **438**, 1083
Siess, L., Dufour, E., Forestini, M., (2000) *Astron.Astroph.* **358**, 593
Skartlien, R., (2001) *Astroph.J.* **554**, 488
Skartlien, R., (2002a) *Astroph.J.* **565**, 1348
Skartlien, R., (2002b) *Astroph.J.* **578**, 621
Smeyers, P., Craeynest, D., Martens, L., (1981) *Astroph. Spa.Sc.* **78**, 483
Smith, M.A., (1977) *Astroph.J.* **215**, 574
Snieder, R., Lomax, A., (1996) *Geophys. J. Int.*, **125**, 796
Soufi, F., Goupil, M.J., Dziembowski, W.A., (1998) *Astron.Astroph.* **334**. 911
Starck, J.-L., Pantin, E., Murtagh, F., (2002) *Pub.Ast.Soc.Pac.* **114**, 1051
Stein, R.F., Georgobiani, D., Trampedach, R., Ludwig, H.-G., Nordlund, Å, (2004) *Solar Phys.* **220**, 229
Stein, R.F., Nordlund, Å, (2000) *Solar Phys.* **192**, 91
Stein, R.F., Nordlund, Å, (2001) *Astroph.J.* **546**, 585
Stellingwerf, R.F., (1978) *Astroph.J.* **224**, 953
Stix, M., (1983) *Astron.Astroph.* **118**, 363
Stoer, J., Bulirsch, R., (2002) *Introduction to Numerical Analysis*, Springer
Stoica, P., Larsson, E.G., Li, J., (2000) *Astron.J.* **120** 2163
Strous, L.H., Goode, P.R., Rimmele, T.R., (2000) *Astroph.J.* **535**, 1000
Struck, C., Smith, D.C., Willson, L.A., Turner, G., Bowen, G.H., (2004) *Mon.Not. Roy.Ast.Soc.* **353**, 559
Talon, S., Charbonnel, C., (1998) *Astron.Astroph.* **335**, 959
Tassoul, J.-L., (2000) 'Stellar Rotation', Cambridge Univ. Press
Tassoul, M., (1980) *Astroph.J. Supp.* **43**, 469
Théado, S., Vauclair, S., Cunha, M.S., (2005) *Astron.Astroph.* **443**, 627
Thompson, M.J., Cunha, M.S., Monteiro, M.J.P.F.G., (Eds.) (2003) *Asteroseismology Across the HR Diagram*, Kluwer
Thompson, M.J., Toomre, J., Anderson, E.R., Antia, H.M., Berthomieu, G., Burtonclay, D., Chitre, S.M., Christensen-Dalsgaard, J., Corbard, T., DeRosa, M., Genovese, C.R., Gough, D.O., Haber, D.A., Harvey, J.W., Hill, F., Howe, R., Korzennik, S.G., Kosovichev, A.G., Leibacher, J.W., Pijpers, F.P., Provost, J., Rhodes Jr., E.J., Schou, J., Sekii, T., Stark, P.B., Wilson, P.R., (1996) *Science* **272**, 1300
Tobias, S.M., Brummell, N.H., Clune, T.L., Toomre, J., (2001) *Astroph.J. Lett.* **502**, L177
Tobias, S.M., Brummell, N.H., Clune, T.L., Toomre, J., (2001) *Astroph.J.* **549**,

1183
Toutain, T., Berthomieu, G., Provost, J., (1990) *Astron.Astroph.* **344**, 188
Toutain, T., Kosovichev, A.G., (2000) *Astroph..J. Lett.* **534**, L211
Townsend, R.H.D., (1997) *Mon.Not. Roy.Ast.Soc.* **284**, 839
Townsend, R.H.D., (2002) *Mon.Not. Roy.Ast.Soc.* **330**, 855
Townsend, R.H.D., (2003) *Mon.Not. Roy.Ast.Soc.* **343**, 125
Tripathy, S.C., Antia, H.M., (1999) *Solar Phys.* **186**, 1
Turcotte, S., Richer, J., Michaud, G., (1998a) *Astroph.J.* **504**, 559
Turcotte, S., Richer, J., Michaud, G., Christensen-Dalsgaard, J., (2000) *Astron.Astroph.* **360**, 603
Turcotte, S., Richer, J., Michaud, G., Iglesias, C.A., Rogers, F.J., (1998b) *Astroph.J.* **504**, 539
Unno, W., Osaki, Y., Ando, H., Saio, H., Shibahashi, H., (1989) *Nonradial Oscillation of stars*, Univ. of Tokyo Press.
VandenBerg, D.A., Stetson, P.B., Bolte, M., (1996) *Ann.Rev. Astron.Astroph.*, **34**, 461
Vaniček, P., (1971) *Astroph. Spa.Sc.* **12**, 10
Ventura, P., D'Antona, F., (2005) *Astron.Astroph.* **439**, 1075
Verwichte, E., Foullon, C., Nakariakov, V.M., (2006) *Astron.Astroph.* **446**, 1139
Vio, R., Strohmer, T., Wamsteker, W., (2000) *Pub.Ast.Soc.Pac.* **112**, 74
Viskum, M., Kjeldsen, H., Bedding, T.R., Dall, T.H., Baldry, I.K., Bruntt, H., Frandsen, S., (1998) *Astron.Astroph.* **335**, 549
Vogt, S.S., Penrod, G.D., (1983) *Astroph.J.* **275**, 661
Wachter, R., Schou, J., Kosovichev, A.G., Scherrer, P.H., (2003) *Astroph.J.* **588**, 1199
Wahba, G., (1977) *SIAM J. Numer. Anal.*, **14**, 651
Wapenaar, K., Thorbecke, J., Draganov, D., (2004) *Geophys. J. Int.*, **156**, 179
Wilcox, J.Z., Wilcox, T.J., (1995) *Astron.Astroph. Supp.* **112**, 395
Wolff, C.L., (1998) *Astroph.J.* **502**, 961
Wolff, C.L., (2002) *Astroph.J. Lett.* **580**, L181
Yecko, P.A., Kolláth, Z., Buchler, J.R., (1998) *Astron.Astroph.* **336**, 553
Young, A.T., Genet, R.M., Boyd, L.J., Borucki, W.J., Lockwood, G.W., Henry, G.W., Hall, D.S., Smith, D.P., Baliumas, S.L., Donahue, R., Epand, D. H., (1991) *Pub.Ast.Soc.Pac.* **103**, 221
Zahn, J.-P., (1992) *Astron.Astroph.* **265**, 115
von Zeipel, H., (1924) *Mon.Not. Roy.Ast.Soc.* **84**, 665
Zhao, L., Jordan, T.H., (1998) *Geophys. J. Int.* **133**, 683
Zhugzhda, Y.D., Dzhalilov, N.S., Staude, J., (1993) *Astron.Astroph. Lett.* **278**, L9
Zwillinger,D., (1989) *Handbook of Differential Equations*, Academic Press

Index

abundances, 237, 239
acoustic cut-off, 25, 72, 80, 101
acoustic depth, 82
acoustic potential, 83
acoustic source, 27, 82, 177
advection, 145
AGB stars, 197
aliasing, 11
anelastic, 154, 253
ARMA, 224

BiSON, 8, 17
buoyancy frequency, 67, 73, 183

closure, 37, 153
completeness, 52, 62, 88
convection, 23, 27, 30, 35–39, 51, 80, 116, 152, 156, 182, 234, 239, 240
COROT, 201, 245
Cowling approximation, 69, 82, 100, 105, 177, 183, 186, 190
Cowling's theorem, 149

degenerate, 51, 56, 58, 60, 79
diffusion, 39, 239
Dirichlet boundary, 56
Dopplergram, 132
dynamical mass, 229
dynamical time scale, 30
dynamo theory, 148, 151

eigenfunction, 17, 42, 52, 56, 59, 61, 62, 67, 73, 88, 89, 271
ensemble averaging, 150, 152
equivalent width, 202
excitation, 82, 196, 242

far field, 161

g-modes, 18, 44, 67, 73, 199
GOLF, 2
GONG, 1, 4, 17, 132

HB stars, 197, 243, 244
helicity, 151, 152
Hermitian, 87

ill-posedness, 52, 86
interferometry, 208, 229
IRIS, 14, 17

JWKB, 74

L-curve, 120
least-squares, 7, 15, 182, 221
limb darkening, 20, 212, 215, 228

mass-luminosity relationship, 233, 235
MDI, 2, 4, 8, 26, 132, 137
mode identification, 192, 208
mode selection, 244
Monte Carlo analysis, 106, 125, 224

Neumann boundary, 56

node, 51, 53, 56, 59, 69, 73, 204
non-adiabatic, 190

opacity, 198, 232, 241
optimal extraction, 205
overshoot, 38, 51, 240

p-modes, 18, 67, 73, 78, 199
perturbation, 29, 61, 69, 92, 182, 185
poloidal vector field, 94, 283
polytropes, 232
propagator, 178
pseudo-modes, 27

r-modes, 252, 253
ray approximation, 134, 166, 171, 189
regularisation, 87, 115, 118, 119, 123
rotational a-coefficients, 110
rotational splitting, 109

self-adjoint, 87, 178
singular value decomposition (SVD), 86, 120, 121, 220
SoHO, 2, 4, 132
spherical harmonic, 17, 59, 69, 89, 158, 180, 188, 215, 282

toroidal vector field, 94, 283
tracking, 138, 166
trade-off, 9, 116, 118, 119
turbulence, 36, 81, 84, 152, 182
turning point, 72

unicity, 52, 62, 69, 86, 88, 92, 235

variational principle, 87, 89, 169
VIRGO, 2, 17–19
virial theorem, 30
vorticity, 151

wavelet, 22, 122, 174, 220
weighted cross validation, 120
weights, 7, 8, 20, 53, 86, 89, 93, 113, 114, 116–119, 139, 203, 222, 255
white dwarfs, 197
window function, 12

ZAMS, 197, 237